HURRICANES

A Reference Handbook
Second Edition

Upcoming Titles in ABC-CLIO's
Contemporary
World Issues
Series

Domestic Violence in America, Margi Laird McCue
Genetic Engineering, second edition, Harry LeVine, III
Global Warming, Gary C. Byner
Gun Control in the United States, Gregg Lee Carter
Human Rights Worldwide, Zehra F. Kabasakal Arat
International Environmental Disputes, Aaron Schwabach
Juvenile Justice, Donald J. Shoemaker and Timothy W. Wolfe
Mental Health in America, Donna R. Kemp
The United Nations System, Chadwick F. Alger
U.S. Hegemony, Raymond Tanter and Clare M. Lopez
U.S. Homeland Security, Howard Ball
Wetlands in America, William M. Lewis
World Population, Geoffrey Gilbert

Books in the Contemporary World Issues series address vital issues in today's society such as genetic engineering, pollution, and biodiversity. Written by professional writers, shcoloars, and nonacademic experts, these books are authoritative, clearly written, up-to-date, and objective. They provide a good starting point for research by high school and college students, scholars, and general readers as well as by legislators, businesspeople, activists, and others.

Each book, carefully organized and easy to use, contains an overview of the subject, a detailed chronology, biographical sketches, facts and data and/or documents and other primary-source material, a directory of organizations and agencies, annotated lists of print and nonprint resources, and an indes.

Readers of books in the Contemporary World Isses series will find the information they need in order to have a better understanding of the social, political environmental, and economic issues facing the world today.

HURRICANES

A Reference Handbook

Second Edition

Patrick J. Fitzpatrick

CONTEMPORARY WORLD ISSUES

Santa Barbara, California
Denver, Colorado
Oxford, England

Copyright © 2006 by Patrick J. Fitzpatrick

All rights reserved. No part of this publication may be reproduced, stored in a retrieval system, or transmitted, in any form or by any means, electronic, mechanical, photocopying, recording, or otherwise, except for the inclusion of brief quotations in a review, without prior permission in writing from the publishers.

Library of Congress Cataloging-in-Publication Data

Fitzpatrick, Patrick J.
  Hurricanes : a reference handbook / Patrick J. Fitzpatrick.—[2nd ed.].
    p. cm. — (Contemporary world issues)
  Rev. ed. of: Natural disasters, hurricanes. c1999.
  Includes bibliographical references and index.
  ISBN 1-85109-647-7 (acid-free paper) — ISBN 1-85109-652-3 (eBook)
  1. Hurricanes—Handbooks, manuals, etc. I. Fitzpatrick, Patrick J. Natural disasters, hurricanes. II. Title. III. Series.
  QC944.F58 2006
  551.55'2—dc22   2005025641

10 09 08 07 06   10 9 8 7 6 5 4 3 2 1

ABC-CLIO, Inc.
130 Cremona Drive, P.O. Box 1911
Santa Barbara, California 93116-1911

This book is printed on acid-free paper ∞

Manufactured in the United States of America

Contents

Preface xiii
Acknowledgments xix

1   **Background on Hurricanes, 1**
    Hurricane Formation and Life Cycle, 2
      Genesis Stage, 3
      Intensification Stage and the Importance of Water
        Temperature, 8
      Weakening Stage and
        Dissipation, 10
      Extratropical Transition, 12
    The Atlantic Hurricane Season,
      13
    Naming Hurricanes, 14
    Hurricane Structure, 15
      Cloud Patterns, 15
      The Eyewall and Eye, 18
      Hurricane Size, 21
      Lightning (Or the Lack Of), 22

2   **Problems, Controversies, and Solutions: Hurricane
    Destruction, Forecasting, Global Warming Controversy,
    Mitigation, and Insurance Procedures, 25**
    Hurricane Destruction, 25
    Forecasting Hurricanes, 35
      Hurricane Opal: A Close Call, 38
      Factors Controlling Hurricane Motion, 39
      Computer Models, 40
      Observation Platforms, 42
      Summary of NHC Forecast Procedures, 44

Society's Preparation for Hurricanes: New Worries, 45
Forecasting Annual Hurricane Activity, 49
Is Global Warming Increasing the Number of
   Hurricanes?, 54
   What Is the Greenhouse Effect?, 54
   What Is Global Warming?, 56
   Has the Number of Tropical Storms Increased, and
      Have Hurricanes Become Stronger Due to Potential
      Global Warming?, 56
   Should Global Warming Occur in the Future, Will
      Hurricanes Increase in Number and Intensity?, 59
Attempts at Hurricane Modification, 60
Mitigation and Insurance Procedures, 64
   Home Protection, 64
   The Roof, 65
   Exterior Doors and Windows, 66
   Storm Shutters, 67
Hurricane Preparedness Steps, 69
   Before Hurricane Season, 69
   When a Hurricane Watch or Warning Is Issued, 72
   If Evacuation Is Necessary, 74
   After the Hurricane Passes, 74
   Make Plans for Your Pets, 77
   Insurance Preparation, 78

3 **Worldwide Perspective, 83**
   International Impact, Names, and Locations
      of Hurricanes, 83
   Hurricane Categories around the World, 93
   Monitoring, Forecasting, and Warning of Tropical Cyclones
   Worldwide, 97
      The World Weather Watch, 97
      Climate Variations in Worldwide Tropical Cyclones
         and Seasonal Forecasting, 100
   Worldwide Naming Conventions, 101
   Storm Size Variations, 103

4 **Chronology, 105**
   Chronology of Weather Advances Related to the
   Study and Forecasting of Hurricanes, 105
   Chronology of Significant Land-falling U.S.
      Hurricanes since 1900, 151

Chronology of Some Hurricanes That Impacted History, 187

5 **Biographical Sketches, 193**

6 **Data, Opinions, and Letters, 231**
Data, 232
Interesting Category 5 Facts, 243
World Records, 243
Deadliest U.S. Tropical Storms and Hurricanes during 1970–1998, 248
Opinions, 249
Excerpts from the 1996 IPCC Report on Whether Global Warming Is Influencing Global Hurricane Activity, 251
Excerpt from "Tropical Cyclones and Global Climate Change: A Post-IPCC Assessment," by Henderson-Sellers et al. 1998, 252
Letters, 253

7 **Directory of Organizations, 259**
American Meteorological Society, 259
American Red Cross and the International Federation of Red Cross and Red Crescent Societies, 262
Cooperative Institute for Meteorological Satellite Studies at the University of Wisconsin-Madison, 265
Federal Emergency Management Agency, 266
Geophysical Fluid Dynamics Laboratory, 269
Hurricane Hunters, 270
Hurricane Research Division, 272
Joint Typhoon Warning Center, 275
National Aeronautics and Space Administration, 276
National Centers for Environmental Prediction, 278
National Climatic Data Center, 279
National Oceanographic and Atmospheric Administration, 280
National Weather Association, 281
National Weather Service, 283
Naval Research Laboratory, 284
NOAA Aircraft Operations Center, 286

Tropical Prediction Center, 290
The Weather Channel, 292
World Meteorological Organization, 293

8 **Print, Nonprint, and Internet Resources, 297**
Young Adult (Elementary) Books on Hurricanes, 297
Mariner Books, 298
General Books on Hurricanes, 299
Popular Weather Magazines, 302
Hurricane Preparation, 303
Societal Impact of Hurricanes, 303
Technical, 304
General Meteorology Books and Textbooks, 306
Encyclopedias and Glossaries on Meteorology, 307
State Books, 308
Individual Hurricanes, 309
Historical Hurricane Tracks, 317
National Hurricane Center Annual Summaries, 318
Joint Typhoon Warning Center Annual Tropical Cyclone Summary, 321
Annual Summary of Australian Cyclones, 322
Annual Summary of Cyclones in the North Indian Ocean, 322
Annual Summary of Tropical Cyclones for All Ocean Basins, 322
Natural Disaster Survey Reports, 322
NCDC *Storm Data* Publication, 323
How to Order a U.S. Government Technical Report, 324
Nonprint Resources, 324
  Hurricane Videos, 324
  Software Tracking, 327
  Emergency Preparedness, 327
  Hurricane and Weather Internet Sites, 328

Epilogue, 335
  Hurricane Katrina, 335

Glossary, 339
  Acronyms, 339
  Abbreviations for Cited Journals, 340
  Terms, 340

Conversion Tables, 350
  Length, 350
  Approximate Conversion of Latitude, Longitude, and Length (Use with Some Caution), 351
  More Precise Distance Calculations between Two Latitude and Two Longitude Points, 351
  Area, 352
  Volume, 352
  Time, 352
  Speed, 353
  Mass, 354
  Pressure, 354
  Converting °C to °F, 354
  Converting °K to °C, 355
  Converting One-Minute Maximum Sustained Winds Speed to Central Pressure in a Tropical Storm or Hurricane , 356
  Computing Wind Gusts from Sustained Wind Speed, 357
Tracking Hurricanes and Understanding National Hurricane Center Forecasts, 363
  Forecast Advisory, 364
  Public Advisory, 366
  Discussion, 368
  Strike Probability, 370
  Final Comments on National Hurricane Center Statements, 373
  Aircraft Reconnaissance Information, 373

References, 379

Index, 403

Preface

The year 2004 was an active Atlantic hurricane season with fifteen total storms—twelve affecting land somewhere—and perhaps a harbinger of the next decade for the U.S. coastline. Five hurricanes made landfall, the most hurricane landfalls since 1985. Two major hurricanes (Charley and Jeanne) hit Florida for the first time since 1950, with a third hurricane (Frances) making landfall in Florida and a fourth (Ivan) in Alabama just west of the Florida Panhandle. Charley is the second costliest U.S. hurricane ($15 billion) on record, while Ivan is the third costliest ($14.2 billion). Total U.S. damage by hurricanes and tropical storms in 2004 is estimated at near $45 billion—the costliest hurricane season on record for the United States. At press time for this book (August 2005), the 2005 Atlantic hurricane season looked prime for another busy season, with an early-season record of ten total storms thus far, two being major hurricanes and two making landfall in the western Florida panhandle again!

As discussed in this book, hurricane activity tends to follow twenty- to thirty-year cycles, so it is anticipated that future hurricane seasons could be just as costly or worse. Noted hurricane expert Dr. Bill Gray has warned for twenty years that Florida and other developed coastal communities, which have experienced decadesof relatively low major hurricane impact, could be in for a rude awakening. In a recent article, Gray writes:

> Florida's four destructive hurricanes fortunately came ashore along coastlines that were not very densely populated. Pensacola, FL was the largest Florida community feeling the direct brunt of one of these four

damaging hurricanes. The coastal and inland areas around Punta Gorda-Port Charlotte (where Charley came ashore), and Stuart (where Frances and Jeanne came ashore) do not have large coastal populations. The three major Florida coastal population concentrations from Tarpon Springs to Sarasota, West Palm Beach to South Miami, and Daytona Beach to Melbourne (and inland to Orlando) were all removed from the direct brunt of these four hurricanes. Economic loss many times greater could have occurred if the center of any one of these four hurricanes had come into one of these more concentrated Florida population areas. For instance, it has been estimated that if Hurricane Andrew (1992) had come inland just 15–20 miles north of its actual landfall near Homestead that it would have caused two to three times the $40 billion dollars (adjusted to 2004 dollars) in property loss that resulted. . . .

It is important that Floridians view this terribly damaging landfall season from a longer period perspective. Overall Florida has been extremely fortunate in recent years. Between 1966–2003 (38 years) the Florida Peninsula has experienced the landfall of only one major hurricane (Andrew, 1992). But in this long major hurricane lull period since the mid-1960s, Florida's population and coastal development has exploded. Few of the new Floridians have experienced a major hurricane hit. Most Floridians were not prepared for this unusual onslaught of four devastating storms in such a short period of time. But old-timers who lived in Florida in the 1930s through the 1950s well remember that Florida used to be hit by many hurricanes. Between 1928–1965 (41 years) the Florida Peninsula experienced 14 major (Cat. 3-4-5) hurricane landfalls (1 per 3 years).

For many years we have been discussing how lucky Florida had been with regards to its few recent landfalling major hurricanes. We said it was inevitable that this period of few major hurricane strikes would end and that the long period climatology would eventually reassert itself. There was no way, however, of knowing that the law of averages would try to catch up to its deficit so rapidly in one year! (Gray and Klotzbach 2004)

The possible return of major hurricane landfalls provides additional dilemmas besides their financial impact. Storms such as Ivan and Frances placed substantial strains on evacuation procedures, exposing serious flaws in communication and traffic operations. During Hurricane Ivan, Alabama residents complained of a lack of shelters and a delayed transition of converting both lanes of I-65 to northbound traffic. In Louisiana, evacuating New Orleans residents sat in a clogged interstate system; it took twelve hours or more to reach Baton Rouge, normally a one-hour trip!

This book, *Hurricanes: A Reference Handbook,* discusses such issues and is the follow-up edition to *Natural Disasters: Hurricanes,* published in 1999. As in the first edition, the purpose of this book is to provide background information on issues, people, organizations, statistics, and publications related to hurricanes and to provide guidance on where additional information can be obtained about a specific topic. Much of the book has been meticulously referenced so that the reader can explore these publications if he or she wishes. Most material is based on peer-review literature and has been updated.

The book follows the ABC-CLIO convention of background and history (chapter 1); problems, controversies, and solutions (chapter 2); worldwide perspective (chapter 3); chronology (chapter 4); biographical sketches (chapter 5); facts and data (chapter 6); organizations (chapter 7); and print/nonprint resources (chapter 8). However, since meteorology and hurricanes are ultimately a scientific topic with societal impacts, some liberties have also been used by the author. For example, "solutions" in the context of hurricanes are interpreted as weather forecasting and mitigation strategies, and the book is written in a readable but technical overtone in some parts.

Chapter 1 from the first book has been divided and expanded into chapters 1 and 2 in the second edition. These chapters focus mostly on the Atlantic Ocean and therefore contain a United States perspective. Chapter 1 describes hurricane formation, their life cycle, storm structure, and naming procedures. Chapter 2 is broader in scope, discussing their destructive behavior, forecasting procedures, societal issues, and annual variation in hurricane activity. The controversial topics of potential global warming issues and weather modification attempts are addressed. Chapter 2 then deviates to hurricane preparation, insurance, and mitigation procedures.

Chapter 3 is a new chapter discussing hurricanes from a global perspective. It describes international naming and classification procedures, global activity, worldwide impact, the World Meteorological Organization's World Weather Watch, and weather features unique to the western Pacific Ocean. A Saffir-Simpson Hurricane Scale, modified for the tropical Pacific Ocean, is also introduced.

Chapter 4 provides a chronology of forecast and scientific advances with regard to hurricanes. Chapter 4 also provides a descriptive timetable of significant U.S. land-falling hurricanes during the twentieth century as well as a listing of hurricanes that have changed history. Comments received by the author from the first edition indicate that this was one of the most popular chapters, providing a unique niche on hurricane history. As such, a considerable amount of effort went into this chapter to expand it and add new, interesting material on individual hurricanes and detailed references. For example, statistics on Hurricane Camille (1969) have been updated and myths exposed about the so-called hurricane party.

Chapter 5 contains biographies of important hurricane scientists and forecasters. Many are colorful people and should make for enjoyable reading. Chapter 6 contains tabular data, opinions on hurricane issues, and interesting letters from hurricane survivors. Chapter 7 describes relevant organizations involved in hurricane forecasting, research, and mitigation. Chapter 8 contains a comprehensive description of publications, electronic, videos, and websites related to hurricanes. Separate sections in the back of the book include acronyms, a glossary, conversion factors, a description of National Hurricane Center forecasts, the Beaufort scale, and directions on how to interpret reconnaissance plane measurements.

The book is not meant to be read from front to back. While chapters 1, 2, and 3 are loosely connected, the real intention is for each chapter to be a unique source of information. This book is a useful reference for students, the general public, emergency managers, public officials, journalists, and meteorologists. Anyone in a weather-sensitive field on the ocean will also find this book practical.

Page space limitations, the transient nature of referenced websites, and new weather information provided considerable challenges in the preparation and freshness of this book. The author has developed a website to tackle these dilemmas, located at http://www.drfitz.net, that will provide additional information not included in this book. Stale links will be updated periodically on this website as well. Finally, this website will serve as an authoritative source for information during hurricane season. Readers are also encouraged to email the author on questions regarding these powerful and fascinating storms.

*Dedicated to my beautiful daughters Megan and Katie.
No matter what the weather, you always bring sunshine.*

Acknowledgments

This book could not have been completed without the assistance and patience of many people. First of all, thanks to my co-workers at Mississippi State University's GeoResources Institute, and to my daughters, parents, and sister for their understanding during this time-consuming but rewarding endeavor. I also thank Lisa Bothman for her kindness and sweet nature during the writing process.

I am indebted to Dr. Chris Landsea for recommending me to ABC-CLIO as an author. Thanks to all the contributors and aides to the previous edition, and to the NOAA library in Silver Spring, Maryland, for new material acquired during my visit there for this edition.

I gratefully acknowledge Dr. Bill Gray for opening many opportunities involving the study of hurricanes while I was a graduate student at Colorado State University, and to Texas A&M University for developing my weather aptitude in the early years of college.

This publication was made possible through support provided by NASA through the University of Mississippi's Enterprise for Innovative Geospatial Solutions under the terms of Agreement No. NAG13–03012. NASA also supported this work with Grant NCC13–99001. The opinions expressed herein are those of the author and do not necessarily reflect the views of NASA, the University of Mississippi, or Mississippi State University.

1

Background on Hurricanes

The term *hurricane*—derived from the Caribbean Indian word *harakan,* meaning "evil spirit and big wind"—is a large rotating system of oceanic tropical origin with sustained surface winds of at least 74 mph somewhere in the storm. Due to the earth's rotation, these storms spin counterclockwise in the Northern Hemisphere and clockwise in the Southern Hemisphere. Both types of hemispheric spins are referred to as *cyclonic rotation,* because the sense of spin about a local vertical axis is the same as the earth's rotation when viewed from above. These storms occur worldwide and are called different names in different locations (see chapter 3 for details). In deference to U.S. readers, and for consistency, storms in this 74-mph or faster category will be called hurricanes in this book except in chapter 3.

Wind and rain, as experienced by most people worldwide, typically occur along the boundary of air masses with different temperature and moisture properties—a boundary known as a *front.* Along a front, warmer air is forced to rise over colder air, the ascending air condenses into clouds, and under favorable conditions the cloud droplets grow large enough so that gravity overcomes the ascending air and they fall as rain. These contrasting air masses also result in winds up to 25 mph. Many other situations can also cause inclement weather associated with local topographical conditions or wind patterns several miles or more above the ground. One unique aspect of hurricanes is that they form over the ocean in the tropics, typically removed from contrasting air masses. It is for this reason that hurricanes are somewhat rare, requiring a confluence of special conditions to form.

A hurricane does not form instantaneously but reaches this status in an incremental process. Initially, such a system begins as a *tropical disturbance* when a mass of organized, oceanic thunderstorms, not associated with a front, persists for 24 hours (NHC 2005). Sometimes partial rotation is observed, but this is not required for a system to be designated a tropical disturbance. The tropical disturbance becomes classified as a *tropical depression* when a closed circulation is first observed and sustained winds are less than 39 mph everywhere in the storm. When these sustained winds increase to 39 mph somewhere in the storm, it is then classified as a *tropical storm* and given a name.

It is important to note that these categories are defined by *sustained winds*, not instantaneous winds. Sustained winds are the average speed over a period of time at 33 feet above the ground. In the Atlantic, this averaging is performed over a 1-minute period (Holland 1993). The actual wind will be faster or slower than the sustained wind at any instantaneous period of time. For example, a hurricane with maximum sustained winds of 90 mph will actually contain gusts of 100 mph or more. Also, one should note that these categories are defined by *maximum winds* somewhere in the storm, almost always near the center, and that winds may be slower in other parts of the storm. For example, maximum sustained winds of 90 mph may only be concentrated in the northeast section near the hurricane center, with the southwest quadrant containing weaker winds.

Hurricane Formation and Life Cycle

Hurricane formation occurs in two distinct phases. The first phase is called the *genesis stage* and includes tropical disturbances and tropical depressions. The second phase includes the tropical storms and hurricanes and is called the *intensification stage*. These phases are separated because most disturbances and a few depressions never reach tropical storm intensity and dissipate. Eventually all storms will weaken in the *weakening stage*. Most dissipate in this stage. However, a decaying hurricane can evolve into a nontropical storm and possibly even reintensify in the *extratropical stage*.

Genesis Stage

Tropical disturbances form in regions where there is a net inflow of air at the surface, known as *convergence*. When convergence occurs at the surface, it must ascend to balance this accumulation of air. As air rises, it will saturate and form the base of a cloud. Once the air is saturated, ascent may be enhanced where the atmosphere is in a state of *static instability*. In a statically unstable atmosphere, saturated air forced upward by convergence is less dense than surrounding unsaturated air. As a result, it accelerates upward because the air is buoyant relative to its environment, forming towering puffy clouds. The concepts of cloud formation and atmospheric instability are beyond the scope of this book, and the reader is referred to other meteorology books on the subject (Ahrens 1994; Nese and Grenci 1998; Danielson, Levin, and Abrams 1998).

While convergence in a statically unstable atmosphere is a first criterion, thunderstorm formation is common in the tropics, and much of the warm, humid tropics are already in a state of static instability. Therefore, these two conditions, while necessary, are only a prerequisite. Several conditions must simultaneously exist for a tropical disturbance to develop complete rotation and become a tropical depression. First, the disturbance must be in a *trough*, defined as an elongated area of relatively low *atmospheric pressure*. Atmospheric pressure is the weight of a column of air on a given area of earth, typically 1 meter squared or 1 centimeter squared. The most frequent unit of pressure is the millibar, or mb, although it is also popularly measured as the height of a column of mercury supported by the atmosphere's weight. Troughs fall in four general categories: equatorial troughs, monsoon troughs, frontal troughs, and surface troughs.

For genesis to occur, troughs must contain a weak, partial cyclonic rotation. However, all troughs at least 5 degrees from the equator will obtain a partial cyclonic spin due to the *Coriolis force*. The Coriolis force results from an apparent twisting of the north-south-east-west coordinate system as the earth spins. If the earth was not rotating, air moving north, south, east, or west would continue in those directions unchanged (known as an inertial, or absolute, coordinate system). However, since the earth is rotating, air motion experiences an apparent deflection since the coordinate system is changing relative to the earth (known as a noninertial, or relative, coordinate system). This effect becomes increasingly influential away from the equator where the earth's curvature increases. To someone living on earth, it appears that the wind is deflected to the right in the Northern Hemisphere and to the left in the Southern Hemisphere. This is important, because without the Coriolis force, areas of low pressure such as a trough would "fill up" as air flows toward it in an attempt to compensate for air pressure differences. Instead, air is deflected around the trough by the Coriolis force, and a balance develops between the pressure differences and the Coriolis force, maintaining the trough's existence. The result is that troughs away from the equator will have a partial cyclonic spin due to the Coriolis force. This also explains why hurricanes do not form on or near the equator.

Troughs do occur near the equator. In some places around the earth, the warmest oceanic regions straddle the equator. As a result, air converges near the equator where persistent Southern Hemisphere and Northern Hemisphere winds (called trade winds) meet, often resulting in towering thunderstorms—a region known as an *equatorial trough* (also called the *InterTropical Convergence Zone,* [ITCZ]). However, since the Coriolis force is too weak, the equatorial trough does not have any cyclonic rotation. Therefore, genesis does not occur in the equatorial trough even though thunderstorm activity is rampant.

In situations where air temperature increases away from the equator, *monsoon troughs* occur. In monsoon troughs, Southern Hemisphere and Northern Hemisphere air converges at 10–20 degrees latitude, and the Coriolis force induces the partial cyclonic spin necessary for genesis. Some monsoon troughs occur where a large land mass exists poleward of an ocean basin; they exist because the land is warmer than the ocean during the warm months. Examples include the Indian monsoon, the Australian monsoon, and the eastern Pacific monsoon. However, they can also occur without any land influence where water is warmer away from the equator, such as in the western North Pacific Ocean. Likewise, monsoon troughs can form entirely over land masses without the nearby presence of water. The most notable example is North Africa's desert, where the Sahara results in considerably warmer temperatures than Central Africa during the summer.

The vast majority of genesis cases are associated with the monsoon trough. However, the dynamics can be complex and are still not well understood. Some disturbances undergo the transition to tropical depression directly inside monsoon troughs. However, others experience this transition as *tropical waves* (called *easterly waves* in the United States) that form when a monsoon trough "breaks down" into a cyclonic wavelike pattern in the wind field and travels westward away from the monsoon trough. This breakdown is associated with another kind of instability called *dynamic instability*, which occurs under certain flow configurations and/or temperature patterns. This instability results because a perturbation in the monsoon trough acquires *kinetic energy* (the energy of motion) from another source and grows with time. This source may be from the kinetic energy of the large-scale wind flow or from *potential energy* (stored energy) due to certain temperature patterns. Often, both sources may be available.

For example, the African monsoon is in the vicinity of a strong easterly wind. This easterly wind can accumulate too much kinetic energy, break down, and transfer its energy into wave formation; the waves then propagate downstream in the easterly current. As the waves move westward, they enter a region where potential energy is available due to the unique temperature pattern in North Africa. The desert air north of the wave is warmer than in Central Africa. The warm air ascends as it flows toward the western edge of the wave, while the cooler southern air sinks on the eastern edge. This lowers the wave's center of mass, releasing stored energy that increases the wave's rotation. In this example, the dynamic instability occurs by the transferring of kinetic energy from the easterly wind into wave growth and the conversion of potential energy (from the temperature field) into kinetic energy for wave growth. While one wave-growth mechanism may dominate, the processes are not always exclusive but may occur simultaneously. Likewise, one process may initiate the growth of a small perturbation in the monsoon trough, and the other may help it amplify afterward, as is the case over Africa.

About 60 percent of the U.S. hurricanes originate from tropical waves that break off from the African monsoon trough and propagate into the Atlantic! Tropical waves are fairly persistent features and can propagate long distances. Some eastern Pacific hurricanes can even be linked to easterly waves that originated in Africa! Dynamic instability can also be initiated in monsoon troughs when very strong thunderstorm activity creates the necessary potential energy conditions. Tropical waves are not unique to the African monsoon, though, and form in many other regions in the Atlantic such as the Caribbean Sea (Molinari et al. 1997) as well as in other ocean basins when dynamic instability is present.

The remaining genesis cases are associated with *frontal troughs* and *surface troughs* away from the equator. A frontal trough is the remnant of a front that has lost its contrasting temperature characteristic and entered the tropics. A surface trough encompasses all other kinds of troughs, such as those associated with a region of moisture contrast or broad thunderstorm complexes. Some surface troughs are even triggered by weather features 40,000 feet aloft.

In summary, *one prerequisite for genesis is a trough at least 5 degrees from the equator where the Coriolis force can induce a partial cyclonic rotation, and these troughs fall into three categories: a monsoon trough, frontal trough, and surface trough. Tropical (easterly) waves sometimes break off from monsoon troughs due to dynamic instability and are also a source of genesis.* The role of dynamic instability for genesis within troughs is still unclear but appears to play a major role in some cases. In the Atlantic about fifty-five to seventy-five tropical waves are observed annually, but only 10–25 percent of these develop into a tropical depression or beyond. Only a small percentage of frontal troughs and surface troughs also become tropical depressions or stronger. A trough 5 degrees or more from the equator is necessary for convergence, partial cyclonic rotation, and sometimes dynamic instability, but clearly genesis requires additional conditions.

The second condition required for genesis is a water temperature of at least 80°F. Heat transferred from the ocean to the air generates and sustains static instability (and therefore thunderstorms) in the disturbance. *The third genesis condition is weak vertical wind shear,* defined as the difference between wind speed and direction at 40,000 feet aloft and the surface. In other words, for genesis to occur the wind must be roughly the same speed and blowing from the same direction at all height levels in the atmosphere. This allows thunderstorms and the wind structure to grow unimpeded (more detail is presented in the intensification section).

The three conditions (warm water, surface trough, and weak vertical wind shear) are apparently necessary but insufficient conditions to develop complete rotation (and, by definition, a tropical depression). In the Atlantic, even when all these conditions appear favorable, sometimes genesis does not occur, and predicting when a tropical depression will form remains a vexing forecast problem.

There is much speculation on what the "missing link" (or links) might be. Studies have shown that subsidence, which suppresses cloud formation by: (1) drying the atmosphere (2) impeding ascending air; and (3) evaporating cloud droplets, stifles genesis (Gray 1968; DeMaria, Knaff, and Connell 2001). Indeed, a general rule of thumb used by forecasters is that genesis does not occur until surface pressure falls below 1010 mb in the Atlantic; pressure above this indicates subsidence. Dust storms accompany the African monsoon and propagate far into the Atlantic, and some forecasters speculate that their accompanying dust, dry air, and wind shear impede genesis (Dunion and Velden 2002; Dunion and Velden 2004). A few researchers theorize that some tropical waves need an additional "kick" by propagating into another dynamically unstable region (Molinari, Moore, and Idone 1997), another monsoon trough (Holland 1995), or a region of enhanced convergence (Ritchie and Holland 1999) for the additional growth necessary to reach tropical depression stage. In some occasions, wind surges from the Southern Hemisphere into a disturbance have been correlated to genesis, perhaps creating these favorable conditions (Love 1985; Zehr 1992; Tomas and Webster 1997).

One interesting observation is that genesis tends to be clustered with time in global basins (Gray 1979). Sometimes, genesis may not occur for several weeks, then suddenly several tropical storms and hurricanes form. Research shows that this clustering is partially due to a sinking/ascending undulation that travels eastward from India to North America and is called the *Madden-Julian Oscillation*. This oscillation, with a repeat time of thirty to sixty days, tends to favor genesis when the ascending branch occurs in the presence of a monsoon trough or tropical wave. The ascending branch is favorable for cloud growth in disturbances and increasing the number of disturbances, while the descending branch suppresses clouds. This phenomenon forms in the near-equatorial Indian Ocean and propagates eastward. Its influence is strongest in the Indian Ocean and western Pacific Ocean, becoming less significant as it propagates eastward; it also has less impact away from the equator. Nevertheless, research shows that it does impact genesis in the eastern Pacific (Maloney and Hartmann 2000a), the Atlantic (Maloney and Hartmann 2000b), and other ocean basins such as Australia (Hall, Matthews, and Karoly 2001). So, tropical genesis is more likely in the fifteen- to thirty-day favorable phase of the Madden-Julian Oscillation. However, most clustering in the Atlantic is likely due to the juxtaposition of favorable conditions, with the Madden-Julian Oscillation playing a minor role.

Once a tropical depression forms, the favorable conditions of wind shear, warm water, and complete cyclonic rotation provide the "ignition" process for further development. The ascending air in the depression stimulates low-level inflow toward the center. This inflow slowly increases the cyclonic circulation of the disturbance. The system starts to develop a warm-core temperature pattern aloft, which lowers surface pressure. As long as the depression remains over warm water in a low vertical wind shear environment and maintains thunderstorm activity near the center, the system will likely develop. Weak wind shear is a crucial factor, because it allows vertical orientation of the thunderstorms, preserves the structure of the system, and maintains the low-level inflow. Should the disturbance move into an environment where wind speed increases dramatically with height (or wind direction changes dramatically with height), the thunderstorms tilt downwind and rotate cyclonically to the left (Corbosiero and Molinari 2002; Black et al. 2002), and the warm core aloft is eroded (Knaff et al. 2004). As a result, strong wind shear disrupts the vertical structure of the system, and inflow weakens. Likewise, dry air intrusion, movement over colder water, or movement over land is detrimental. Should adverse conditions occur, the depression weakens and eventually dissipates.

Typically, the genesis time frame of both disturbance and depression lasts for several days or longer. However, under ideal conditions, a disturbance or depression can evolve much quicker. When the cyclonic sustained winds increase to 39 mph somewhere in the depression, the system is upgraded to a tropical storm. At this point, the intensification stage begins.

Intensification Stage and the Importance of Water Temperature

For tropical storm intensification into a hurricane, the same conditions that allowed its initial development (warm water, moist air, and weak wind shear) must continue. When favorable environmental factors persist, the rate of development increases compared to the genesis stage. This is because as the wind increases, more moisture is transferred from the ocean to the air, and when this moisture changes from a gas to a liquid stage during cloud formation, latent heat associated with this phase change is released into the vortex. Furthermore, because the system has complete rotation, a larger percentage of this latent heat is retained in the storm (unlike in a nonrotating thunderstorm, in which all the latent heat released by the clouds just propagates away). The column of air begins to warm, which lowers surface pressure. More air will flow toward the lower surface pressure, trying to redistribute the atmosphere's weight, resulting in faster winds. The faster cyclonic winds also enhance convergence. Both factors increase thunderstorm production and low-level inflow. A feedback mechanism now occurs in which faster cyclonic winds breed more potent thunderstorms, dropping central surface pressure more and creating stronger inflow, which breeds faster cyclonic winds, etc. Under favorable environmental conditions, a tropical storm can "spin up" rather quickly, with winds increasing as much as 50 mph or more in a day. When sustained winds reach 74 mph somewhere in the storm, it is classified as a hurricane.

Water temperature is unquestionably linked to the development of these storms. Hurricanes rarely form over water colder than 80°F (although exceptions do occur, for reasons discussed in chapter 3). They also weaken dramatically if a mature system moves over water colder than 80°F or if they make landfall, since their heat and moisture source has been removed. For a hurricane to maintain thunderstorms through static instability, warm, moist surface air is required near the low-pressure center. This warmth is provided by sensible heat transfer from warm ocean water, because otherwise air flowing toward lower pressure would expand and cool. (To convince yourself of this, let air out of a tire and feel how cool it is. When one lets air out of a high-pressure tire, the air expands as it enters the lower pressure environment and cools. This cooling occurs because the motion of gas molecules slows down as the air expands, and temperature is essentially a measurement of molecular motion). In other words, sensible heat flux from the warm water compensates for expansional cooling due to lower pressure, maintaining warm surface air near the storm center required to maintain the thunderstorms. This sensible heat flux at the surface also increases with wind speed, providing another positive feedback.

The warmer the water, the greater are the chances for genesis, the faster is the rate of development, and the stronger these storms can become. Under conditions of prolonged weak wind shear and water temperature greater than 85°F, sustained winds may reach almost 200 mph. Table 1.1 shows the potential intensity that a tropical storm or hurricane can achieve for a given water temperature (DeMaria and Kaplan 1994).

Weakening Stage and Dissipation

Fortunately, due to some inhibiting factor, few hurricanes reach their maximum potential. Conditions that stop intensification include wind shear, landfall, dry air intrusion, storm-induced ocean cooling by mixing or upwelling colder water beneath the warm surface water, and movement over colder water. Temporary occurrence of any (or a combination) of these influences will stall development or cause weakening.

TABLE 1.1

Potential intensity (as measured by sustained winds in mph) of a mature hurricane for a given water temperature (in °F).

Water temperature (°F)	Potential sustained winds for a mature tropical storm or hurricane once it is past the genesis stage (in mph)
70	101
71	104
72	107
73	110
74	114
75	118
76	122
77	127
78	132
79	138
80	145
81	152
82	160
83	169
84	178
85	189
86	201

This table is not valid for the genesis stage since hurricanes do not form over water colder than 80°F. However, it is valid for a mature hurricane moving over colder water. Adapted from DeMaria and Kaplan (1994).

When hurricanes significantly weaken over warm water, there are two possible culprits. The primary cause is usually vertical wind shear. Wind shear is the result of environmental wind direction changing with height, the environmental wind increasing by 20 mph or more with height, or a combination of both. Wind shear disrupts the vertical structure of the hurricane. Large hurricanes can withstand wind shear better than small hurricanes (Wong and Chan 2004), making prediction of the impact of wind shear on hurricanes sometimes difficult. The other weakening factor is sometimes dry air intrusion into the hurricane's inner core, which disrupts its inner-core thunderstorms.

Even when no inhibiting factors are evident over warm water, hurricanes that reach their potential intensity rarely maintain their intensity for any appreciable period. Apparently, the internal physics of a hurricane preclude a steady-state storm. Instead, strong wind conditions promote interior adjustments near the storm's center. These inner processes are discussed in the "Hurricane Structure" section later in this chapter.

Persistent occurrence of any or several inhibiting factors will cause disintegration of the hurricane. Of these possibilities, most dissipating cases occur due to landfall or movement over colder water. When a hurricane moves over cold water, expansional cooling dominates that stabilizes the atmosphere, disintegrates the thunderstorms, and weakens the hurricane. In addition, while warm water is significant, it is just as important that the warm water be at least 100 feet deep. The upper ocean layer under a hurricane can cool due to enhanced air-sea exchanges, mixing of the layer by wind forcing, and mixing by ocean currents (Shay et al. 1998). A thick layer of warm water is required to reduce or offset these effects. In general, mixing results in 1–3°F cooling of the ocean under a hurricane (Cione and Uhlhorn 2003). In addition, a sluggish process occurs in slow-moving hurricanes where ocean water is transported away from the storm's center; this is known as *upwelling* (Black 1983). Water from below is required to replenish the lost surface water. If the warm ocean layer is too thin, cold water from below the layer will be upwelled to the surface, cutting off the storm's warm-water energy supply and weakening the storm. In fact, sometimes a hurricane will kill itself when it becomes stationary for an extended period of time. Should a hurricane stop moving for several days, it can mix and upwell the ocean significantly, replacing all the warm water with cold water, and the hurricane will dissipate. An example is Hurricane Roxanne in 1995 when it became stationary in the Bay of Campeche.

When a hurricane moves over land, the weakening occurs even faster because not only has the surface heat flux been lost but also the moisture source for cloud formation. Since land has more friction than water, this also weakens land-falling hurricanes, but it's a minimal influence compared to the loss of sensible heat and moisture flux. As a result, hurricanes making landfall experience rapid decay. If the storm remains over land, its maximum sustained winds will decrease on average 45 mph per day, and the rate of dissipation is even faster for initially strong storms (Kaplan and DeMaria 1995; Emanuel 2000). Thirty-six hours after landfall, inland storms rarely contain winds above tropical depression strength.

Extratropical Transition

Often, public interest in a hurricane diminishes when a hurricane moves to higher latitudes or makes landfall, weakens, and begins to lose its tropical characteristics. However, a decaying storm can evolve into a fast-moving and occasionally rapidly developing *extratropical cyclone* that produces intense rainfall, very large waves, and even hurricane-force winds. Such a situation poses a significant risk to midlatitude locations during the summer and fall, demanding public attention.

Extratropical transition is a gradual process whereby a hurricane moves northward into a nontropical environment with significant temperature and moisture contrasts, wind shear, and colder water temperature. As discussed above, these situations are destructive to tropical systems, and about half completely decay. However, extratropical low-pressure systems (the ones with cold fronts and warm fronts that affect the U.S. year-round) thrive in such an environment, and the remaining 50 percent of weakening hurricanes experience an extratropical transition (Hart and Evans 2001). In particular, if the hurricane's remnants interact with an upper-level trough, a preexisting extratropical cyclone, or a region of large temperature contrasts, it may transform itself into an extratropical system.

During extratropical transformation, the inner core of the hurricane loses its symmetric cloud appearance with diminishing thunderstorm activity. The nearly axisymmetric wind and rain distributions evolve into broad asymmetric patterns that expand greatly in area (Jones et al. 2003). A cold front to the south and a warm front to the east form on the system, resembling a comma-shaped cloud pattern. The storm motion increases dramatically, with translation speeds greater than 45 mph, further contributing to the asymmetric structure of the storm, expanding the area of gale-force winds, and posing serious forecast difficulties. If the storm is over the ocean, the fast motion and expanded winds generate large waves and swell. The extratropical cyclone itself may gradually decay, but about half reintensify, with a few experiencing rapid development with winds up to hurricane strength (Hart and Evans 2001). In either situation (decay or reintensification), substantial rain can occur, resulting in severe flooding. During extratropical transition, precipitation expands poleward of the storm center, with rainfall maximums on the left of the storm track (Jones et al. 2003).

The Atlantic Hurricane Season

Hurricane season is limited to the warm seasons. In the Atlantic, the official hurricane season begins June 1 and ends November 30, although activity has been observed outside this time frame on rare occasions. However, hurricanes are most numerous and strongest in late summer and early fall. This is because the favorable conditions—deep warm water, moist air, weak wind shear, and cyclonic disturbances—are optimum in late summer. In particular, water's temperature peaks in late summer. This seems paradoxical since the longest day is in June. However, days are still longer than nights until fall; therefore, the water is still accumulating heat into late summer. The monsoon troughs are most active in late summer as well, and large-scale circulation patterns favor weak wind shear in late summer. Exceptions to this late-summer/early-fall peak in hurricane activity occur in certain parts of the world such as India (see chapter 3). In addition, while activity does peak in late summer, the western North Pacific hurricane season lasts all year.

Genesis locations also vary during the hurricane season (Neumann et al. 1999). For example, genesis patterns in the Atlantic and Gulf of Mexico can be categorized into three time periods: early season (June 1–July 15), midseason (July 16–September 20), and late season (September 21–November 30). Early-season storms mostly originate in the western Caribbean

Sea and the Gulf of Mexico. Midseason storms form in the main basin of the tropical Atlantic Ocean. Genesis still occurs in the Gulf of Mexico, but not in the majority of cases, and is virtually nonexistent in the Caribbean Sea. Genesis in the tropical Atlantic Ocean peaks in the midseason (during the so-called Cape Verde season) because water temperatures are warm enough to immediately impact tropical waves propagating off the African continent. The late season witnesses a more gradual change in which genesis in the main basin of the tropical Atlantic Ocean quickly declines and the Gulf of Mexico slowly declines; however, the Caribbean Sea experiences a revival of storm formation. The midsummer genesis lull in the Caribbean Sea is possibly due to local enhancement of trade winds mixing the ocean in that area (Inoue, Handoh, and Bigg 2002) or because conditions favorable for dynamic instability only occur in the early season and late season (Molinari et al. 1997).

Naming Hurricanes

When a tropical depression is upgraded to a tropical storm, it is assigned a name. Before this practice was started, tropical storms and hurricanes were identified in many confounding ways. Some legendary storms have been inconsistently named for the holiday they occurred on (Labor Day Hurricane of 1935), the nearest saint's day (Hurricane Santa Anna of 1825), the area of landfall (Galveston Hurricane of 1900), or even for a ship (Racer's Storm in 1837). However, most storms before the 1940s never received any kind of designation.

During the early 1940s, storm identification remained perplexing. At first, forecasters used cumbersome latitude-longitude identifications. This naming convention was too long and was confusing when multiple storms were present in the same ocean basin. Right before World War II, this procedure was changed to a letter designation (e.g., A-1943).

The naming of hurricanes was pioneered by the Australian forecaster Clement Wragge (Holland 1993), who used letters of the Greek alphabet, characters from Greek and Roman mythology, and, amusingly, the names of politicians with whom he was displeased. Wragge is apparently also the first person to give hurricanes female names. In 1941, George Stewart published the novel *Storm* in which an intense snowstorm on the U.S. West

Coast was given the female name "Maria" (Stewart 1941). As a result, Stewart is sometimes given credit for the notion of naming U.S. storms, although in this case it was a winter storm. However, the naming convention began in earnest during World War II when western North Pacific forecasters began informally naming tropical storms after their girlfriends and wives. This practice became entrenched in the system, and beginning in 1945 Northwest Pacific storms officially were given female names. Atlantic storms were officially given names starting in 1950 that were radio code words (e.g., Able, Baker, Charlie). Hurricane Easy and Hurricane King in 1950 are well-known Florida storms that had phonetic names. In 1953 the U.S. Weather Bureau switched to a female list of names, with male names added in 1979.

Today, six lists of names for Atlantic storms are used. Table 1.2 shows a list of Atlantic Ocean names for the 2005–2010 seasons. Each list is used alphabetically for a particular year. Then, should an Atlantic storm be very destructive or should it have a noteworthy impact on human lives or the economy, its name is retired (table 1.3) and replaced by a new name beginning with the same letter (LePore 1996). The letters "Q," "U," "X," "Y," and "Z" are not used by the National Hurricane Center (NHC), leaving twenty-one names on the list. It has never happened in the historical database, but should more than twenty-one Atlantic storms occur in a year, they would be given Greek letter names (Alpha, Beta, etc.). Also, NHC names *subtropical cyclones* even though they have different characteristics compared to tropical storms and hurricanes. More information on hurricane naming procedures from the international perspective and on subtropical cyclones is provided in chapter 3.

Hurricane Structure

Cloud Patterns

The structure of a hurricane is certainly one of the most fascinating, awesome, and bizarre features in meteorology. Distinct cloud patterns exist for each stage of a hurricane's life cycle. These patterns are so unique that a meteorologist can estimate the intensity of a depression, tropical storm, or hurricane solely based on cloud organization and cloud height using a methodology known as the *Dvorak technique* (Dvorak 1975). Dvorak identified

TABLE 1.2
Tropical storm and hurricane names for the Atlantic Ocean from 2005 to 2010.

2005	2006	2007	2008	2009	2010
Arlene	Alberto	Andrea	Arthur	Ana	Alex
Bret	Beryl	Barry	Bertha	Bill	Bonnie
Cindy	Chris	Chantal	Cristobal	Claudette	Colin
Dennis	Debby	Dean	Dolly	Danny	Danielle
Emily	Ernesto	Erin	Edouard	Erika	Earl
Franklin	Florence	Felix	Fay	Fred	Fiona
Gert	Gordon	Gabrielle	Gustav	Grace	Gaston
Harvey	Helene	Humberto	Hanna	Henri	Hermine
Irene	Isaac	Ingrid	Ike	Ida	Igor
Jose	Joyce	Jerry	Josephine	Joaquin	Julia
Katrina	Kirk	Karen	Kyle	Kate	Karl
Lee	Leslie	Lorenzo	Laura	Larry	Lisa
Maria	Michael	Melissa	Marco	Mindy	Matthew
Nate	Nadine	Noel	Nana	Nicholas	Nicole
Ophelia	Oscar	Olga	Omar	Odette	Otto
Philippe	Patty	Pablo	Paloma	Peter	Paula
Rita	Rafael	Rebekah	Rene	Rose	Richard
Stan	Sandy	Sebastien	Sally	Sam	Shary
Tammy	Tony	Tanya	Teddy	Teresa	Tomas
Vince	Valerie	Van	Vicky	Victor	Virginie
Wilma	William	Wendy	Wilfred	Wanda	Walter

Unless a hurricane name is retired and replaced by a new name, the list is recycled every six years (i.e., the names used in 2005 will be used again in 2011, the names used in 2006 will be used again in 2012, etc.). From NHC (2005).

five patterns related to a storm's development stage and environmental influences. By picking which pattern conforms to the situation, combined with a set of rules, one uses a flowchart to estimate a storm's intensity. Different flowcharts exist for visible, infrared, and enhanced infrared satellite imagery. These five patterns are the curved band pattern, the shear pattern, the eye pattern, the central dense overcast pattern, and the embedded center pattern.

One of the most common types of cloud evolution is the curved band pattern, which occurs when wind shear is weak. During genesis, typically a mass of thunderstorms with a weak rotation is first observed, known as stage 1 of genesis. Usually these thunderstorms will temporarily dissipate or weaken, leaving a residual circulation, although in the case of a tropical wave sometimes a cloud pattern resembling an upside-down "V" is observed. Often, no further development occurs, and the distur-

TABLE 1.3
Retired Atlantic Hurricane names and the year of their occurrence in alphabetical order.

A	Audrey (1957)	Agnes (1972)	Anita (1977)	Allen (1980)	Alicia (1983)
	Andrew (1992)	Allison (2001)			
B	Betsy (1965)	Beulah (1967)	Bob (1991)		
C	Connie (1955)	Carla (1961)	Cleo (1964)	Carol (1965)	Camille (1969)
	Celia (1970)	Carmen (1974)	Cesar (1996)	Charley (2004)	
D	Diane (1955)	Donna (1960)	Dora (1964)	David (1979)	Diana (1990)
E	Edna (1968)	Eloise (1975)	Elena (1985)		
F	Flora (1963)	Fifi (1974)	Frederic (1979)	Fran (1996)	Floyd (1999)
	Fabian (2003)	Frances (2004)			
G	Gracie (1959)	Gloria (1985)	Gilbert (1988)	Georges (1998)	
H	Hazel (1954)	Hattie (1961)	Hilda (1964)	Hugo (1989)	Hortense (1996)
I	Ione (1955)	Inez (1966)	Iris (2001)	Isidore (2002)	Isabel (2003)
	Ivan (2004)				
J	Janet (1955)	Joan (1988)	Juan (2003)	Jeanne (2004)	
K	Klaus (1990)	Keith (2000)			
L	Luis (1995)	Lenny (1999)	Lili (2002)		
M	Marilyn (1995)	Mitch (1998)	Michelle (2001)		
O	Opal (1995)				
R	Roxanne (1995)				

Before retiring practices were established, some names were simply not used anymore. For example, in 1966 "Fern" was substituted for "Frieda" and no reason was given. Also, Carol was used in 1965, but was probably retired retrospectively for the damage a 1954 storm of the same name caused. From NHC (2005).

bance is merely a maritime nuisance. Should strong thunderstorms return (sometimes within twelve hours, often taking several days) and a complete circulation forms, stage 2 of genesis, which coincides with the depression stage, commences (Zehr 1992). A dominant band of clouds gradually takes on more curvature around a cloud-minimum center. When the band curves at least one-half distance around the storm center, typically tropical storm intensity has been achieved.

This curved band consists of a 100-mile-wide region of clouds with embedded potent thunderstorm bands reaching 50,000 feet in height; the cloud shield extends another 400 miles farther out with less potent thunderstorm squall lines. These thunderstorm bands, known as *spiral bands,* are 5–30 miles wide and 50–200 miles long. In between the bands is light to moderate rain or areas of sinking air (downdrafts). Spiral band evolution and motion can be explained by a special class of wave solutions called *vortex Rossby waves* (Montgomery and Kallenbach 1997; Chen, Brunet, and Yau 2003), discussed in the chapter 5 biography on Carl-Gustav Rossby. In the periphery of a tropical storm, wind speed will fluctuate, with fastest sustained winds of 30–40 mph in the spiral bands. As one approaches the center of a tropical storm, winds will consistently increase, with the strongest winds close to the center.

As the tropical storm strengthens, the dominant cloud band continues coiling around the storm center. When the band completely coils around the center, hurricane intensity usually has been reached. At this point, a clear region devoid of clouds forms in the center known as the *eye,* surrounded by a ring of thunderstorms known as the *eyewall.*

The Eyewall and Eye

The eyewall contains the fiercest winds and often the heaviest rainfall, making this feature the most dangerous part of a hurricane. The eyewall slants outward with height, giving the eye a "coliseum" appearance, as if one is in a giant football stadium made of clouds. In the eye, winds become weak, even calm! This transition from hurricane force winds to calm is rather sudden (often within minutes) and is truly a bizarre weather feature. The average eye-size diameter is between 20 and 40 miles. Typically an eye starts at about 35 miles wide during the transition from tropical storm to hurricane. As the hurricane intensifies, usually the eye contracts. Small eyes are correlated to very intense hurricanes, with diameters as little as 9 miles. However, intense hurricanes with large eyes also occur.

The cause of eye formation is still not understood, but there is some consensus among meteorologists that near 74 mph the strong rotation impedes inflow to the center, causing air to instead ascend 10–20 miles from the center. Consider what happens as one drives at fast speeds in a vehicle. As the driver makes a sharp turn, an "invisible" force makes the driver lean outward. This outward-directed force, called the *centrifugal force,* occurs because the driver's momentum wants to remain in a straight line, and since the car is turning, there is a tugging sensation outward. The sharper the curvature or the faster the rotation, the stronger is the centrifugal force.

As rotation increases in a developing hurricane, air is subjected to these outward accelerations that counteract inflowing air. The centrifugal force eventually dominates near hurricane strength, causing inflowing air to not reach the center and instead ascend about 15 miles from the center (Zhang, Liu, and Yau 2001). This strong rotation also creates a vacuum of air at the center, causing some of the air flowing out the top of the eyewall to turn inward and gently sink to balance this loss of mass. This subsidence suppresses cloud formation, creating a pocket of generally clear air in the center (although low-level short clouds or high-level overcast skies may still exist). People experiencing an eye passage at night often see stars. Trapped birds are sometimes seen circling in the eye, and ships trapped in a hurricane eye report hundreds of exhausted birds resting on their decks. The landfall of Hurricane Gloria (1995) on southern New England was accompanied by thousands of birds in the eye.

The sudden change of violent winds to a calm state is a dangerous situation for people ignorant about a hurricane's structure. Those experiencing the calm of an eye may think that the hurricane has passed, when in fact the storm is only half over with dangerous eyewall winds returning from the opposite direction typically within 20 minutes or less, although sometimes the pause is longer. Marjory Stoneman Douglas, the famous Florida Everglades activist, writes in the well-known book *The Everglades: River of Grass* of the eye passage experienced by survivors in the 1926 Miami hurricane:

> Late that night, in absolute darkness, it hit, with the far shrieking scream, the queer rumbling of a vast and irresistible freight train. The wind instruments blew away at a hundred twenty-five miles [per hour]. The leaves went, branches, the bark off the trees. In the slashing assault people found their roofs had blown off, unheard in the tumult. The water of the bay was lifted and blown inland, in streaming sheets of salt, with boats . . . , coconuts, debris of all sorts, up on the highest ridge of the mainland. . . .
>
> At eight o'clock next morning the gray light lifted. The roaring stopped. There was no wind. Blue sky stood overhead. People opened their doors and ran, still a little dazed, into the ruined streets. . . . Only a few remembered or had ever heard that in the center of a spinning hurricane there is that bright deathly stillness.

It passed. The light darkened. The high shrieking came from the other direction as the opposite whirling thickness of the cyclonic cone moved on over the darkened city. (Douglas 1997, 339–341)

In fact, many killed in the Labor Day Hurricane had wandered outside during the passage of the eye. In addition to the return of fierce winds, sometimes the highest storm surge (to be discussed in chapter 2) occurs on the back side of the eye.

The first appearance of an eye on a satellite picture is also an indicator that rapid intensification could occur. Because air sinks in the eye, it compresses and warms, enhancing the warm core of air aloft. This upper-air warmth reduces the weight of the air column and causes surface pressure to drop in the eye dramatically, producing large pressure differences between the storm and its environment. In response to this pressure gradient, winds increase substantially. The positive-feedback process between winds, surface fluxes of heat and moisture, convergence, thunderstorms, and eye warming accelerates. Satellite imagery shows that as a hurricane intensifies, the eye becomes warmer and the eyewall clouds grow taller. If conditions remain favorable (weak wind shear and warm water temperature), the positive relationship among all these processes can lead to rapid intensification where maximum sustained winds increase by 40–50 mph in one day (Fitzpatrick 1997).

Sometimes a second eyewall forms outside the original eyewall about 40–60 miles from the center. This outer eyewall "cuts off" the inflow to the inner eyewall, causing the inner one to weaken and dissipate. Because the eye is wider, temporary weakening occurs. The outer eyewall will begin to contract inward to replace the inner eyewall, and about twelve to twenty-four hours later intensification resumes. This internal adjustment process, known as the *concentric eyewall cycle* (Willoughby, Clos, and Shoeibah 1982), is one reason strong hurricanes experience intensity fluctuations even in otherwise favorable conditions. On rare occasions, the second eyewall forms far from the center, resulting in a gigantic eye with a diameter greater than 150 miles (Lander 1999).

Another factor that causes minor intensity fluctuations is diurnal temperature swings (Browner, Woodley, and Griffith 1977). Apparently, daytime heating and nighttime cooling affects a storm cloud's top temperature, enhancing or reducing static stability. The result is considerable variability in cloud patterns. In addition, air outside tropical thunderstorms sinks more at night, possibly enhancing convergence into the hurricane (Gray and Jacobson 1977). As a consequence, thunderstorms tend to "pulsate," being strongest at sunrise and weakest at sunset (all other environmental factors being equal), causing intensity bursts at these periods. Diurnal oscillations are also obvious in the formative stages of hurricanes.

Once a hurricane reaches a mature state and exists in a favorable environment of warm water temperature, weak wind shear, and moist air, it obtains the classic appearance of a nearly circular eye, surrounded by a nearly uniform ring of thunderstorms in the eyewall region with little deep thunderstorm activity outside this ring. This type of symmetric storm is called an *annular hurricane* and tends to experience little intensity change (unless a concentric eyewall cycle occurs) until the environment changes or it makes landfall (Knaff, Kossin, and DeMaria 2003).

Hurricane Size

Outside the eyewall region, the weaker spiral bands accompanying the hurricane typically affect a large area. On average, the width of a hurricane's cloud shield is about 500 miles, but it may vary tremendously. For example, the cloud shields of Hurricane Gilbert (1988) and Hurricane Allen (1980) covered an impressive one-third of the Gulf of Mexico. Some are giants, such as Supertyphoon Tip (1979) that was twice the size of Gilbert or Allen!

However, cloud size is not an accurate indication of hurricane size. Hurricane size is typically categorized by the radial extent of gale-force winds (32–54 mph). Mariners and the navy typically avoid winds stronger than gale-force. Furthermore, many hurricane preparedness exercises are required to be completed before gale-force winds begin, and often bridges are closed when gale-force winds ensue. It should be noted that other more sophisticated means exist to quantify hurricane size (Liu and Chan 1999, 2002; Carr and Elsberry 1997), but gale-force winds are the most relevant. Therefore, using winds of 35 mph as the gale-force criteria, the average Atlantic hurricane size has a radius of 150 miles, or equivalently a diameter of 300 miles (Liu and Chan 1999).

There are seasonal, latitudinal, and life cycle variations of hurricane size in the Atlantic (Merrill 1984). Small hurricanes are more common early in the season, and large ones are more common late in the season. Hurricanes tend to also be larger farther north. As a tropical depression intensifies, the size tends to contract, while during the mature and weakening stages, the radius of gale-force winds tends to expand. As will be discussed in chapter 2, hurricane size also plays a small role in its own movement. It is sufficient to state here that larger hurricanes influence their own motion more than smaller hurricanes (Carr and Elsberry 1997).

However, storm size does not correlate with storm intensity. Both small and large hurricanes can have sustained eyewall winds in excess of 100 mph and even 150 mph. Examples of small category 5 hurricanes include Camille (1969) and Andrew (1992). Likewise, gale-force winds in tropical storms or category 1 hurricanes can extend over a large region.

Very small hurricanes are called *midgets*. Hurricane-force winds are typically confined to a region 25 miles wide, and gale-force winds may only extend 40 miles from the center. However, again, there is no correlation between size and intensity. While many midgets are tropical storms or minimal hurricanes, the disastrous Labor Day Hurricane of 1935 was a midget, with hurricane-force winds confined within 17 miles of the storm center. Likewise, Tropical Cyclone Tracy that hit Australia in 1974 was extremely devastating but very small. Gale-force winds only extended 25 miles from Tracy's center, with hurricane-force winds extending 13 miles. Midget hurricanes are very rare in the Atlantic, but a little more frequent elsewhere in the world.

Lightning (Or the Lack Of)

One would assume that the most dangerous storm on earth contains abundant lightning. However, lightning is surprisingly infrequent in hurricanes within 60 miles of the storm center (Molinari, Moore, and Idone 1999). Lightning is a discharge of electricity and requires two regions containing opposite charges. To understand why hurricanes lack electrical activity, a discussion on lightning processes is required first.

Charge separation can only occur in strong thunderstorm updrafts—in other words, where abundant buoyancy is available—so that air can accelerate upward at speeds of 15 mph or more. Strong updrafts transport cloud water droplets aloft into regions where the temperature is below freezing; some droplets become ice, and the rest remain in a liquid state (known as *supercooled water*). The stronger the updraft, the more plentiful are the supercooled droplets. When supercooled droplets are abundant, ice crystals collide with the supercooled droplets, freeze on contact, and stick together, a process called *riming*. The icy matter (rime) that forms is called *graupel*. In this situation, a mixture of three substances now exists—graupel, smaller ice crystals, and supercooled water droplets.

Current theory suggests that charge separation involves collisions between the graupel and small ice particles in the presence of supercooled droplets (Illingworth 1985; Saunders 1993). The key is the existence of supercooled droplets; without them, no charge separation occurs, and there is no lightning. Laboratory experiments show that, in the presence of supercooled water, graupel colliding with ice crystals acquire a positive charge when temperatures are colder than $-4°F$; for temperatures between $-4°F$ and $32°F$, charging is reversed, with graupel acquiring a negative charge. Some studies show that this charge reversal may be related to relative humidity distribution in clouds (Berdeklis and List 2001), but this process requires more research. Nevertheless, cloud electrification studies show that typically thunderstorms have a positive charge at the top and a negative charge at the bottom. Because of this charge difference, *intracloud lightning*, which is the most common type of lightning and looks like a diffuse brightening that flickers, can occur. *Intercloud lightning* is also possible when two clouds exist next to each other with a charge difference. Finally, because the earth's surface becomes positively charged beneath a thunderstorm from a process known as *inducement*, cloud-to-ground lightning also occurs.

The reason hurricanes contain little lightning activity is relatively weak updrafts. A hurricane contains a warm-core center aloft, and furthermore the ocean often lacks sustained surface heating. As a result, less buoyancy is available for the inner-core thunderstorms. Because the updrafts are weak, fewer water droplets are transported above the freezing level. Hurricanes also experience much horizontal mixing in the rapidly rotating core, splintering ice crystals in the below $32°F$ layer, effectively creating an abundance of "baby" ice crystals. What little supercooled water exists in the subfreezing region is quickly diffused onto these ice splinters. Consequently, hurricanes lack supercooled water, and without this key agent, charge separation cannot occur (Black and Hallet 1986; Cecil and Zipser 2002). Therefore, most hurricanes lack lightning activity.

A dozen or less cloud-to-ground strikes per hour occur within the storm's eyewall. Lightning is more common in the outer core of hurricanes, with flash rates on the order of hundreds per hour where the thunderstorms are farther removed from the warm core and cloud buoyancy is less impeded, allowing stronger updrafts. In contrast, a typical (land-based) thunderstorm may have lightning flash rates greater than 1,000 per hour.

Exceptions are observed where a hurricane or tropical storm's inner core experiences a burst of electrical activity, and these have been correlated to storm intensification. Examples include Hurricanes Diana (1984), Elena (1985), Florence (1988), Bob (1991), and Andrew (1992) (Molinari, Moore, and Idone 1999). Since lightning is dependent on updraft strength, this suggests that unusually vigorous eyewall thunderstorms are associated with sudden storm intensification. It also shows that the National Lightning Detection Network, consisting of 100 magnetic direction finders across the U.S. that sense lightning strike location, may provide forecasters a tool for anticipating rapid intensification in hurricanes near the coast.

2

Problems, Controversies, and Solutions

Hurricane Destruction

Coastal communities devastated by strong hurricanes usually take years to recover. Many forces of nature contribute to the destruction. Obviously, hurricane winds are a source of structural damage. As winds increase, pressure against objects increases at a disproportionate rate, roughly the square of the wind speed. For example, a 50-mph wind causes a pressure of 5.5 pounds per square foot. In 100-mph winds, that pressure becomes 30 pounds per square foot. When the wind exceeds design specifications, structural failure occurs. Debris is also propelled by strong winds, compounding the damage. Other concerns include downed trees and power poles, causing power outages sometimes for weeks.

Isolated pockets of enhanced winds also occur in hurricanes. Accompanying the steady winds will be several seconds of wind gusts that can amplify or initiate destruction. More powerful wind entities also occur in isolated regions. As hurricanes make landfall, interactions with the thunderstorms form columns of rapidly rotating air in contact with the ground, known as *tornadoes*. Also accompanying the thunderstorms are areas where heavy rainfall accelerates air to the ground, known as *downbursts,* and spreads out at speeds greater than 100 mph. In addition, another phenomena called *mesoscale vortices*, or sometimes *mesovortices*, was documented in Hurricane Hugo (1989) and Hurricane Andrew (1992)

(Willoughby and Black 1996; Fujita 1993). These are whirling vortices that form at the boundary of the eyewall and eye where there is a tremendous change in wind speed. The updrafts in the eyewall stretch the vortices vertically, making them spin faster with winds up to 200 mph. Damage by these three wind phenomena occurs in narrow swaths inland, although sometimes it is difficult to discern during the poststorm analysis which wind event caused a particular swath's destruction.

Flooding produced by the rainfall can also be quite destructive and is currently the leading cause of hurricane-related fatalities in the United States. A majority (57 percent) of the 600 U.S. deaths between 1970 and 1999 due to hurricanes or their remnants was associated with inland flooding (Rappaport 2000). Most of these flood-related deaths occur at night to people in cars. Typically, they result from flash floods in which rivers and streams run out of their banks with strong currents, and these currents are particularly dangerous when they cross a roadway. Moving water just 2 inches deep can cause a driver to lose control of the car and crash, and 1 foot of water will carry most cars off the road and drown the occupants. Other drivers drown when they drive into a flooded road, not realizing the water is several feet deep, or into a canal or stream because the road is obscured with water. Nonvehicular drownings occur when people get caught in currents; just 6 inches of fast-moving water will sweep people off their feet. Poor vision at night, especially when power is out, enhances these problems.

Hurricanes average 6–12 inches of rain at landfall regions, but this amount varies tremendously. Hurricane Floyd (1999) dumped 15–20 inches of rain over portions of eastern North Carolina and Virginia; 12–14 inches over portions of Maryland, Delaware, and New Jersey; 4–7 inches over eastern Pennsylvania and southeastern New York; and up to 11 inches over portions of New England. Since the soil was already saturated from Hurricane Dennis (1999) and other recent rain events, an inland flooding disaster ensued that drowned fifty people.

Flooding potential is not proportional to intensity. In fact, weaker tropical storms often produce greater amounts of rain, such as a U.S.-record 42 inches of rain in twenty-four hours by Tropical Storm Claudette (1979) near Houston, Texas, causing $400 million in damage. Rainfall accumulation generally increases for slow-moving storms. Hurricane Danny (1997) sat over

Mobile Bay for almost one day before moving inland, dumping at least 37 inches of rain in coastal Alabama in thirty-six hours, of which 26 inches fell in seven hours! More recently, Tropical Storm Allison's (2001) heavy rains of 20–35 inches produced catastrophic flooding over portions of the upper Texas coastal area (especially Houston) and significant flooding along the remainder of its track through Louisiana, Mississippi, and Alabama. Forty-one deaths are attributed to Allison's heavy rain, flooding, tornadoes, and high surf.

Heavy rainfall is not just confined to the coast. The remnants of hurricanes can bring heavy rain far inland, particularly dangerous in hills and mountains where acute concentrations of rain turn tranquil streams into raging rivers in a matter of minutes. In addition, mountains "lift" air in hurricanes, increasing cloud formation and rainfall. Rainfall rates of 1–2 feet per day are not uncommon in mountainous regions when hurricanes pass through. In fact, the highest hurricane rainfall amounts have occurred in the mountains of La Reunion Island (see chapter 4 for details).

Some examples regarding rainfall damage and fatalities follow. Hurricane Camille (1969), which made landfall in Mississippi, dumped 30 inches of rain in six hours in the Blue Ridge Mountains, triggering flash floods and mud slides that killed 114 people in Virginia and 2 in West Virginia. One of the most widespread floods in U.S. history was caused by Hurricane Agnes (1972), resulting in 188 deaths and $2.1 billion in property damage along most of the Eastern United States. Mud slides are also often a problem in hilly terrain, burying homes and destroying property particularly in underdeveloped mountainous countries where the warning system, hurricane preparedness, and infrastructure are insufficient. Many of these countries have high terrain near the coast where mud slides and floods can occur with tragic results. Eastern Hemisphere countries sustain the worse fatalities, since more storms occur there. A few of many examples include the Philippines, where Typhoon Kelly (1981) killed 140 and Tropical Storm Thelma (1991) killed 3,000; China, where Typhoon Peggy (1986) killed 170, Typhoon Herb (1996) killed 779, and Typhoon Nina (1975) killed at least 10,000; and Korea, where Typhoon Thelma (1987) resulted in 368 killed or missing. However, Western Hemisphere countries also suffer sizable casualties due to flooding and mud slides, such as in the Caribbean nations, where Hurricane Fifi (1974) killed 8,000–10,000; Hurricane Flora (1963) killed 8,000, and Hurricane Gordon (1994) killed 1,145; Mexico, where Hurricane Gilbert (1988) killed 202 and Hurricane Pauline (1997) killed 230; and Central America, where Hurricane Mitch (1998) killed 11,000–18,000.

When it comes to flooding fatalities, Hurricane Mitch (1998) is the most deadly hurricane to strike the Western Hemisphere in the last 200 years. Mitch produced an estimated 75 inches of rainfall in the mountainous regions of Central America, resulting in floods and mud slides that destroyed the entire infrastructure of Honduras and devastated parts of Nicaragua, Guatemala, Belize, and El Salvador. Whole villages and their inhabitants were swept away in the torrents of floodwater and deep mud that came rushing down the mountainsides, destroying hundreds of thousands of homes. The president of Honduras, Carlos Flores Facusse, claimed that the storm destroyed fifty years of progress. Typically in all the examples listed above, the number of homeless is in the 10,000–100,000+ range.

Although all these elements (wind, rain, floods, and mud slides) are obviously dangerous, historically most people have been killed in the *storm surge*, defined as an abnormal rise of the sea along the shore. The storm surge, which can reach heights of 20 feet or more, is caused by the winds pushing water toward the coast. As the transported water reaches shallow coastlines, bottom friction slows the motion, causing water to pile up. Ocean waters begin to rise gradually, then quite quickly as the storm makes landfall (storm surge does *not* occur as a tidal wave, as depicted in at least one Hollywood movie).

Some factors that determine a storm surge's height include storm intensity, storm size, storm speed, and the angle at which the hurricane makes landfall. The storm surge increases with storm intensity and size. Slow-moving storms allow more time for water to pile up on the coast, whereas fast-moving storms cause the surge to spike over a few hours with an overall lower surge. The storm surge is greater when landfall is perpendicular to the coastline, since some of the surge will be deflected offshore when storms land at a sideways angle. The shape of the coastal estuary is another important component, since coastal points and channels tend to enhance the surge in certain regions. One more element is the proximity of shallow water near the coast. Low-lying regions adjacent to shallow seas (such as the Gulf of Mexico in the southern United States and the Bay of Bengal bordering

Bangladesh and India) are particularly vulnerable to the storm surge, since this allows more water to pile up before inundating the coast. Another minor contribution to the storm surge is the low pressure of a hurricane, which allows water to expand (known as the *inverse barometer effect*). For every 10-mb pressure drop, water expands 3.9 inches. The storm surge is always highest on the side of the eye corresponding to onshore winds, which is usually the right side of the point of landfall in the Northern Hemisphere, called the *right front quadrant.* Winds are also fastest in the right front quadrant because storm motion (which averages about 10 mph but varies substantially) is added to the hurricane's winds. Because winds spiral inward, the storm surge is greatest along the eyewall.

The total elevated water includes three additional components—the astronomical tide, the steric effect, and ocean waves. The astronomical tide results from gravitational interactions between the earth and the moon and sun, generally producing two high and two low oceanic tides per day in most locations. Should the storm surge coincide with the high astronomical tide, additional feet will be added to the water level; the highest oceanic tides occur when the earth, sun, and moon are aligned (known as *syzygy*). The total water elevation due to the storm surge, astronomical tides, and wave setup (see next paragraph) is known as the *storm tide.* Therefore, storm surge is officially defined as the difference between the actual water level under the hurricane's influence and the level due to astronomical tides and wave setup. In practice, water-level observations during posthurricane surveys are always storm tides, and it is difficult to distinguish between storm surge and storm tide elevations. Therefore, the two terms are used interchangeably.

Waves are another important contributor. Under normal conditions, waves that reach the coast break, and the water sinks and flows back out to the sea under the next incoming wave. However, in hurricane conditions, excessive incoming water is not balanced by outgoing water. As a result, water levels increase due to *wave setup.* This process is most pronounced when deep water is near the shore, because in shallow water waves break far offshore. The final contributor is water temperature. Because warm water expands, water levels are naturally highest in the summer. This process, known as the *steric effect,* is always present during the hot hurricane season.

Water is very powerful, weighing some 1,700 pounds per cubic yard, and therefore most inundated structures pounded by waves and the storm surge will be demolished. Ocean currents set up by the surge, combined with the waves, can severely erode beaches, islands, and highways. Most people caught in a storm surge will be killed by injuries sustained during structural collapse or by drowning. Death tolls for unevacuated coastal regions can be terrible. The worst natural disaster in U.S. history occurred in 1900 when a hurricane-related 8–15-foot storm tide inundated the island city of Galveston, Texas, and claimed more than 6,000 lives. In 1893, nearly 2,000 were killed in Louisiana and 1,000 in South Carolina by two separate hurricanes.

Hurricane Camille (1969), with sustained winds of at least 180 mph, produced a storm tide of 23 feet in Pass Christian, Mississippi. As the storm tide penetrated far inland, Camille killed 172 people in Mississippi and 9 in Louisiana. In Louisiana, Camille produced a storm tide that pushed water over both levees near the mouth of the Mississippi River, "removing almost all traces of civilization," as one U.S. Department of Commerce (1969) report states (ESSA 1969). Floods from the storm surge penetrated 8 miles inland in the Waveland-Bay St. Louis region, and in river estuaries the flood extended 20–30 miles upstream. When combined with the flood deaths in Virginia (114) and West Virginia (2), Camille killed a total of 297 people. A total of $1.5 billion in property damage was incurred, with total devastation on the immediate coastline and severe damage further inland.

However, these statistics pale in comparison to the lives taken in coastal India and surrounding countries by storm tides. The most vulnerable area is coastal Bangladesh, a huge river delta fertile enough to support large numbers of farmers and fishermen (about 1,500 per square mile). Many are essentially nomads, staking claims on frequently changing temporary islands. The geography of this area favors large storm surges, because the narrow inlet of the bay funnels large amounts of water inland and because shallow water extends 60 miles offshore. Storm surge warnings are issued more than a day in advance, but communicating this information is difficult due to the rural and nomadic nature of the population. Furthermore, many choose to ignore the warnings or find it too difficult to evacuate since no transportation infrastructure exists (Rosenfeld 1997). The results have been the worst storm surge fatalities in history.

Six hurricanes have hit Bangladesh since 1960, killing at least 10,000 each time (Rosenfeld 1997). The most tragic incident occurred on November 12, 1970, when a hurricane with 125-mph winds caused a 20-foot storm tide that killed at least 300,000 people. This storm triggered a revolution against Pakistan that brought independence for Bangladesh. In 1991, 139,000 people perished in another Bangladesh hurricane. In general, this area has a history of hurricane-induced fatalities: 300,000 people were killed near Calcutta, India, in 1737; 20,000 were killed in Coringa, India, in 1881; and 200,000 were killed in Chittagong, Bangladesh, in 1876. Unfortunately, evacuation procedures and public response has not changed much since 1991, so the chance of similar tragedies remains high. China, Thailand, and the Philippines have also seen smaller but significant losses in the tens to hundreds of people by a single storm in recent years due to storm surges.

The combination of heavy rainfall and a storm surge can be particularly devastating, especially along coastal streams. Under these conditions, residents will first experience the surge, which typically last for one day, followed by flooding from runoff that persists for weeks. For example, coastal homes in Pascagoula, Mississippi, were devastated by slow-moving Hurricane Georges (1998) because the storm surge first inundated the region, followed by flooding from the Pascagoula River.

The expected level of damage for a given hurricane intensity is described by the *Saffir-Simpson Hurricane Scale* (Simpson 1974). It was devised in 1971 by Herbert Saffir, an engineer in Miami, for the World Meteorological Organization, and was given to the National Hurricane Center (NHC). Robert Simpson, the director of NHC, then added the storm surge portion. This scale, only valid for the Atlantic Ocean, classifies hurricanes into five categories according to central pressure, maximum sustained winds, storm surge, and expected damage (table 2.1). An alternative scale for the tropical Pacific Ocean is shown in table 3.1. Although all categories are dangerous, categories 3, 4, and 5 are considered *major hurricanes,* with the potential for widespread devastation and loss of life. Whereas only 21 percent of U.S. land-falling tropical systems are major hurricanes, they historically account for 83 percent of the damage (Pielke and Landsea 1998). On average, the Atlantic has two major hurricanes per year (see table 3.2). Fortunately, category 5 hurricanes are infrequent in the Atlantic Ocean and seldom sustain themselves at such intensities for very long before weakening

TABLE 2.1
The Saffir-Simpson Hurricane Scale for Atlantic Hurricanes (Simpson 1974)

Category	Central pressure in mb inches (approximate)	Maximum sustained winds in mph	Storm surge in feet (approximate)	Potential damage scale	Damage
1 (Minimal)	< 979 > 28.91	74–95	4–5	1	Damage primarily to shrubbery, trees, foliage, and unanchored mobile homes. No real damage to building structures. Low-lying coastal roads inundated, minor pier damage, some small craft in exposed anchorages torn from moorings.
2 (Moderate)	965–979 28.50–28.91	96–110	6–8	10	Considerable damage to shrubbery and tree foliage; some trees blown down, and major damage to exposed mobile homes. Some damage to roofing, windows, and doors of buildings. Coastal roads and low-lying escape routes inland cut by rising water two to four hours before arrival of hurricane center. Considerable pier damage, marinas flooded, small craft torn from moorings. Evacuation of shoreline residences and low-lying island areas required.
3 (Extensive)	945–964 27.91–28.47	111–130	9–12	50	Large trees blown down. Foliage removed from trees. Structural damage to small buildings; mobile homes destroyed. Serious flooding at coast, and many smaller coastal structures destroyed. Larger coastal structures damaged by battering waves and floating debris. Low-lying inland escape routes cut by rising waters three to five hours before arrival of hurricane center. Low-lying areas flooded 8 miles or more inland. Evacuation of low-lying structures within several blocks of shoreline possibly required.

4 (Extreme)	920–944	27.17–27.88	131–155	13–18	250	All signs blown down. Extensive damage to roofing, windows, and doors. Complete failure of roofs on smaller buildings. Flat terrain 10 feet or less above sea level flooded as far as 6 miles inland. Major damage to lower floors of coastal buildings from flooding, battering waves, and floating debris. Major erosion of beaches. Massive evacuation: all residences within 500 yards of shore and single-story residences on low ground within 2 miles of shore.
5 (Catastrophic)	<920	<27.17	>155	>18	500	Severe and extensive damage to residences and buildings. Small buildings overturned or blown away. Severe damage to windows and doors; complete roof failure on homes and industrial buildings. Major damage to lower floors of all structures less than 15 feet above sea level. Flooding inland as far as 10 miles. Inland escape routes cut three to five hours before arrival of storm center. Massive evacuation of residential areas on low ground within 5–10 miles of shore.

In practice, the maximum wind speed determines the category. Many factors affect central pressure and storm surge, so these values are only estimates for a particular category. In fact, the storm surge may vary by a factor of 2 depending on the coastline's proximity to deep or shallow water. This scale is not valid for other ocean basins, since some countries (e.g., Australia and India) use different definitions of sustained winds, contain different types of foliage with different damage thresholds than U.S. foliage, have different building construction standards (some of which may be further weakened by termites), and have coral reefs that modify storm surge damage. An alternative scale for the tropical Pacific Ocean is shown in table 3.1 "Potential Damage Scale" provides a scale relative to a category 1 hurricane, where a category 1 hurricane is scaled as "1" (Pielke and Landsea 1998). For example, a category 3 hurricane typically causes fifty times as much damage as a category 1 hurricane.

to a lower category. Only three recorded category 5 hurricanes have made landfall in the United States (the Labor Day Hurricane of 1935 in the Florida Keys, Hurricane Camille in 1969, and Hurricane Andrew in 1992).

Another danger from hurricanes is extreme ocean currents. In calm conditions, when waves strike the beach at an angle, the leading edge of the wave hits the shallow water sooner than the rest of the wave front and slows down, bending the wave as it moves ashore. This results in a current with a net flow parallel to the coast, known as a *longshore current*. A hurricane can disrupt (or "rip") the longshore current with its large incoming waves, causing a narrow band to be deflected offshore, known as a *rip current*. (This is also called a *riptide*, a misleading name since there is no tidal component, or incorrectly an *undertow*, because rip currents do not pull you under the water.) Rip currents occur even if the hurricane is far away or making landfall elsewhere since their waves can travel far from the storm, and this is why they are an unexpected threat to beach swimmers. This powerful current pulls people away from the shore, and a number of swimmers drown each year from it. Hurricanes enhance rip current activity, but rip currents are always a threat when wave activity is strong and account for 80 percent of lifeguard rescues. Swimmers caught in a rip current should try to swim parallel to shore—rather than directly into the current—until out of the rip (about 10 to 20 yards sideways on average), then swim back to the beach. Rip currents appear as a plume of dirty water (from stirred-up sediment) moving away from shore—often carrying foam, seaweed, or debris—with a break in incoming waves. Rip currents are also common near jetties and piers even in calm conditions, so these are areas swimmers should always avoid.

While hurricanes cause much misery, they can also be beneficial. Hurricanes often provide much-needed rain to drought-stricken coastlines. Their ocean interactions can flush bays of pollutants, restoring the ecosystem's vitality. After the record rainfall from Hurricane Claudette (1979) in Texas, fish were being caught in the northern industrialized reaches of Galveston Bay that had vanished from there for several years. Finally, in cruel Darwinism fashion, weak sea life and plants will perish, leaving only the strong to survive and reproduce.

Another positive consequence is that sometimes hurricanes "correct" man's mistakes. For example, in the early 1900s nonnative foliage such as Australian pine trees had been planted on the tip of Key Biscayne, Florida (now the Bill Baggs State Park). These nonnative plants had few natural enemies in their new environment and quickly dominated plant life, resulting in a loss of natural habitat. However, these Australian nonnatives lacked the ability to withstand hurricane-force winds, and Hurricane Andrew destroyed them all in 1992. Park officials seized the opportunity to replant the park with native foliage.

Forecasting Hurricanes

Hurricanes are one of the most difficult phenomena to forecast in meteorology. A forecaster must understand all facets of meteorology for hurricane prediction since these storms encompass all weather processes, from individual thunderstorms to rainband physics to air-sea coupling to interaction of the hurricane itself with the surrounding atmosphere. These large and small-scale weather features interact with each other in complicated ways. Only a computer model can predict these processes, and sometimes with limited success. Even worse, since hurricanes occur over the data-sparse ocean, forecasters have few observations for seeing the current state of a hurricane or inputting into the computers. Without knowing what the storm is doing now, how can one anticipate its future? Forecasting hurricanes is indeed a challenge.

For evacuation and emergency preparedness purposes, forecasters are most concerned with predicting where the storm will go and, should the hurricane threaten land, where the eye will make landfall. Of nearly equal importance is forecasting hurricane intensity and the width of the tropical- and hurricane-force winds around the hurricane. Forecast statements are also issued for expected rainfall and storm surge. A *tornado watch*, which means conditions are favorable for tornado development, is always issued by the Storm Prediction Center in Norman, Oklahoma, preceding hurricane landfall. When a tornado has been observed by spotters (trained volunteers in the SKYWARN program) or inferred by an instrument called Doppler radar (which can detect areas of rapid rotation in a thunderstorm), a *tornado warning* is issued by the local National Weather Service office.

Inland flooding is a big concern with hurricanes, as it causes enormous economic damage, social disruption, and the largest loss of lives. These floods can occur far from the coast anywhere in the path of a hurricane. Therefore, in recent years the National Weather Service has increased its focus on predicting hurricane rainfall. Hurricanes undergoing extratropical transition (see chapter 1) or interacting with other weather features are particularly susceptible to creating floods. Other factors that enhance flooding potential are storm speed, because the slower the system moves, the more time for rain to fall over a location; orography, in which geographical barriers such as hills and mountains promote lifting air; and antecedent conditions (soil wetness and preexisting water-levels). The wetter the soil, the greater the chance of runoff. Also, the higher the water level in rivers, lakes, and reservoirs, the greater the chance of flooding when heavy rain falls.

Uprooted trees are also a problem in hurricanes. A wet soil makes trees more susceptible to uprooting. If the soil is saturated, 30–40 mph winds will blow trees down. If the soil is dry, trees will stay upright until the wind reaches 60 mph. The type of tree is also important, since shallow-rooted trees are more susceptible to uprooting, all other conditions being equal. Common shallow-rooted trees susceptible to hurricane winds along the coast are dogwood, water oak, pecan, sweet bay, and red maple. Common deep-rooted trees that can withstand strong winds are live oak, longleaf pine, pond cypress, and bald cypress. The type of soil also matters, as trees growing in sandy soils are more deeply rooted than trees growing in soils with an inhibiting clay layer or a high water table. Another factor is the height of the tree. The taller the tree, the greater is its chance of breaking, especially if the trunk's diameter decreases little with height. For this reason, tall, slim slash and longleaf pines are extremely vulnerable. Open-crowned and lacy-foliaged trees, such as cypress and mimosa, offer less resistance to the wind and thus are better able to survive. On the other hand, magnolia trees with their heavy, wind-catching foliage are more susceptible to damage than their root system and trunk structure would indicate. Palm trees offer little surface to the wind because they have almost no laterally extended crown and branches. This characteristic makes them fairly wind-firm, despite their limited root systems.

Predicting heavy rain events is the responsibility of the Hydrometeorological Prediction Center (HPC). The HPC interacts with the NHC, discussing the potential for flooding due to precipitation. A crude technique for predicting total rainfall in inches is to divide 100 by the forward speed of the hurricane in miles per hour; for example, a hurricane moving 10 mph would produce 10 inches of rain. But rainfall can vary, so this is only a rule of thumb. Of particular concern are slow-moving storms, which can dump rain for days. In addition, rivers can swell from copious rain. The National Weather Service has twelve specialized centers at some local offices, called River Forecast Centers, that generate forecasts for U.S. rivers and streams.

While storm surge is always a concern, generally evacuations mandate that people leave areas susceptible to the surge. Therefore, storm surge fatalities in the United States have decreased immensely since Hurricane Camille (1969). The United States has two tools for locating storm surge-prone areas. First, for several decades, the Army Corps of Engineers has surveyed areas hit by a storm surge. Therefore, the United States has an excellent storm surge historical record for decision making. Second, using a storm surge computer model called the Sea, Lake, and Overland Surges from Hurricanes (SLOSH), hundreds of simulations can be conducted to explore hurricane scenarios for each local region. These scenarios include different storm paths, intensities, size, translation speed, and landfall locations. These calculations are then composited to determine the highest storm surge value possible for a category 1–5 hurricane based on a particular family of tracks, known as Maximum Envelop of Water (MEOW) maps. Emergency planners and local government officials design evacuation plans based on the MEOW maps generated by SLOSH. These maps are also particularly important for studying the scenarios in which the hurricane storm surge could topple local levees, such as in the New Orleans area. For a "second opinion," the Army Corps of Engineers also uses a separate storm surge model, called the ADvanced CIRCulation (ADCIRC) model, for its own studies.

Hurricanes have a reputation for being unpredictable at times. They can suddenly turn, speed up, slow down, stall, or loop. They can also reform their center of rotation when thunderstorms are not uniformly distributed around the center, making the storm suddenly "jump" from one spot to the next (an example is Hurricane Earl in 1998). Track forecasts have improved considerably the past twenty years, but errors still occur. Intensity forecasts, however, currently exhibit little skill. The process of anticipating whether a hurricane will strengthen, weaken, or not change in intensity—and predicting how quickly any intensity change may occur—is still an unsolved forecasting problem. Such uncertainties are fraught with potential disasters, such as occurred in the "near-miss" with Hurricane Opal (1995).

Hurricane Opal: A Close Call

Hurricane Opal formed in the southern extent of the Gulf of Mexico off the Yucatán peninsula. Initial computer models forecasted that it would accelerate toward Florida. However, it only slowly drifted northward. Therefore, NHC forecasters became skeptical as each subsequent model-run projected a fast northward motion that did not materialize. Opal was also a category 1 hurricane and was not expected to intensify past category 2. Based on Opal's slow motion and the fact that it was a minimal hurricane, at 5 P.M. on October 3, 1995, NHC decided that a hurricane warning was not necessary until the following morning for the Florida panhandle.

Late that night, Opal intensified explosively to a category 4 with winds of 150 mph, just shy of category 5 status. (Later analysis attributed this rapid development to movement over an isolated pool of very warm water. Interactions with environmental 40,000-foot winds may have also promoted the intensification.) Even worse, Opal finally began accelerating toward Florida. Since most people were sleeping, it was difficult to alert the public about this new, unanticipated danger. Many people near Pensacola Beach awoke the next morning with a near-category 5 hurricane just offshore. The last-minute evacuation procedures clogged the roads and Interstate 10, creating traffic jams that moved less than 10 mph. It was NHC's worst nightmare coming true.

Fortunately, Opal weakened before landfall from a category 4 to category 3 hurricane. While still dangerous and destructive, this was a bit of a reprieve. Furthermore, while many residents were unable to use the roads and had to stay home during the storm, most residents on the immediate coast did manage to evacuate. Opal caused $3 billion in damage and killed nine people in Florida, Alabama, Georgia, and North Carolina, but none of these deaths were from the storm surge. The people of Florida had dodged a bullet.

Opal demonstrated that large track forecast errors still occur, and when combined with unexpected intensification, many congested coastal regions are today toying with human catastrophe since seashore development has proceeded without seemingly any considerations for speedy evacuation. These issues will be discussed later, but first a background on storm motion and its prediction is necessary.

Factors Controlling Hurricane Motion

To understand hurricane motion, it is helpful to use the analogy of a wide river with a small eddy rotating in it. To a first approximation, the river transports the eddy downstream. However, the eddy will not necessarily move straight because the speed of the current varies horizontally; for instance, it may be faster in the center and slow down toward the riverbanks. As a result, the eddy may wiggle off a little to the left or right as it moves downstream and, depending on the situation, may speed up or slow down. Furthermore, this eddy's rotation may alter the current in its vicinity, which in turn will alter the motion of the eddy.

Likewise, one may think of a hurricane as a vortex embedded in a river of air. The orientation and strength of large-scale pressure patterns basically dictate the hurricane's motion, except that the steering depends on both the horizontal and vertical wind distribution. In general, hurricanes are steered by the deep-layer mean of the wind averaging 500–700 miles around the storm (Dong and Neumann 1986). However, there is a tendency for stronger hurricanes to be steered by winds higher aloft because the storm has more vertical development (Velden and Leslie 1991), thus requiring the forecaster to identify the best "steering height" for a particular storm. To make a track forecast, one must first predict the large-scale flow within which the hurricane is embedded, and this can be difficult. How does one separate the hurricane winds from the environmental winds, and how does one know what height best represents the steering flow?

Forecast errors also occur when the steering current is weak and ill-defined, thus causing the hurricane to stall or slowly drift. However, the largest errors are associated with rapid changes in the strength and orientation of the steering current. Under these situations, a hurricane may accelerate its motion and/or turn. For example, sometimes forecasters are faced with the dilemma of a hurricane moving toward the U.S. East Coast, not knowing for sure whether it will hit land before being turned back to the right by westerly steering currents poleward of the hurricane.

Studies show that the hurricane motion deviates a little from the environmental steering flow. This is because a hurricane can internally change its course as well. Just as in the river example, the hurricane can interact with surrounding pressure fields, thus altering its own steering current. Research has also shown that the earth's rotation interacts with the hurricane's rotation, inducing a weak poleward and westward drift, known as the *beta drift*. The beta drift results in a 1–2 mph northwest motion in small storms (for the Northern Hemisphere), while in large storms the motion is 3–4 mph (Carr and Elsberry 1997).

On rare occasions when two hurricanes get too close to each other—especially in the Pacific and Southern Hemisphere oceans—they may begin to move toward each other and rotate cyclonically about a common midpoint between them, known as the *Fujiwhara effect*. This consists of three phases: (1) *approach and capture*, when the two hurricane tracks become impacted by each other rather than by steering currents, a process in which the two hurricanes usually begin to feel each other's influences at a distance of 1,300 miles, with approach and capture dominating at a distance of 850 miles or less; (2) *mutual orbit*, which is the actual mutual rotation about a common center; and (3) *cessation of orbit*, either by a rapid escape of one hurricane from another or, on rare occasions, when one of the systems is destroyed by merging into the other (Lander and Holland 1993). Since the beginning (capture) and end (release) of the mutual cyclonic rotation are relatively rapid, anticipation of these events is critical to the track forecast.

Even when a hurricane moves relatively straight, a detailed analysis shows small oscillations about the mean path, called *trochoidal motion*. These small oscillations are due to a variety of factors, which include a nonsymmetrical distribution of thunderstorms surrounding the center, shifting the center of mass (Willoughby 1988); mesoscale vortices in the eye inducing a track meander (Smith, Ulrich, and Dietachmayer 1990), as well as other eye and eyewall dynamics; and a tilted storm, causing the upper- and lower-level centers to rotate about a midlevel center (Wang and Holland 1996).

Computer Models

Until the late 1970s, NHC forecasters relied on meteorological intuition, statistical schemes, and the storm's past trend to make track predictions. They would also look at past tracks from other years to see which way similar storms moved. These techniques are still used today.

Problems, Controversies, and Solutions 41

However, forecasters began to realize that such difficult predictions required the use of *computer models*. Computer models ingest current weather observations and approximate solutions to complicated equations for future atmospheric values such as wind, temperature, and moisture. The NHC uses a suite of models that differ in their mathematical assumptions and complexities in describing atmospheric processes. Some of these differences exist because certain atmospheric features have been removed to make the computer program run faster. Other differences exist because meteorologists are still uncertain about how to formulate certain weather features such as cloud process on a computer. The most complex models must be run on the fastest computers in the world, known as supercomputers. Model predictions can vary because of these mathematical differences. Model forecasts also contain small errors because the mathematical solutions contain a small amount of uncertainty, which accumulates with time; therefore, all forecast errors grow with time. Models will also produce incorrect forecasts if they are initialized with bad data, which is a frequent problem over the data-sparse ocean (as the saying goes, "garbage in, garbage out"). Even where observations exist, measurements contain small errors that cause forecast inaccuracies to grow with time. The result is that even a slight difference in measurements will cause diverging long-range forecasts. Ultimately, even if the model was perfect in a utopian world, these measurement errors cause long-range weather forecasts to be unpredictable beyond 5–7 days. This is why the NHC only issues hurricane forecasts covering five days. A historical perspective on the science of predictability limits is presented in chapter 4.

Surprisingly, it is possible to manipulate these uncertainties to a forecaster's advantage. In theory, by generating multiple forecasts with slightly different weather observations that represent the probable error distribution of the measurements, the true weather forecast will emerge within these computer solutions. This technique is called *ensemble forecasts* (Toth and Kalnay 1993). Generally, the mean of the forecast ensemble represents the most accurate forecast. Also, many of the forecasts will cluster together around the true prediction. Furthermore, the forecast spread indicates the possible forecast error and forecast uncertainty. When the ensemble technique indicates widely varying forecasts, one knows there is considerable uncertainty in the hurricane prediction; likewise, when the spread is small, the forecaster will have confidence in the prediction. Again, this technique is only useful for up to five days, at which point all the forecasts begin to widely diverge.

In addition, because hurricanes exist in data-sparse regions, one can locate the most useful region to take plane measurements by using a complicated mathematical technique known as a *breeding cycle*. The breeding technique simulates the development of growing forecast error (Toth and Kalnay 1997). The region where observations benefit computer models the most will be located where these breeding modes grow with time (Aberson 2003). Such "targeted observations" allow reconnaissance planes to maximize the positive impact of additional observations on track forecasts.

Computer models have revolutionized all aspects of meteorology, including hurricane track forecasts, and they are improving each year. The NHC has access to multiple models from the United States (including the U.S. Navy), Britain, Canada, Japan, and Europe. Just as in ensemble forecasts, often the best forecast is an average of these different models, called a *consensus forecast* (Goerss 2000). However, in general the most reliable models, which use sophisticated hurricane wind-profile software, are the Geophysical Fluid Dynamics Laboratory (GFDL), developed specifically for hurricanes, and the Global Forecast System (GFS), developed by the National Centers for Environmental Prediction (NCEP) (Elsberry 2005). The NHC leans toward these two models more than other models. The history and recent improvements to the GFDL model are listed in chapter 4. Unfortunately, computer guidance for intensity forecasts is still unreliable. Forecasters use statistical schemes and intuition to make their intensity predictions.

Observation Platforms

Hurricane forecasters monitor the latest observations from satellites, ships, radar, buoys, oil rigs, and other sources to assess any track or intensity changes. When a hurricane is far offshore in data-void regions, the intensity is estimated from satellite cloud patterns using the *Dvorak technique* (see chapter 1 for details). While generally robust, the Dvorak technique can produce wrong intensity values, as in the case of Typhoon Omar (1992). The Dvorak technique estimated Omar to be a category 1 storm, but it was discovered to actually be a category 2 storm hours before hitting Guam, surprising the island's inhabitants and perhaps amplifying damage due to lack of preparation.

Therefore, the most important data platform is *reconnaissance planes* that fly into the hurricane's eye and take critical meteorological measurements. The precise information required for obtaining hurricane structure, location, and intensity cannot be obtained in any other way and is crucially important for forecasts and evacuation procedures. Forecasters credit a reconnaissance plane for discovering that Hurricane Camille (1969) had strengthened from sustained winds of 115 mph to 150 mph on August 16, 1969, at 5 p.m. CDT, thirty hours before landfall. Based on this new information, additional evacuations were prompted that may have saved up to 10,000 lives by one estimate. A reconnaissance plane also measured Hurricane Opal's (1992) unexpected rapid intensification. Reconnaissance flights are expensive and only occur in the Atlantic Ocean since just the U.S. government has the financial capability to fund them. At one time, Congress had considered halting reconnaissance flights. However, once their forecast and evacuation impacts were made clear by scientists and evacuation managers, not only was funding preserved but a line item was added to the U.S. budget explicitly setting aside funds for Atlantic reconnaissance flights. At one time, U.S. Air Force reconnaissance flights also occurred in the western Pacific Ocean, but budget restrictions forced those flight missions to end in 1987. Occasional surprises such as Typhoon Omar are the result.

Continuous reconnaissance flights begin once a tropical system moves close enough to land. The 53rd Weather Reconnaissance Squadron (also known as the "Hurricane Hunters") at Keesler Air Force Base near Biloxi, Mississippi, takes most of these measurements on WC-130s. Two additional P-3 Orion planes with more sophisticated instruments (including a radar) are also available by the National Oceanographic and Atmospheric Administration (NOAA) Aircraft Operations Center at MacDill Air Force Base in Tampa Bay, Florida. The P-3s are deployed on less routine flights to perform analysis of hurricane structure in conjunction with scientists at the Miami NOAA Hurricane Research Division (HRD). The P-3s also transmit data to the NHC.

Obviously, reconnaissance flights carry an element of risk. In September 1955, a navy plane and its crew of nine plus two Canadian newsmen were lost in the Caribbean Sea while flying in Hurricane Janet. Three air force aircraft have been lost flying in typhoons in the Pacific. Fortunately, no planes have been lost in the last twenty-five years, but scary moments still happen. In one extreme circumstance during Hurricane Hugo (1989), one of the

P-3 planes encountered a severe mesoscale vortex that damaged the plane, causing it to sink and then regaining altitude only moments before it would have crashed into the ocean. Most flights are less eventful but do contain some "bumps" between thunderstorm updrafts and downdrafts, and there are occasional roller-coasterlike drops (or ascents) of 3,000 feet in less than a minute due to strong updrafts and downdrafts. Since hurricane winds are strongest just above the earth's surface (about 1,000 feet) and weaken with height, reconnaissance flights are performed at 5,000 or 10,000 feet. During the hurricane penetration, information about the horizontal wind and temperature structure is transmitted to the NHC. Once the plane enters the eye, it deploys a tube of instruments (called a *dropsonde*) that parachutes downward from flight level to the sea, sending valuable intensity measurements back to the NHC.

Summary of NHC Forecast Procedures

Forecasters at the NHC critically evaluate each individual computer prediction, as well as forecast ensembles, for reasonableness and consistency with other forecasts, observed data, statistical projections, the historical record, and extrapolated positions. They also monitor the latest observations and satellite imagery for last-minute adjustments. Based on all of this information, every six hours the forecasters produce a new forecast of the storm's projected center location, intensity (defined as maximum one-minute sustained winds at 33 feet elevation), and storm size for the next five days in several text and graphical packages. Examples of these NHC forecast products are provided in the glossary.

When hurricane conditions on the coast are possible within thirty-six hours, a *hurricane watch* is issued. When hurricane conditions are likely within twenty-four hours, a *hurricane warning* is issued. These watches and warnings refer to the arrival of hurricane-force winds of 74 mph, not eye landfall that generally occurs a few hours later.

Track forecasts have significantly improved in the last thirty years, especially the two- and three-day forecasts (Franklin, McAdie, and Lawrence 2003). In 1970, the forty-eight- and seventy-two-hour track forecast errors were 300 and 480 miles, respectively; today they are 170 and 250 miles, respectively. The twenty-four-hour error has decreased more slowly, from 140 miles in 1970 to 90 miles today. The annual rate of improvement, on average, is 1.0 percent, 1.7 percent, and 1.9 percent for the twenty-four-, forty-eight-, and seventy-two-hour track forecasts, respectively. These improvements are average errors, so individual predictions can be substantially better or worse. The four- and five-day forecasts, which were issued starting in 2003, contain significant uncertainty with average errors of 290 and 370 miles, respectively, and should only be used as general guidance. Because of these errors, the predicted track is shown graphically on the NHC's website and on most TV stations as a widening cone showing the uncertainties, with the track forecast of the storm center usually in the middle of this cone. Generally, the edge of this cone represents the most deviating models.

The NHC also predicts storm size (radius of strong wind values), such as the radius of tropical storm force (39 mph) winds. Storm size forecasts can also contain errors, so areas not in the path of the storm center still need to be on guard. Typically, the twenty-four-, thirty-six-, and forty-eight-hour error for storm size averages 25, 30, and 35 miles, respectively. The NHC also predicts intensity, but this is a tough challenge. There is much about how the hurricane core interacts with the ocean and atmosphere that is not understood. As a result, intensity forecasts are often significantly in error and are a priority research topic. The twenty-four-, forty-eight-, and seventy-two-hour intensity error averages 10, 17, and 22 mph, respectively. As illustrated in the Hurricane Opal example, rapid intensity swings are often unpredictable as well.

Because of these potential track, storm size, and intensity prediction errors, it is important that all communities in the hurricane-warning area take the threat seriously, even though the landfall forecast may be 50 miles or more away. Never was this more apparent than for Hurricane Charley (2004), which was forecast to hit Tampa Bay as a category 3 but hit 100 miles south in Port Charlotte as a category 4, surprising many residents. Port Charlotte was in the hurricane-warning region, but many residents confidently believed that the storm was going to Tampa and did not expect the slight track turn or the rapid intensification.

Society's Preparation for Hurricanes: New Worries

Since Hurricane Camille's storm surge killed 181 people in 1969 along the coast of Mississippi and Louisiana, U.S. hurricane-related storm surge fatalities have dropped dramatically as storm surveillance, evacuation procedures, public awareness, and forecasts have improved. However, new concerns have emerged as coastal population growth and property development have exploded. In fact, U.S. population increases are currently largest in coastal communities. The population has grown on average by 3–4 percent a year in hurricane-prone regions, and in some states such as Florida the growth is greater. As a consequence, property damage costs due to hurricanes skyrocketed in the 1990s (table 2.2).

This trend is disturbing in other ways. First of all, during this period of higher property costs, the number of major hurricanes (category 3 or better) actually *decreased* between 1970 and 1995 (Landsea et al. 1996). This decrease follows an active period of intense hurricanes during the 1940s to 1960s. Such shifts in weather activity are known as *multidecadal changes.* This change is likely due to a cooling in the Atlantic Ocean water temperature and may also be related to a drought that started in 1970 in Africa, where many tropical waves form before propagating out into the Atlantic.

Second, the number of major hurricanes making landfall on the U.S. East Coast dramatically decreased between 1965 and 1995. During the period 1944–1964, seventeen major hurricanes hit the East Coast (Landsea 2000). In contrast, only five major hurricanes made landfall on the East Coast between 1965 and 1995 (with a period between 1965 and 1983 where *no* East Coast landfall occurred). It is during this period that much population growth and property development occurred on the East Coast,

TABLE 2.2
Hurricane-induced property damage costs by decade, adjusted to 1990 dollars

Decade	Property damage
1900–1909	$1 billion
1910–1919	$2.5 billion
1920–1929	$2 billion
1930–1939	$4.5 billion
1940–1949	$4.4 billion
1950–1959	$10 billion
1960–1969	$18 billion
1970–1979	$14 billion
1980–1989	$15 billion
1990–1995	$32.5 billion

with people apparently ignoring or not knowing what happened during 1944–1964.

Some evidence suggests that major hurricane activity occurs in forty- to sixty-year cycles, where twenty to thirty years of active Atlantic hurricane seasons will be followed by twenty to thirty years of relatively quite hurricane seasons (Gray, Sheaffer, and Landsea 1996). In fact, it appears that an active period began in the mid-1990s. The year 1995 was extraordinarily active, with nineteen total storms; eleven reached hurricane status, and five became intense hurricanes. The years 1996 and 1998 were also above normal. In fact, the four-year period of 1995–1998 had a total of thirty-three hurricanes—an all-time record in the Atlantic Ocean. And in 2004, three major hurricanes made devastating landfalls in the Florida peninsula, and a fourth major hurricane, Ivan, demolished Gulf Shores and Pensacola! Global weather patterns are also emerging that are favorable for more active hurricane seasons in the future, such as a return of warm Atlantic water temperatures (Gray et al. 1998). When the more active hurricane phase returns (if it hasn't already), property damage costs could easily exceed $50 billion in one year, especially should a major city be hit. Regardless of trends in Atlantic hurricane season activity, each year the United States has at least a one in six chance of experiencing hurricane-related damage of at least $10 billion (Pielke and Landsea 1998).

Even more disturbing is the sharp population increase. In Florida alone, the population has increased from 500,000 in 1930 to 13 million today. Some coastal regions are so congested now that it is becoming more difficult to perform a timely evacuation. As discussed earlier, track forecasts have improved by 1–2 percent per year (Franklin, McAdie, and Lawrence 2003), but the 3–4 percent population growth could overwhelm these better predictions, resulting in higher casualties again. Even worse, should a hurricane unexpectedly change course or accelerate its motion, there may not be enough lead time for orderly evacuation of intensively developed coastal regions, trapping many residents when the hurricane makes landfall.

In addition to potential casualties, hurricanes are expensive. A typical hurricane warning costs an estimated $192 million due to preparation, evacuation, and lost commerce. At least $300,000 of business losses and hurricane-preparation costs are incurred per day for every mile of coastline evacuated. Some experts claim that the cost is even larger, perhaps $1 million per mile. Therefore, a top priority has become accelerating the improvements in forecasting accuracy through expanded and more accurate reconnaissance observations. In the late 1990s, HRD modernized the dropsonde using Global Positioning System (GPS) technology, which improved hurricane observations with unprecedented accuracy. In addition, the new dropsondes have the ability to take observations below 1,500 feet—something the older dropsondes could not do. A new, faster plane has also been added to the fleet—the Gulfstream IV. Since the Gulfstream is a jet, it can reach altitudes of 45,000 feet, unlike older propeller-driven planes belonging to the Hurricane Hunters and to the Aircraft Operations Center, which can only reach 25,000 feet at best. The Gulfstream data, combined with new data assimilation techniques in models, has improved track forecasts by 15–35 percent.

If this track improvement can be sustained, it may allow the NHC to shrink the warning zone by 50–80 miles, which would save millions in evacuation costs, especially if a major city is excluded from the warning area. It would also reduce traffic congestion, speeding up evacuation from the most threatened region. Furthermore, if homeowners take advantage of more accurate forecasts by seriously protecting their homes, it is estimated that another 10–15 percent in costs would be saved. And, any lives saved by the improved forecasts would be priceless.

Other exciting technological developments have occurred. In 1999 the Hurricane Hunters replaced their WC-130H aircraft with WC-130J planes, which are faster, more fuel-efficient (resulting in longer flights), and, most importantly, can travel at 37,000 feet or more. This augments operational measurements critical to hurricane forecasting. In addition, NASA scientists, NOAA scientists, and universities have collaborated on several field programs to collect data at all levels of hurricanes using multiple aircraft, remote sensing technologies, and portable surface-measuring platforms (see chapter 4 for details about the Third Convection and Moisture Experiments, the China LAndfalling Typhoon EXperiment, and The Coupled Boundary Layer Air Sea Transfer). The data sets produced by this and future missions will be unprecedented in their comprehensiveness, providing researchers with new information toward understanding hurricanes.

In the next few decades, another new type of plane may be taking oceanic observations. Scientists in Australia and Canada are experimenting with small, pilotless aircraft, typically called *drones* or sometimes *aerosondes*. These drones theoretically can be programmed to fly a fixed path for more than twenty-four hours, continuously ingesting weather observations that would improve forecasts. The first successful cross-Atlantic flight by an unmanned aerosonde was accomplished on August 22, 1998. The unique aerodynamic structure and relative weightlessness of drones also allows them to theoretically withstand hurricane winds and strong updrafts in thunderstorms, thereby providing timely and continuous crucial weather data at a cheaper cost than airplane reconnaissance.

Forecasting Annual Hurricane Activity

In the early 1980s, Dr. William Gray at Colorado State University asked the question, "Why do some Atlantic hurricane seasons have many storms, and why are other seasons inactive?" He further wondered if one could predict, months in advance, the number of tropical storms and hurricanes for the upcoming Atlantic tropical season. Based on his research, in 1984 Gray began to publicly predict how active the Atlantic hurricane season would be before it starts. His predictions have been remarkably accurate at times and have proven to be skillful (Owens and Landsea 2003). His forecasts, which are well publicized in the media, are issued every December, April, June, August, and September. His forecasts are also available at http://tropical.atmos.colostate.edu.

Gray and his students have discovered several surprising global signals that affect Atlantic hurricane activity (Gray et al. 1998). The reasons for some of these associations remain unclear and are still under research. These include:

1. *El Niño-Southern Oscillation* (ENSO), also known as *El Niño*. El Niño occurs when warm water in the equatorial western Pacific Ocean shifts east, which occurs about every three to seven years. What does the Pacific Ocean have to do with Atlantic hurricanes, you ask? This change in the Pacific Ocean alters global wind patterns and modifies many weather elements, including reducing hurricane activity. In El Niño years, wind shear is enhanced in the Atlantic Ocean that is destructive to tropical storm formation. An El Niño officially ends when the warm eastern Pacific waters cool to average temperatures; however, usually the cooling continues, which is called

La Niña. La Niña years are associated with weaker than average Atlantic wind shear and enhanced Atlantic hurricane activity. The occurrence of an El Niño or La Niña is one of the most important modulators of Atlantic hurricane activity.

2. *African rainfall.* There is a strong correlation between rainy years in the western Sahel of Africa and major (category 3, 4, or 5) hurricane activity. Likewise, during drought years in Africa, major hurricane activity is reduced. Most major hurricanes come from tropical waves that originate over Africa. It has been hypothesized that during rainy years, tropical waves are stronger and more conducive to genesis as they move westward off the African coast. However, Goldenberg and Shapiro (1996) show that years with strong (weak) wind shear are associated with droughts (rain) in Africa. Rainfall is also likely related to changes in water temperature in the Atlantic Ocean, where years with warm (cold) water are correlated with wet (dry) years in Africa and more (fewer) strong hurricanes (C. Landsea, personal communication, 1998).

3. *Pressure and temperature difference between the western African coast and the Sahel region during the previous February–May period.* When these differences are conducive to onshore flow that transports moisture over Africa, hurricane activity is often enhanced; when it is conducive to offshore flow, hurricane activity is often suppressed.

4. *Caribbean Sea surface pressure.* When pressure is lower (higher) than normal in the Caribbean Sea, Atlantic hurricane activity is enhanced (suppressed). High pressure indicates sinking air, which suppresses thunderstorm formation in easterly waves and other tropical disturbances. Knaff (1997) also showed that higher pressure is correlated with strong wind shear.

5. *Quasi-Biennial Oscillation* (QBO). The QBO is an oscillation of equatorial winds between 13 and 15 miles aloft. These winds change direction between westerly and easterly every twelve to sixteen months. Westerly winds are associated with more hurricanes than easterly winds, especially when the wind is westerly at both 13 and 15 miles aloft. It has been postulated that a westerly QBO reduces wind shear, but the reason for this hurricane association remains unclear.

6. *Caribbean wind shear.* During seasons when shear is greater than normal in the Caribbean Sea, hurricane activity is suppressed, and vice versa for seasons with weak wind shear.
7. *Atlantic Ocean water temperature.* Obviously, warmer than normal sea-surface temperature is favorable for active hurricane years, and vice versa for colder than normal water.
8. *Strength of Azores high-pressure system.* The Azores High is a permanent feature located 20–30° west in the Atlantic Ocean. When pressure is higher than normal in the Atlantic, tropical northeast winds tend to be stronger than normal. This, in turn, upwells cold water to the surface off the northwest African coast, thereby reducing the chance of tropical genesis as tropical waves propagate off Africa. There also is some long-term feedback, such that a strong Azores High in the fall results in high pressure in the Caribbean Sea the following year (Gray et al. 1998; Knaff 1997). Therefore, when the Azores high is stronger (weaker) than normal in the central Atlantic, hurricane activity is enhanced (suppressed).

In general, when more of these predictors are favorable for hurricane activity than unfavorable, Gray predicts an above-average hurricane season, and when most are unfavorable, a quiet hurricane season is predicted. When the same number of predictors have positive and negative influences, an average hurricane season is predicted—about nine named storms, six of which will become hurricanes and two of these becoming intense (table 2.3). However, caution must be advised here since some predictors are more important than others. For instance, despite the fact that most factors were favorable for an active hurricane season in 1997, a record El Niño that year dominated the weather, resulting in a quiet hurricane season. Also, one should be careful about attributing Atlantic hurricane activity to one feature in any year, because many of these features are interrelated; for example, an El Niño tends to be associated with strong Caribbean wind shear.

Gray quantitatively predicts several parameters using statistical techniques and intuition, such as the number of named storms (which includes tropical storms and hurricanes), the number of hurricanes, the number of days with tropical storms or hurricanes, etc. He also forecasts the number of major hurricanes,

TABLE 2.3
Number of named storms (including both tropical storms and hurricanes) predicted by Dr. Bill Gray each year in August, versus the number actually observed

Year	Number of named storms forecasted	Number of named storms observed
1984	10	12
1985	10	11
1986	7	6
1987	7	7
1988	11	12
1989	9	11
1990	11	14
1991	7	8
1992	8	6
1993	10	8
1994	7	7
1995	16	19
1996	11	13
1997	11	7
1998	10	14
1999	14	12
2000	11	14
2001	12	15
2002	9	12
2003	8	7
2004	13	14

which includes category 3, 4, or 5 hurricanes, since these are the storms that cause about 83 percent of total hurricane damage (Pielke and Landsea 1998). Table 2.3 summarizes Gray's forecast of named storms versus the observed number.

Note in table 2.3 that in sixteen of the twenty-one forecasts, Dr. Gray correctly predicted whether the observed number of storms would be above or below the average of nine named storms. However, years that were incorrect (1989, 1993, 1997, 1998, and 2002) are instructive because they lead to new insights on how to improve the forecasts. For example, after analyzing the 1989 underforecast of storm activity, the importance of African rain was discovered, since 1989 was the only nondrought year that decade. Starting in 1990, African rainfall was included in Gray's seasonal forecasts. Also, since the predictions are based on statistical probabilities, the forecast will fail in some years (just as local National Weather Service statistical forecasts of rain and temperature will occasionally fail). Gray has to also make a correct prediction of parameters such as African rain and El Niño for an accurate seasonal hurricane forecast. If he forecasts a basic predictor incorrectly (such as not predicting the record El Niño of 1997), the seasonal forecast will probably be wrong. Starting in 1998, Gray also began issuing forecasts for hurricane landfall probability based on Atlantic sea surface temperature patterns.

These seasonal forecasts draw immense public interest each year. In addition, insurance companies find them useful in preparation for the upcoming hurricane season. Insurance companies also use this information to buy insurance for themselves, called *reinsurance*. A reinsurance program is divided into layers with different reinsurers assuming the risk of different layers (Murnane 2004). For example, a direct insurer may be responsible for the first $50 million of losses, then reinsurer A assumes 90 percent of the next $50 million, reinsurer B assumes 90 percent of the subsequent $75 million, and reinsurers C and D split 90 percent of the next $50 million. The direct insurer is then responsible for losses in excess of $250 million. In this way, reinsurers retain a fraction of the losses to help avoid excessive claims. The top property catastrophe loss is due to hurricanes (see chapter 6). Therefore, knowing the activity of the upcoming hurricane season five months or more in advance allows reinsurers to adjust their exposure during annual negotiations. Direct insurers benefit because they can enlarge their underwriting capacity and reduce the probability of financial ruin.

Gray's pioneering work has inspired other universities and organizations, such as Dr. Jim Elsner at Florida State University and the Tropical Storm Risk group at University College London, to issue seasonal forecasts using different statistical methodologies. It also motivated NOAA's Climate Prediction Center (CPC) to start issuing seasonal hurricane forecasts in 2002. Using basically the same predictor's as Gray, the CPC issues a more general outlook in terms of probability. For example, the August 2003 forecast predicted a high likelihood (60 percent) of an above-normal season, a 35 percent probability of a normal season, and only a 5 percent chance of a below-normal season. The CPC then give a range of predicted named storms, hurricanes, and major hurricanes (i.e., twelve to fifteen named storms) rather than a specific number. These numbers are clumped together into an Accumulated Cyclone Energy (ACI) index, a measure of overall activity. The goal is to actually predict a correct range of the ACI index, with generally two of the three predictors in the correct range. These forecasts are issued in May with an update in August and are available at http://www.cpc.noaa.gov.

All hurricane seasons should be taken seriously by coastal residents, even if the seasonal forecast is "below normal." Many devastating hurricanes have occurred in otherwise inactive seasons, such as Hurricane Andrew (1992), Hurricane Alicia (1983), and Hurricane Allen (1980).

Is Global Warming Increasing the Number of Hurricanes?

Some environmental groups claim that hurricanes are becoming stronger and more numerous (Leggett 1994) due to a reputed phenomena known as *global warming*. Media outlets also have made such assertions (Anon. 1996). First of all, the topic of global warming is itself very political and controversial. Although a majority of scientists believe that global warming is occurring, many scientists still discredit these allegations with solid arguments. This book will not discuss the global warming controversy in any detail, since many other books are devoted to the subject, including the ABC-CLIO book *Global Warming* (Newton 1993).

However, we will discuss two issues concerning any links to possible global warming and hurricanes. The first question is: Have hurricanes increased in numbers or intensity in the past two decades due to possible global warming? The second question is: Will hurricane activity increase in the future due to possible global warming?

In short, the answer to the first question is "no." The answer to the second question is "maybe," although reductions in hurricane activity are also possible. Before addressing these issues in detail, a review of the greenhouse effect and global warming is required.

What Is the Greenhouse Effect?

A balance between incoming solar radiation from the sun and outgoing radiation from the earth primarily regulates the atmosphere's temperature. This is a complicated process requiring some explanation. When an object becomes warmer than $-273.15°C$ (the temperature where all molecular motion ceases, also called absolute zero on the Kelvin temperature scale, and equivalent to -459.67°F), it begins to emit electromagnetic radiation. Therefore, as the sun warms the earth environment, the earth begins to emit its own radiation. However, radiation is dependent on temperature. The hotter an object is, the shorter the wavelength of peak emission. The much hotter sun emits most of its radiation as visible light. The cooler earth emits its energy at a much longer wavelength, which human eyes cannot see, called the *infrared spectrum*.

It turns out that molecules in the atmosphere behave differently with regard to the passage of infrared radiation and visible radiation through them. It's a complex process: the end result is that some air molecules allow visible radiation to pass through them but obstruct infrared radiation. With this background, we can now discuss the heat balance of the earth and its atmosphere.

As solar radiation flows from the sun toward the earth, it passes freely through the atmosphere and warms the earth. However, as the earth tries to emit this heat back to space in the infrared spectrum, some of it becomes "trapped" by the atmosphere, thus warming the air. A portion of this energy is radiated back to the earth, which warms the surface. The earth, in turn, reradiates this infrared energy upward, where it is again absorbed and warms the lower atmosphere some more. In this way, the atmosphere acts as an insulating layer, keeping part of the infrared radiation from escaping rapidly into space. Consequently, the earth's surface and lower atmosphere are much warmer than they would be without this selective absorption of infrared radiation. This process is often called the *greenhouse effect* since it is analogous to a glass building that allows visible light inside but prevents some infrared radiation from leaving, thus keeping the plants inside warm (even in winter). It's important to realize that the atmosphere's greenhouse effect is a natural process, and without it the earth would be a much colder, unlivable planet with an average surface temperature of 0°F.

There is a distinction between the atmosphere and a greenhouse: in a greenhouse, the glass allows visible light to pass through but restricts the passage of infrared, while in the atmosphere molecules are differentiating between visible and infrared radiation. Also, a greenhouse warms quickly because glass is a physical barrier to air movement, while in the atmosphere mixing occurs freely between warm air at the ground and cooler air aloft, thus slowing down warming. Nevertheless, the phrase "greenhouse effect" is used throughout the media and meteorology, so the nomenclature will be used in this book as well.

Certain molecules are more effective at trapping infrared radiation than others. The most important greenhouse gas is water vapor, because it strongly absorbs a portion of outgoing infrared radiation and is a plentiful gas. The other greenhouse gas is carbon dioxide CO_2, which is equally as absorptive but far less plentiful than water vapor and therefore plays a much smaller role in the greenhouse effect. CO_2 is a natural component of the atmosphere, produced mainly by the decay of vegetation.

What Is Global Warming?

However, CO_2 is also produced by the burning of fossil fuels such as coal, oil, gasoline, and natural gas. Observations show that CO_2 has increased by more than 10 percent since 1958, coinciding with increases in fossil fuel emissions from automobiles, factories, and other power sources. *Global warming theory* states that the greenhouse effect will be enhanced as CO_2 increases, because more outgoing infrared radiation will be absorbed. There is also the possibility that increasing the temperature will increase water vapor concentrations (the major greenhouse gas) due to increased evaporation rates, further enhancing warming prospects. Some scientists contend that global warming will increase the earth's average surface temperature by 2–8°F. However, other scientists contend that feedback processes involving clouds and the ocean may cancel that potential warming. Some scientists, based on decades of observations, also say that global warming is already occurring. However, other scientists using different observation techniques show that no true warming has happened yet; still others attribute any perceived warming to natural climate variability. The honest truth is, no reputable scientist is 100 percent confident that global warming has occurred or will ever occur. But let's play devil's advocate for the moment and discuss how these scenarios could change hurricane activity.

Has the Number of Tropical Storms Increased, and Have Hurricanes Become Stronger Due to Potential Global Warming?

The global average of eighty-nine tropical storms (see table 3.2) has not changed in the last few decades, which suggests that even if global warming is occurring, it hasn't altered the number of these tropical systems (Landsea 2000). Furthermore, while annual and decadal variations in hurricane numbers occur in regional ocean basins, no long-term trends are observed. Indeed, some ocean basins experience ten- to twenty-year cycles where the total number of storms changes. For example, since 1980 the number of tropical storms has increased in the Northwest Pacific, but this increase was preceded by a nearly identical decrease from about 1960 to 1980. A recent analysis also shows a slight decreasing trend in hurricanes in the Northwest Pacific, even though water temperature has shown a slight upward trend during 1960–2003 (Chan and Liu 2004). A downward trend in tropical storm number since the mid-1980s for the Australian region has also been observed, but this is probably an artificial decrease due to a change in tropical cyclone wind designation by that country in the mid-1980s.

As discussed earlier, the number of Atlantic tropical storms has substantial year-to-year variability. However, no significant trend in tropical storm number has been observed since 1944 (Landsea et al. 1996). In contrast, the number of major hurricanes (category 3 or better) has shown a significant *downward* trend during this period, corresponding with an active period of intense hurricanes during the 1940s to 1960s that suddenly changed to a quiet period of intense hurricanes during the 1970s to 1990s. (Such shifts in weather activity are known as *multidecadal changes*). This change is likely due to a cooling in the Atlantic Ocean water temperature during this period and may also be related to a drought that started in 1970 in Africa, where many tropical waves form before propagating out into the Atlantic.

The early 1990s was a particularly inactive period in the Atlantic. No hurricanes were observed over the Caribbean Sea during the years 1990–1994—the longest period of lack of hurricanes in the area since 1899. The period 1991–1994 is the quietest four-year period on record since 1944 in terms of frequency of total storms (7.5 per year), hurricanes (3.8 per year), and major hurricanes (1.0 per year). However, one major hurricane that made landfall during this period was Andrew (1992), which caused record amounts of damage in Miami.

Because Andrew, a category 5 hurricane, hit a metropolitan area, it attracted much attention, including claims that global warming must have caused such a powerful storm because of the unprecedented damage (Leggett 1994; *Newsweek* 1996; U.S. Senate Bipartisan Task Force on Funding Disaster Relief 1995; Dlugolecki et al. 1996). However, it is normal to have two major hurricanes each year (see table 3.2). In fact, it is unusual to have only one major hurricane per year during a four-year period! Andrew just happened to hit a highly developed coastal region.

Therefore, claims of global warming increasing hurricane activity or intensity has little supporting evidence thus far, because a significant *decrease* occurred between 1970 and 1995 during this period of increased fossil fuel emissions. The long-term climate record does not support possible global warming contributing to increased hurricane activity.

The moral is that a reader should view most sensational science reports with some skepticism. Stories about global warming (and most sensational science stories in general) need not necessarily be dismissed as incorrect, but they should be carefully scrutinized. Questions one should ask when reading a science report or trying to learn about a particular science topic are:

1. *Who is the source for the report?* For example, is it in a scientific magazine written by an author with a science degree from college or in a newspaper/news/TV report by an author with no background in the sciences? Just because an article is in a popular media outlet doesn't mean it's shoddy, but it does depend on the quality of the journalist, his or her interview procedures, and scientific aptitude. Some writers work hard to produce an accurate, informative article. On the other hand, many scientists can describe instances where a journalist inaccurately reported their information or took a quote out of context.
2. *Was the report peer-reviewed?* The most reputable reports are in journals that undergo a *peer-review* process. The peer-review process involves rigorous commentary by several experts in the field who either accept or reject the article based on its scientific merit. Usually several rewrites are needed. Examples of general science journals available at bookstores and newsstands that at least undergo some minimal peer review are *Scientific American, Nature,* and *Science.* Examples of peer-reviewed journals in meteorology are *Monthly Weather Review, Weather and Forecasting, Journal of Climate,* and the *Bulletin of the American Meteorological Society.* The latter are all published by the American Meteorological Society (see chapter 7). Other clues that a publication is from a journal are the usage of footnotes and references, few to no advertisements, quarterly or monthly publications, and a tech-

nical format written for experts. In contrast, magazines and newspaper articles are written by journalists or laypersons, have few footnotes or references, are written for the general public, and appear weekly or daily.
3. *Is it a well-balanced report presenting all the issues, or is it strongly biased toward one point of view?* Often a journalist or a magazine funded by a particular special interest group or political party, will strongly skew a report toward the individual's or group's point of view. These reports may have legitimate information, but one should investigate other sources before jumping to conclusions. To assess whether a report is biased, simply look at other reports by the same reporter or magazine to see if similar topics are reported in the same context. Even peer-reviewed articles can contain biases.

In summary, one should be skeptical and inquisitive about most controversial or sensational scientific information they encounter and should always obtain their information from more than one source.

Should Global Warming Occur in the Future, Will Hurricanes Increase in Number and Intensity?

Research has shown that global warming could increase sea surface temperature, humidity, and rainfall in the tropics. Based on this research, some have suggested that hurricanes may increase in frequency, area of occurrence, intensity, and rainfall amounts (Ryan, Watterson, and Evans 1992; Emanuel 1987; Tuleya and Knutson 2002). A recent study suggests global warming has increased hurricane intensity and storm lifetime in the Atlantic and North Pacific (Emanuel 2005). However, any changes in hurricane activity will also be associated with large-scale changes in the tropical atmosphere (Landsea 2000; Henderson-Sellers et al. 1998; Chan and Liu 2004). For example, the static instability threshold necessary for thunderstorm maintenance could change, since temperature may increase throughout the atmosphere, not just at the surface.

In addition, any changes in global wind patterns will profoundly affect hurricane activity. For instance, if global warming increases wind shear, one would see a significant decrease in hurricane activity. Likewise, a reduction in wind shear would dramatically increase hurricane activity. If monsoon activity increases, then tropical cyclogenesis will increase, and vice versa for weakened monsoon troughs. Another wildcard is how global warming would change El Niño activity—if more El Niños occur, hurricane activity would be reduced, and vice versa for fewer El Niños. In summary, the combined changes in water temperature, wind shear, monsoon activity, atmospheric instability, tropical moisture, and El Niño will dictate how hurricane activity changes. Therefore, it is difficult to assess how potential global warming could alter hurricane activity. Besides, there is considerable motivation for society to better prepare for hurricanes independent of global warming concerns!

Attempts at Hurricane Modification

Because of the destructive and life-threatening nature of hurricanes, experimental attempts have been made to weaken them. The main hypothesis involved converting liquid cloud water to ice just outside the eyewall. Water gives off enormous quantities of stored heat (also called *latent heat*) when it changes phase from liquid to ice. Many clouds in the atmosphere exist in a *supercooled* state, which means that the temperature of the liquid droplets is below freezing (32°F), but the droplets lack a "triggering" mechanism to turn to ice. For this conversion to occur, supercooled liquid water needs to attach to a floating aerosol with a molecular structure similar to ice, known as *ice nuclei*. However, ice nuclei are often sparse in the atmosphere, and many supercooled droplets are never converted to ice. If somehow one could introduce artificial ice nuclei (such as silver iodide) into a supercooled cloud, the water would be converted to ice, thus releasing lots of heat into the air and causing the cloud to grow. This premise, known as *cloud seeding*, has been attempted to increase rainfall in drought-stricken regions (with unproven results) and snowpack in the mountains (with successful results).

The seeding process to theoretically weaken a hurricane is more complicated. At first scientists thought they could seed the eyewall and perturb the hurricane's wind outward, thus weakening the hurricane (Anthes 1982; Willoughby et al. 1985). By the mid-1960s, scientists realized that this theory was flawed. A revised theory involved seeding the clouds just outside the eyewall to stimulate cloud growth away from the eyewall. The new outer eyewall would grow, depriving inflow into the older inner eyewall. The result is a weakening inner eyewall, resulting in less subsidence in the eye and a rise in central pressure. If the pressure increases, inflowing air is unable to penetrate to as small a radius, and most of the new ascent occurs at the new outer eyewall. Eventually the new eyewall would replace the old eyewall, but at a larger distance from the center. Just as ice-skaters slow their rotation when their arms are spread out, a larger eyewall radius would cause a reduction in wind speed.

Cloud seeding was first tested when several U.S. government agencies collaborated in a pioneering weather modification effort known as *Project Cirrus* (Willoughby et al. 1985). Among other notable firsts was the cloud seeding of a hurricane. On October 13, 1947, a plane dropped silver iodide into a hurricane moving to the northeast. Observers on the plane noted changes in the visual appearance of the cloud but could not demonstrate any changes in structure or intensity. However, shortly afterward the hurricane reversed course to the west, making landfall on the coasts of Georgia and South Carolina. It is extremely unlikely that seeding altered the course, since hurricanes are mostly guided by constantly shifting large-scale atmospheric currents. However, the political and legal implications taught scientists to be more careful about where they conducted their hurricane seeding experiments. Future attempts at hurricane modification occurred only in hurricanes that were far from all land masses and unlikely to make landfall within twenty-four hours.

The next experiment occurred on September 16, 1961, when a naval aircraft dropped eight canisters of silver iodide into Hurricane Esther. Esther, which had been intensifying, stopped strengthening and the eyewall's distance from the center increased. The next day, a second seeding attempt was made, but the canisters missed the eyewall and the hurricane's intensity did not change. At the time, the experiment was considered successful, although we now know that the eyewall-seeding hypothesis was flawed and that Esther's changes must have occurred naturally. The encouraging results from Esther led to the formal establishment of a hurricane modification program known as *Project STORMFURY* in 1962 (Willoughby et al. 1985).

STORMFURY was a collaboration between NOAA and the U.S. Navy and was directed by the National Hurricane Research Project (now called the Hurricane Research Division). STORMFURY started out apparently well when Hurricane Beulah's eyewall was seeded on August 24, 1963. The eyewall disintegrated, followed by formation of a new eyewall at a larger distance from the eye. The maximum winds decreased by 20 percent and moved farther away from the center. STORMFURY seemed to have a promising beginning.

Four more years were to pass before the next modification experiment. The years 1964–1968 were generally inactive hurricane seasons, and the hurricanes that did occur were either too close to land or out of flight range. Scientists also realized that the eyewall-seeding hypothesis was incorrect, resulting in the revised hypothesis of seeding outside the eyewall. The modified hypothesis was tested on Hurricane Debbie on August 18 and August 20, 1969, when more than 1,000 seedings of silver iodide were made each day. The eyewall shifted outward each day, and the winds decreased by 31 percent and 15 percent, respectively.

Ironically, this was essentially the end of Project STORMFURY. The 1970 hurricane season yielded no suitable candidates for seeding. In 1971, the only eligible storm was Hurricane Ginger, a late-season, diffuse system. It was seeded twice but was a poor candidate because it lacked a small, well-defined eye; the seeding had no effect on Ginger. In 1972, all the storms were too weak, too close to land, or out of flight range. In fact, in general the 1970s were a period of below-average hurricane activity, especially for intense hurricanes. Other difficulties also plagued STORMFURY. The U.S. Navy ended its support to pursue goals more closely related to national defense. Several of the aircraft had become too old for reliable use anymore. Permissions for seeding from Caribbean countries and Mexico became more difficult to obtain, and the State Department increased the time restriction for seeding before landfall (Posey 1994). Eventually, only a narrow zone north of Puerto Rico in which hurricanes were at least thirty-six hours from landfall was allowed for seeding. Unfortunately, no storms ever passed through the small permissible trapezoid of ocean north of Puerto Rico between 1973 and 1979. STORMFURY scientists attempted to move the project into the Pacific where storms are more numerous, but Japan and Australia blocked that move. The Chinese government also was not in favor of the project. Chinese officials were convinced that seeding during the Vietnam War had been successful in muddying the Ho Chi Minh Trail, making it impassible, and were concerned that seeding hurricanes offshore would remove beneficial rainfall to their continent (List 2004). Finally, weather modification experiments in general were falling into public disfavor (Pielke and Pielke 1997). In 1983, Project STORMFURY was terminated.

Also, from the very beginning of STORMFURY and throughout the project's lifetime, several scientists expressed concerns about the seeding hypothesis as well as the interpretation of STORMFURY's experiments. These doubts became substantiated by additional reconnaissance observations in the 1970s and 1980s and by increased knowledge of hurricane structure and evolution during this period. These findings (Willoughby et al. 1985), which refute the basic premise of project STORMFURY, are summarized below:

1. Observational evidence shows that hurricanes actually contain too little supercooled water and too much natural ice for seeding to be effective.
2. Hurricanes go through a natural (but temporary) weakening process in which a new eyewall forms outside the original eyewall. The outer eyewall "chokes off" inflow to the inner eyewall, causing it to dissipate. The outer eyewall then propagates inward, replacing the original eyewall. This process, known as the *concentric eyewall cycle*, lasts twelve to thirty-six hours and is associated with temporary weakening (Willoughby, Clos, and Shoreibah 1982). Therefore, much of the observed weakening in Esther, Beulah, and Debbie may be the result of natural internal evolution rather than seeding. Hurricane intensity is also controlled by external influences, such as water temperature and wind shear. For example, some evidence exists that Debbie moved into a strong wind shear environment conducive to weakening (P. Black, personal communication, 1994). Therefore, the expected results of seeding are indistinguishable from naturally occurring intensity changes.

Project STORMFURY should not be viewed as a failure, however. Even though only a few seeding experiments were performed, many reconnaissance flights were conducted into hurricanes to understand their formation, structure, and evolution. The legacy of Project STORMFURY is a wealth of observations that has augmented our understanding of hurricanes and improved hurricane forecasts. For more information about Project STORMFURY, see Posey (1994), Willoughby et al. (1985), and chapter 6.

The notion of hurricane modification still remains, though (Posey 1994). Occasionally, someone will ask, "Why don't we just nuke a hurricane and destroy it?" Radioactivity aside, such a question demonstrates an extreme underestimate of hurricane power. For example, Hurricane Andrew (1992) generated the equivalent energy of a 10-megaton bomb continuously during its existence—not in a split second as in a bomb explosion! It is doubtful that a hurricane would even be affected by a nuclear blast; the storm would simply spew radioactive fallout throughout the region. Another hypothesis suggests that future earth-orbiting solar power stations could beam microwaves around a hurricane at a frequency where water vapor would absorb it, altering the temperature and pressure fields and ultimately the storm path (Hoffman 2004). Other ideas involve removing a hurricane's energy source. Some commercial entrepreneurs have proposed covering the ocean ahead of a hurricane with an impermeable chemical film or biodegradable powder that would impede evaporation from the sea, thus weakening the hurricane as it moves into that region (Posey 1994). This is unrealistic, though, because it is unlikely that any surface film or powder could withstand the 30- to 50-foot ocean waves in a hurricane, and covering hundreds of square miles of ocean with a substance is a formidable task. Others have promoted stimulating cloud growth in the outer core of a hurricane by increasing the surface temperature with carbon black, which possibly would restrict inflow to the hurricane's inner core (Gray and Frank 1993); however, such a scheme would suffer the same obstacles as Project STORMFURY. Chemical films and carbon black also pose environmental problems. The wisest course of action is to reduce societal exposure to hurricane impacts, not use modification attempts. Pielke and Pielke (1997) offer several recommendations to improve hurricane preparedness policies.

Mitigation and Insurance Procedures

Proper preparation for hurricanes is of paramount importance. Dealing with insurance companies after a hurricane damage likewise can be a harrowing experience. This section discusses procedures for mitigation and insurance. The Federal Emergency Management Agency (FEMA) produces authoritative documents on these issues. The following is based on FEMA's guidelines, but with significant additions from other sources.

Home Protection

Hurricane-force winds are destructive to homes. In extreme storms, such as Hurricane Andrew, the force of the wind alone can cause weak places in your home to fail. Flying debris also breaks windows and doors, allowing high winds inside the home. After Hurricane Andrew, a team of experts examined homes that had failed and ones that had survived. They found four areas that should be checked for weakness: roof, windows, doors, and garage door.

The Roof

Wind damage begins with items such as television antennas, satellite dishes, unanchored air conditioners, wooden fences, gutters, storage sheds, carports, and yard items. As winds increase, the home becomes vulnerable. Wind interacting with a building is deflected over and around it. Positive (inward) pressure occurs on the windward walls and tries to push the building off its foundation. Therefore, the building needs to be anchored properly to its foundation to resist these lateral forces. Negative (outward) pressure occurs to the side and leeward walls. The resulting suction tries to peel away siding. In addition, wind flowing over a roof creates lift similar to that on an airplane wing, especially along windward eaves, roof corners, and leeward ridges. Because wind pressure increases with height, building damage typically begins at roof level and progresses downward and inward. Therefore, roof protection is of critical importance.

Damage depends on the type of roof. Homes with gabled roofs are more likely to suffer damage during a hurricane. A gabled roof looks like an "A" on the ends, with the outside wall going to the top of the roof. The end wall of a home with a gabled roof takes a beating during a hurricane, and those that are not properly braced can collapse, causing major damage to the roof.

In most homes, gabled roofs are built using manufactured trusses. Sheets of roof sheathing, often plywood, are fastened to the trusses with nails or staples, and roofing material is fastened to the sheathing. In many cases, the only thing holding the trusses in place is the plywood on top. This may not be enough to hold the roof in place during a hurricane. Installing additional truss bracing makes the roof's truss system much stronger. If most of the large nails or staples coming through the sheathing have missed the trusses, then the sheathing is not properly installed.

In gabled roofs, truss bracing usually consists of 2×4s that run the length of the roof. The 2×4s should extend the length of your roof, overlapping the ends of the 2×4s across two trusses. Braces should be installed 18 inches from the ridge, in the center span, and at the base, with 8–10 feet between the braces. Two 3-inch, 14-gauge wood screws or two 16d (16 penny) galvanized common nails should be used at each truss. Because space in attics is limited, 3-inch, 14-gauge screws can be used instead; they are easier to install but are discouraged.

Gable end bracing consists of 2×4s placed in an "X" pattern from the top center of the gable to the bottom center brace of the fourth truss, and from the bottom center of the gable to the top center brace of the fourth truss. Most hurricane building codes require two 16d galvanized common nails to attach 2×4s to the gable and to each of the four trusses. If 2×6s are required, then four nails must be used.

Regardless of the roof type, hurricane straps consisting of galvanized metal are designed to help hold the roof to the walls. Hurricane straps may be difficult for homeowners to install and so require a professional. Also, in some communities, the law requires hurricane straps.

Even if the roof survives, shingles typically are blown away. Furthermore, many shingles simply shed water and are not waterproof. As a result, wind-driven rainwater can enter the home because of lost shingles, flapping loose shingles, or through cracks, damaging the ceiling and other household items. To minimize this damage, aged shingles should be replaced, and all shingles should be inspected for looseness before a hurricane. Hurricane building codes require that shingles be sealed with wet/dry cement or class 3 silicon with at least six nails; the nails should go cleanly into the truss. If excessive numbers of shingles are loose or if the roof is more than ten years old, it is best to replace the roof, preferably with a roof covering and sheathing designed to withstand high winds. However, high-wind proof doesn't mean hurricane proof—many manufacturers do not include a hurricane warranty since the designs are for short-term fast winds, not the sustained strong winds of hurricanes. Because shingles are high maintenance, some people forego shingles and have a metal roof installed instead.

Exterior Doors and Windows

The exterior walls, doors, and windows are the protective shell of your home. If the home's protective shell is broken, high winds can enter and put pressure on the roof and walls, causing damage. Therefore, the doors and windows should be strengthened or protected.

Many homes have a double entry door, with an active and an inactive (fixed) door. The bolts or pins that secure most doors are not strong enough for hurricane-force winds. Some door manufacturers provide reinforcing bolt kits made specifically for their doors. The door bolt materials should cost from $10 to $40, depending on the type and finish. Doors with windows will need additional protection from flying debris. The section on storm shutters describes how to protect windows.

Double-wide (two-car) garage doors can pose a problem during hurricanes because they are so large that they wobble as the high winds blow and can pull out of their tracks or collapse from wind pressure. If garage doors fail, high winds can enter the home through the garage and blow out doors, windows, walls, and even the roof.

Certain parts of the country have building codes requiring garage doors to withstand high winds. Some garage doors can be strengthened with retrofit kits, usually costing $70 to $150. Many garage doors can be reinforced at their weakest points. Retrofitting the garage doors involves installing horizontal bracing onto each panel. This horizontal bracing can be part of a kit from the garage door manufacturer. Garage doors may also need heavier hinges and stronger center and end supports. The garage door track should also be checked. If the track is loose or can be twisted, then a stronger track should be installed.

After the garage door is retrofitted, it may not be balanced. Lower the door about halfway and let go. If it goes up or down, the springs will need adjusting. The springs are dangerous and should be adjusted by a professional.

If the garage door cannot be retrofitted, a specially reinforced garage door can be purchased that is designed to withstand winds of up to 120 miles per hour. These doors can cost from $400 to $450 (excluding labor) and should be installed by a professional.

Storm Shutters

Installing storm shutters over all exposed windows and other glass surfaces is one of the easiest and most effective ways to protect your home. All windows, French doors, sliding glass doors, and skylights should be covered. Before installing a shutter, check with a local building official to find out if a building permit is required.

Plywood shutters can offer a high level of protection from flying debris during a hurricane. After measuring each window and each door that has glass, 8 inches should be added to both the height and width to provide a 4-inch overlap on each side of the window or door. Sheets of plywood are generally 4 × 8 feet. Plywood shutter installation requires bolts, wood or masonry anchors, large washers, and ⅝-inch exterior-grade plywood. Windows 3 × 4 feet or smaller installed on a wood-frame house can use ¼-inch lag bolts and plastic-coated permanent anchors. The lag bolts should penetrate the wall and frame surrounding the window at least 1¾ inches. For larger windows, ⅜-inch lag bolts should be used that penetrate the wall and frame surrounding the window at least 2½ inches. Windows 3 × 4 feet or smaller installed on a masonry house can use ¼-inch expansion bolts and galvanized permanent expansion anchors. The expansion bolt should penetrate the wall at least 1½ inches. For larger windows, ⅜-inch expansion bolts should be used that penetrate the wall at least 1½ inches. Required tools include a circular or hand saw, a drill with the appropriately sized bits, a hammer, and a wrench to fit the bolts. Eye protection and work gloves are needed for safety.

After cutting the plywood to the measurements for each opening, holes should be drilled 2½ inches from the outside edge of the plywood at each corner and at 12-inch intervals. You should also drill four holes in the center area of the plywood to relieve pressure during a hurricane. Each hole position needs to be marked on the outside wall. On wood-frame houses, make sure that the anchors are secured into the solid wood that frames the door or window and not into the siding or trim. Each shutter should be marked to indicate where it is to be installed. The shutters and bolts then need to be stored in an accessible place.

Obviously, protecting windows with plywood can be inconvenient, and wood is not the best protection against strong winds. Plywood shutters, if not adequately fastened, will be ripped off in a hurricane. Furthermore, wood can rot and warp with time and may be useless when it is needed the most. As a result, some homeowners have elected to use aluminum shutters instead. They are considerably more expensive but are effective against hurricane winds. Different types include storm panels, Bahama or awning, accordion, and rolldown. These shutters meet the tough building code standards of Miami Dade County and are therefore sufficient for all hurricane-threatened regions.

The cheapest and most popular shutters are storm panels, which require a preinstalled steel channel on the top and bottom of the window in which the panels are slid into and must be anchored prior to installation of the shutters. They require some physical effort to install and may cost from $7 to $15 per square foot of coverage. They need to be stored when not in use, but because they overlap they use little space. Bahama and awning shutters are mounted above the window. They provide shade when not in use but are lowered and fastened to the wall when a hurricane threatens. Prices are $15 to $20 per square foot. Accordion shutters are made of aluminum blades connected to each other vertically, moving horizontally between an upper and lower track. They fold away on each end of the window when not in use and then are pulled out during a hurricane threat (hence the name "accordion"). They cost from $15 to $25 per square foot. The rolldown shutters are motorized and can be lowered with the flip of a switch. In addition to protecting against hurricanes, they provide effective security against theft. They are very expensive, costing from $20 to $35 per square foot of window area, but are the easiest to use.

A final suggestion is to install impact-resistant windows in addition to a storm shutter. This type of window can provide sufficient protection in weaker storms, thereby avoiding the trouble of using shutters except in strong hurricanes. As a weak alternative, you can also use safety film on windows that will hold glass shards together if the window is broken; however, safety film is generally not adequate protection for windows in hurricanes.

Source: Against the Wind: Protecting Your Home from Hurricane Wind Damage, FEMA, 7 pp. (also available at http://www.fema.gov); *Hurricane Shutter Information,* Hurricane Research Division, http://www.aoml.noaa.gov/hrd/shutters/; *Rollac Shutter of Texas, Inc.,* http://www.rollac.com.

Hurricane Preparedness Steps

The following is a list of hurricane preparedness steps.

Before Hurricane Season

- Be familiar with escape routes, with several routes for different contingencies. Also know the location of the nearest shelter.
- Know the storm surge history and elevation of area and home.
- Post emergency phone numbers (fire, police, and ambulance) by the telephone.

- Teach children how to call 911 and the emergency numbers for help.
- Teach your family basic safety measures, such as CPR and first aid.
- Know how to turn off utilities.
- Identify family meeting places in case you are separated. Choose a place in a building or park outside your neighborhood. Everyone should be clear about this location.
- Develop an emergency communication plan. Identify a dependable out-of-state contact and provide this phone number to all family members.
- Check your home for loose and clogged gutters and downspouts.
- Test smoke detectors monthly and change batteries twice a year.
- Keep trees and shrubbery trimmed. Cut weak branches and trees that could brush against the house.
- Determine where to move cars and boats in case of floods or evacuation.
- Purchase and learn how to use a generator. A 3,500-watt generator can run a refrigerator, window air conditioner, TV, and a few lights. A 5,000-watt generator provides more flexibility for running additional appliances and is preferred for backup power. Run it monthly to be certain it is operational.
- Also purchase an inverter, which converts direct current from a car battery to alternate current and can be used to run a few small appliances such as a light, fan, and TV. A $70 model can power up to 800 watts. By idling the car outside (not in the garage, which can cause carbon monoxide poisoning), the battery will be continuously charged by an alternator, and small appliances hooked to an extension cord can be used. A car is actually quieter and more fuel-efficient than a generator!
- A rechargeable portable power unit and a 40-watt solar panel are recommended in addition to an inverter and generator. This setup also can provide light and electricity to small appliances, and the power unit can be recharged in eight hours using the solar panel.
- Make a homemade air conditioner, which consists of an ice chest, a styrofoam lid to cover the cooler, and a battery-powered 6-inch fan. Make a 1-inch hole on one end of the lid and a 6-inch hole on the other end where the fan is

placed on its side, pointing upward. Fill the cooler with ice. With the fan on, warm air enters the small hole and cold 50°F air comes out the fan. This will cool a very small room for a few hours and then will require fresh ice. Commercial ice chest air conditioners are also available at a cost of about $40. This is a decent alternative when not enough power is available for a conventional air conditioner.
- Prepare a disaster supply kit, which consists of:

  Flashlight and extra batteries
  Portable battery-operated radio and extra batteries
  Deluxe first-aid kit
  Nonelectric can opener
  Extra cash and credit cards
  Extra keys
  Sturdy shoes, rubber boots, a change of clothing, and rain gear
  Portable cooler
  Matches or butane lighters
  Water purification pills (available at sporting good stores and drugstores)
  Pure, unscented liquid chlorine bleach for water purification
  Fire extinguisher
  Rope
  Duct tape
  Tarp
  Baby food, diapers, and formula
  Soap, shampoo, toiletries, and hygiene items
  Insect repellent and mosquito netting
  Heavy-duty trash bags
  Alarm clock
  Blankets and pillows
  Eating and cooking utensils
  Emergency cooking facilities (grill or camp stove)
  Gloves and goggles
  Pots and pans
  Sleeping bags
  Sponges and paper towels
  Toys (for children)
  Lanterns and extra fuel
  Sunscreen
  Signal flare

72 Problems, Controversies, and Solutions

Paper and pencil
Cards, games, and books
Waterproof containers

- Assemble clean-up and repair supplies:

Hand ax
Wrecking and crow bars
Corner braces
Brooms and mops
Caulk and caulking gun
Steel chain
Chain saw and handsaw
Cleaning supplies and disinfectant
Dust pan
Extension cords
Hammers and hatchets
Ladders
Paint
Plastic sheeting
Rakes and shovels
Wheelbarrow

When a Hurricane Watch or Warning Is Issued

- Listen to radio or television for hurricane progress reports.
- Store at least three days' worth of nonperishable (canned or packages) food and beverages. This includes staples (sugar, salt, and pepper), high-energy (carbohydrate) foods and protein bars, vitamins, and comfort/stress foods (cookies and candy). Two weeks' worth is recommended.
- Check the disaster supply kit.
- Fill bathtub and large containers with water for sanitary purposes.
- Bring in outdoor objects such as lawn furniture, toys, and garden tools. Anchor objects that cannot be brought inside but that could be wind-tossed. Remove outdoor antennas, if possible.
- Secure home by closing or installing window shutters.
- Turn the refrigerator and freezer to the coldest setting. Open only when necessary and close quickly if power is lost.

- Unplug small appliances.
- Turn off propane tanks.
- Fuel and service car.
- Moor boat securely with storm lines or move it to a designated safe place. However, many insurance companies will not cover a boat outside a flood protection levee system, so it's best to move the boat at least inside the levees. Also, boat owners who use stackable storage marinas should move their boats, as stackable systems can collapse in hurricanes and marinas typically absolve themselves of responsibility in their contracts. Fill the boat with fuel and charge the battery in case the boat is needed during a flood. Remove loose items from trailored boats, and also remove the drain plug. Loosen the trailer straps and tie a bowline to a sturdy object that will hold the boat in case of high floodwaters.
- Store valuable items and irreplaceable items (such as photos) that must be left behind in lockable plastic storage containers or waterproof containers.
- Gather vital documents such as deeds, insurance papers, birth certificates, stocks and bonds, legal papers, credit card and bank account numbers, and immunization records, and keep these in a mobile waterproof container.
- Purchase a two-week supply of essential medicines, especially prescription drugs.
- Mobile home owners: Mobile homes are particularly vulnerable to high winds. Anchor the mobile home with frame ties and leave.
- High-rise occupants: Hurricane winds are faster at higher elevation, and residents should plan to leave.
- Pool owners: Drop water level 1–2 feet. Turn off pool pump. Add extra chlorine to compensate for heavy rains.
- Identify a safe room in case of high winds. Close all interior doors. Secure and brace external doors. Also be wary of tornadoes spawned in a hurricane. The safest room will be an interior room on the first floor with small or few windows. Lie on the floor under a table or another sturdy object. Wait for the all clear on the radio; don't be fooled by the calm winds in the eye.
- Sewage could be interrupted. Keep 5-gallon buckets for use as emergency toilets. Line each bucket with a heavy-duty plastic trash bag.

- Have camera ready for insurance documentation.
- If you have a gas grill, fill the propane tank for cooking after the storm if the power is lost.
- Put vehicles in the garage.

If Evacuation Is Necessary

- Unplug appliances and turn off electricity and the main water valve.
- Elevate furniture to protect it from flooding or move it to a higher floor.
- Take disaster supply kit.
- Contact family members to coordinate evacuation efforts.
- Leave early, preferably during daylight. In urban areas, take a back road instead of an interstate highway, since these can become jammed.
- Eat before you leave. Meals may be difficult to obtain during the evacuation.
- Take important documents and insurance papers. Bring identification. These are needed for a variety of reasons. In addition, they show proof of residence when you return after the storm.
- Back up computer on a portable media and bring it with you.

After the Hurricane Passes

- Return home only after authorities say it is safe.
- Beware of downed or loose power lines. Report them immediately to the power company, police, or fire department.
- Enter your home with caution. Open windows and doors to ventilate or dry your home.
- Check for gas leaks. If you smell gas or hear a blowing or hissing noise, quickly leave the building and leave the doors open. Call the gas company.
- Look for electrical system damage. If you see sparks or frayed wires, turn off electricity at the main switch box. If you have to step in water to reach the electric box, call an electrician for advice.
- Check for sewage- and water-line damage. If you suspect there is such damage, call the water company and avoid using water or toilets until repairs are made.

- Take pictures of the damage for insurance. Make a list of damage to property.
- If the water supply becomes exhausted, water purification procedures are necessary. First, let any suspended particles settle to the bottom or strain the water through layers of clean cloth. Then, purify by boiling (safest way), using liquid chlorine bleach or purification tablets. The water heater also is a source of water.
- Be wary of unlicensed contractors, con artists, and door-to-door repairmen who will offer to do the work quickly but at inflated prices. Insist on written estimates from several companies, and be sure a contract or business card has an address, a telephone number, and a license number. Contact the Better Business Bureau to do a background check. Don't be pressured into quick decisions. Use contractors known in the community, and check their references. Make sure the contractor can furnish a license and has general liability and workers' compensation insurance. The website http://www.contractors-license.org may be useful. Insist on start and completion dates, and limit down payments to one-third of the total price. Do not pay the balance until the work is completed to your satisfaction and you have proof that any subcontractors have been paid (subcontractors can file a lien against you if the contractor does not pay them).
- Beware of price gouging. Most states have laws against this practice, and some have hotlines to report price gouging.
- The storm surge can drive poisonous snakes from their usual habitats. Be on the lookout for them.
- Avoid using candles and other open flames indoors.
- Do not plug generator directly into house system. Plug appliances directly into the generator.
- Hurricanes moving inland can cause severe flooding. Stay away from riverbanks and streams until all potential flooding is past.
- The IRS is usually generous about extending filing deadlines to October 15, and estimated tax payments deadlines are also usually extended. In addition, an unusual tax break is available to those who suffer a casualty loss in regions the president declares as federal disaster areas. Details are in IRS Publication 547 ("Casualties, Disasters, and Thefts"). Here are a few key points. One has to itemize

the deductions to claim a loss for personal-use property. Also, one can deduct personal-property losses not covered by insurance or reimbursements. However, there are some important, steep limits. First, subtract $100 for each "casualty event." Then, the total casualty (and theft losses) must exceed 10 percent of your adjusted gross income (AGI). For example, suppose a household with $70,000 adjusted gross income experiences $10,000 uninsured storm damage to the house. The casualty loss deduction can be computed by subtracting $100 from the $10,000 ($9,900), minus 10 percent of the $70,000 AGI ($7,000), resulting in a deduction of $2,900. Additional uninsured casualty events can also be added, minus $100.

- A common question is whether evacuation costs such as gasoline, hotel room, meal, and snack expenses are tax deductible. Unfortunately, these are considered personal expenses and are not tax deductible.
- If you are short on cash, you can borrow from your individual retirement account without penalty if you repay the money within sixty days. This should be done with extreme caution and only as a last resort. If you miss the deadline, you will pay a 10 percent penalty as well as taxes on the amount withdrawn.
- Another option for retrieving cash is to amend your previous year's tax return with damages incurred from the current year.
- Several key FEMA disaster air programs exist in the event of a presidential major disaster declaration. This includes rental payments for temporary housing; grants for home repairs and replacement of essential household items not covered by insurance; grants to replace personal property and help meet medical, dental, funeral, transportation, and other serious disaster-related needs not covered by insurance or other federal, state, and charitable aid programs; unemployment payments up to twenty-six weeks for workers who temporarily lose jobs because of the disaster and who do not qualify for state benefits, such as self-employed individuals; low-interest loans to cover residential losses not fully compensated by insurance; loans up to $1.5 million for small businesses that have suffered disaster-related cash flow problems and need funds for working capital to recover from the disaster's

adverse economic impact; loans up to $500,000 for farmers, ranchers, and aquaculture operators to cover production and property losses, excluding primary residence; crisis counseling for those traumatized by the disaster; income tax assistance for filing casualty losses; and advisory assistance for legal, veteran benefits, and social security matters. To see if you qualify, contact FEMA at 1-800-621-FEMA (3362).

- The Red Cross will provide shelter, food, and health and mental health services as well as help individuals and families to resume their normal daily activities independently. This may include a referral or a way to pay for what is needed most: groceries, new clothes, rent, emergency home repairs, transportation, household items, medicines, materials for short-term repairs, and occupational tools. The Red Cross may also help those needing long-term recovery assistance when all other available resources, including insurance, government, private, and community assistance, are either unavailable or inadequate to meet the needs. All assistance is based on verified disaster-caused needs, and all assistance is free—literally a gift as a result of the generous support of the American people. The Red Cross also feeds disaster victims and emergency workers, handles inquiries from concerned immediate family members outside the disaster-affected area, provides blood and blood products to disaster victims, and links disaster victims to other available resources.

Make Plans for Your Pets

- If you evacuate your home, do not leave pets behind.
- Be sure you have up-to-date identification tags, a pet carrier, and a leash for your pets.
- Most emergency shelters will not accept pets. In the event of evacuation, make alternative arrangements for pets, such as with family, friends, veterinarians, or kennels in safe locations. Send medicine, food, feeding information, water, and other supplies with them that will last for two weeks. Be sure to provide food, a water bowl, blankets, your pet's toys, a brush, a nonelectric can opener, newspapers, cat litter, litter box, scoop, and plastic bags for handling waste. Ask if a crate should be provided.

- While hotels generally do not allow pets, some will make exceptions for those leaving evacuation areas. If you are staying at a hotel, be sure to ask if they will allow pets in this circumstance.
- Store a copy of current vaccination records in a plastic bag for safekeeping.

Insurance Preparation

Homeowners insurance should be evaluated before each hurricane season. In particular, homeowners should check the hurricane or windstorm deductible. Most insurance companies require a 2 percent deductible of the value of the house (or policy). This means that for a $100,000 policy, the homeowner has to pay the first $2,000 in damages from a tropical storm or hurricane. Some policies can be as high as 5, 10, or 15 percent. Many homeowners, used to paying a $500 or $1,000 deductible for other damage, are not prepared for such costs. The hurricane deductible was added in 1996 to enable insurance companies, scared by the $20 billion damage caused by Hurricane Andrew in 1992, to continue selling policies in coastal states. Therefore, homeowners should set money aside each year in an emergency fund. Note that this applies to each storm as well. Florida residents suffering damage from both Hurricane Charley and Frances had to pay two deductibles!

You should also check the type of coverage. Replacement cost coverage, which will pay the cost of replacing the home at today's costs up to the limits of coverage on the policy, is recommended. A minimum 80 percent coverage of the current house value (or the maximum amount of insurance available) is required or there will be a penalty applied to reimbursements, so it is important that the replacement cost policy stays current with any house value appreciation. Some policies instead use actual cash value, which is replacement cost minus depreciation. This is a bad deal, because repairs for a ten-year-old home will be covered less than for a one-year-old home. A third alternative, if it's available, is guaranteed replacement cost, which means that the insurer will pay the cost of repairs even if it's greater than the insured amount.

Sometimes obtaining homeowner's insurance can be difficult. For example, insurance companies have reduced their exposure in the coastal regions of Florida and Louisiana. Homeowners who make three or more claims in three years, are identified as high risk and may lose coverage. Other factors, such as bad credit and even making insurance inquiries about home damage but not filing a claim (which some insurance companies will count as unpaid losses!), can count against the homeowner. For those who lose or can't obtain coverage, state-sponsored programs exist for homeowners in hurricane-prone regions. Typically, these rates are higher than standard rates.

Also, when a hurricane is threatening the region, insurance companies will not issue any new policies. This region can be rather large. For example, once a hurricane enters the Gulf of Mexico, no insurance companies will issue policies along the entire Gulf Coast.

Homeowners policies do not cover damage from rising waters, and therefore flood insurance is highly recommended (and often required) for a home. Flood insurance is backed by the federal government and is available to any homeowner, renter, or business owner whose property is in a community that participates in the National Flood Insurance Program (NFIP). The NFIP is paid by policyholder premiums, not taxpayer dollars. In order to participate, the community must adopt and enforce local floodplain management ordinances designed to reduce the risk of future flood losses. If you live in such a community, you can purchase flood insurance from any licensed agent or company—the same one, for example, who handles your homeowners or automobile insurance. The NFIP can also be contacted at 1-800-638-6620. The rates are set by the federal program, and do not vary from insurer to insurer. The rates do vary on whether you are in a flood zone or nonflood zone, the selected deductible, amount of coverage, type of building, and the elevation of the base of the home. Flood insurance covers structural damage, flood debris cleanup, and floor surfaces such as carpeting and tile. However, the standard version of flood insurance does not cover house contents, so make sure this is added to the policy as well (some states require this). Renters can buy flood insurance to cover their belongings. It is important to remember that no claims can be made unless the policy has been in effect for thirty days.

Some people resist buying flood insurance in the faulty belief that if flooding occurs, the government will bail them out. Federal disaster assistance is available only if a flood (or other disaster) is so large and widespread that it warrants a major disaster declaration from the president. More than 90 percent of flood events are not presidentially declared (although generally hurricanes are declared). Furthermore, even for floods that are declared major disasters, the aid available is limited. Most assistance is in the form of loans that must be repaid, with interest, in addition to the mortgage loan that still has to be paid.

Unfortunately, the NFIP is not a perfect system. The NFIP was created in 1968 in response to large-scale flooding of the Mississippi River. When insurance companies threatened to stop coverage because of the escalating costs, the government decided to offer insurance in exchange for local government's adopting strict mitigation standards. However, with time, the NFIP has become a complex system of public agencies and private corporations, further complicated in 1983 when the program authorized private insurers to sell the service. The NFIP has also issued a myriad of policy interpretations over the years. The result has been some insurers not understanding flood insurance policies and making unfair settlement offers. The problems became apparent after Tropical Storm Isabel (2003) caused extensive storm surge damage in Maryland. Many settlement offers were far below (by tens of thousands of dollars) the rebuilding costs. Other policyholders were also unaware that they needed a separate policy for household contents. The State of Maryland eventually had to provide help with a low-interest loan program. Complaints were so numerous that congressional hearings on the NFIP were held, and reforms as of this writing are currently being pursued in Congress. Recommendations include establishing an appeal process through FEMA, clearer paperwork for filing claims and to understand what is covered, and better training of insurance agents. Currently, Congress is also investigating whether the NFIP is sufficient to restore policyholders to their predisaster states as well as investigating its practices and the fairness of insurance adjusters. Despite these problems, the NFIP has been successful in obtaining a large pool of policyholders (4 million) in 20,000 communities and is better than no flood insurance.

Here are some tips regarding insurance:

- Document your belongings. If possible, compile a room-by-room inventory with manufacturers' names, dates, and places of purchases, serial numbers, and prices, especially for major appliances. Use a camera to record images of your belongings. Also, save any appraisals.

- Check the amount of coverage. If you have upgraded the house or bought expensive furnishings, increase the coverage.
- Don't make claims unless there is significant damage. Insurance companies may drop a policy based on the number of claims, and not the total cost of the claim.
- Gazebos, piers, and landscaping generally are not covered by either flood insurance or homeowners insurance.

Here are some tips regarding insurance claims:
- If your property is damaged, contact the agent who sold you the policy. Provide her or him with a phone number where you can be reached. A cell phone number is recommended if the cell phone network is functioning.
- Report claims as soon as possible.
- While your first instinct is to make major repairs, it's important to wait until the insurance adjuster visits the property. However, do take actions to prevent further damage, such as putting a tarp over roof damage to stop further water damage. If you must hire repair crews, make sure all work is documented. Keep receipts for any related expenditures.
- Be patient and expect delays having an adjuster view your property. It can take weeks or months after a major disaster. Try to schedule an appointment if possible. The wait could be worthwhile, because the adjuster could make allowances for damage you did not know were covered by the policy. When the adjuster arrives, present her or him with a copy of the policy, pictures, and a list of possessions. Pictures of the property before and after the storm will be beneficial as well. Separate the damaged from the undamaged property. Take good notes on all conversations with adjusters. Check to see if your insurance company has a mobile catastrophe center where a settlement can be obtained with receipts and pictures.
- The first check is often an advance, not a final payment.
- On-the-spot settlement checks can be accepted right away. If other damage is found, a claim can be reopened within one year of the disaster.
- If both the home's structure and personal belongings are damaged, generally two checks are sent. A separate check may also be sent for living expenses.

- If the home is mortgaged, the check will be made out to both you and the mortgage lender. The lender gets equal rights to the insurance check to ensure that the necessary repairs are made. Therefore, the mortgage company will have to endorse the check. Lenders generally put the money in an escrow account and pay for the repairs as the work is completed. Unfortunately, the mortgage company may withhold funds until the work is completed, forcing the homeowner to pay ahead of time with her or his own cash to get the work completed.
- If possessions are insured and you have a replacement cost policy, normally the items have to be replaced before the insurance company will pay for the replacement cost. If you decide not to replace some items, you will be paid the actual cash value. Generally, the insurance company allows several months to replace items.
- If the home is destroyed and you decide not to rebuild, the settlement amount depends on state law, court policies, and the insurance policy.
- Compare your list of damages to the proposed settlement offer. Don't settle until you understand what is covered.
- If you disagree with an insurance settlement, several options exist. First, talk to your agent or claims manager to explain your situation, and provide supporting documents. If no success, then call the National Insurance Consumer Helpline at 1-800-942-4242. Also, contact your state insurance department, who will try to resolve the matter. If the matter involves flood insurance, appeal the process through the NFIP. If still no success, hire a public adjuster or attorney and provide all relevant paperwork, the insurance policy, and settlement offers. Usually the public adjuster will get 10–15 percent of the settlement offer, and an attorney will get 30 percent.

Source: *Surviving the Storms: A Guide to Hurricane Preparedness*, FEMA, 6 pp. (available from http://www.fema.gov); New Orleans Channel 6 WDSU, http://www.theneworleanschannel.com; Mary Judice, "It's Time to Check Insurance Coverage," *Times-Picayune*, June 8, 2003; Andrew Green, "Panel OKs Flood Insurance Reforms," *Baltimore Sun*, March 31, 2004.

3

Worldwide Perspective

International Impact, Names, and Locations of Hurricanes

Every tropical ocean except the South Atlantic and eastern South Pacific contains tropical depressions, tropical storms, and hurricanes. The international generic term for such storms is *tropical cyclones*, and this designation will be used throughout this chapter when not referring to a storm-intensity category. Tropical cyclones occur in the tropical North Atlantic Ocean (including the Caribbean Sea and Gulf of Mexico); the Eastern North Pacific (off the west coast of Mexico); the Central North Pacific (near Hawaii); the western North Pacific (including the China Sea, Philippine Sea, and Sea of Japan); the North Indian Ocean including the Bay of Bengal and the Arabian Sea; the Southwest Indian Ocean off the coasts of Madagascar and extending almost to Australia; the Southeast Indian Ocean off the northwest coast of Australia; and the Southwest Pacific Ocean from the east coast of Australia to about 140°W.

Regardless of location, tropical cyclones pose the same triple threat of powerful winds, storm surge, and heavy rains throughout the world. The societal impacts depend on storm intensity, the topography of the area, the localand regional economy, the state of development in the country, community demographics, and the status of physical and social infrastructure. A World Meteorological Organization (WMO) survey revealed the following annual statistics: 1,145 dead or missing, 700,000 houses destroyed,

1.5 million people left homeless, and financial losses of $3.6 billion. Land-falling tropical cyclones can devastate regions and cause situations where authorities can no longer provide the basic necessities of food, safe drinking water, adequate health care, and electricity, resulting in social upheaval. Medium- and long-range consequences include the physical and emotional drain on people, breakdown of family and community support systems, destruction of crops and livestock, loss of business revenues, communication failures, and transportation problems.

These problems are compounded in less-developed, highly populated countries. As discussed in chapter 2, India and Bangladesh have historically suffered astounding fatalities—often in the tens of thousands and sometimes in the hundreds of thousands—from storm surge in the Bay of Bengal. The shallow water and the bay's funnel configuration, combined with large astronomical tides, result in storm surges of 20 feet or more. Unfortunately, the societal structure makes evacuations problematic. The delta region is heavily cultivated and densely settled with poor people living in flimsy housing, and many are resistant to evacuation because they don't want to leave their land and meager possessions. The remoteness also makes evacuation difficult. Charitable organizations such as the World Bank and the Red Crescent Society have funded approximately 2,000 shelters, but each can only accommodate 1,000 people. The result is a situation prime for exorbitant fatalities. Since each storm also devastates their livelihood based on agriculture and cattle, the economic hardship magnifies the situation.

Latin America and the Caribbean are also susceptible to large death tolls, as illustrated by Hurricane Mitch (1998) that killed more than 10,000 people, affected 6.7 million, and caused $8.5 billion in damage. Mitch produced the equivalent of a year's worth of rain in some areas of Honduras and Nicaragua, resulting in flash floods and landslides that killed thousands in exposed areas. El Salvador, Guatemala, Belize, and Costa Rica were also impacted. Other countries such as China and Cuba experience similar problems in remote regions (with fortunately lower fatalities).

Pielke et al. (2003) identify that hurricane disasters are correlated to increasing populations, widespread poverty, lack of access to adequate land, deforestation, and urbanization. Population growth increases vulnerability, because more people can be impacted and more people settle in exposed areas. Impoverished people often lack access to favorable land, since much of it is owned by a few large farms, businesses, etc. Others seek employment in cities but can't afford housing. The result is that some live on steep hillsides and flood-prone areas. Agricultural, farming, mining, and logging practices cause deforestation, which increases vulnerability to soil erosion, landslides, and flooding. In addition, population growth is related to poverty, because some of the poorest countries have the highest fertility rates. Because of these factors, international calamities in poorly developed countries are still a major concern.

Aspects of the innovative Saffir-Simpson Hurricane Scale (see chapter 2) do not apply to the tropical Pacific. The scale does not consider: (1) tropical storm-force winds than can do considerable destruction to agriculture and poorly built structures, (2) the effects of termites and wood rot on weakening wood structures, (3) differing building practices compared to the United States, (4) Pacific vegetation; (5) the detrimental effects of salt spray on metals and vegetation, and (6) the effects of coral reefs on storm surge and waves. To address these differences, Guard and Lander (1993) developed a Saffir-Simpson scale for the tropical Pacific, shown in table 3.1.

A variety of regional or local names exists for tropical cyclones throughout the world (Jennings 1970). Probably the best known is *typhoon* (derived from the Chinese word *ty-fung*, meaning "great wind"), which is the equivalent to the Atlantic hurricane, used for storms over the Western Pacific Ocean west of 180°E. In the Philippines, tropical cyclones are called *"chubasco"* or *"baguio."* Japan calls them *"reppu,"* while Mexico calls storms in the eastern North Pacific *"cordonazo."* Some Australians give intense hurricanes the colorful name "cock-eyed Bob."

It is a popular belief that modern Australians call tropical cyclones "willy-willies." The word was coined by a colonial settler sometime in the early 1900s and was only used for tropical cyclones in Northwest Australia. However, this practice ended more than fifty years ago. Further confusing matters is that Southeast Australians use the same term (willy-willy) to describe dust devils. A dust devil is a small, short-lived whirl of dust, sand, or debris in rising columns of air created by strong solar heating near the ground; only occurs over land; and is a different phenomenon than tropical cyclones.

Many U.S. residents perceive the North Atlantic Ocean basin as a proliferate producer of tropical cyclones due to the publicity these storms generate. In reality, the North Atlantic is a marginal

TABLE 3.1
The modified Saffir-Simpson Tropical Cyclone Scale for the tropical Pacific

Category	Maximum sustained winds in mph	Coastal inundation and wave action	Damage
Category A: Weak Tropical Storm	30–49 (peak gusts 40–64)	On windward coasts, sea level rise of less than 2 ft above normal in open bays and inlets due to storm surge and wind-driven waves; breaking waves inside bays can reach 2–3 ft; less than 1 ft over reefs. Rough surf at reef margin with moderately strong longshore currents and rip currents inside reefs.	Damage done to only the flimsiest lean-to type structures. Unsecured light signs blown down. Minor damage to banana trees and near coastal agriculture, primarily from salt spray. Some small dead limbs, ripe coconuts, and dead palm fronds blown from trees. Some fragile and tender green leaves blown from trees such as papaya and fleshy broad leaf plants.
Category B: Severe Tropical Storm	50–73 (peak gusts 65–94)	On windward coasts, sea level rise of 2–4 ft above normal in open bays and inlets due to storm surge and wind-driven waves; breaking waves inside bays can reach 4–6 ft; 1–2 ft over reefs. Very rough surf at reef margin with strong longshore currents and rip currents inside reefs.	Minor damage to buildings of light material; major damage to huts made of thatch or loosely attached corrugated sheet metal or plywood. Unattached corrugated sheet metal and plywood may become airborne. Wooden signs not supported with guy wires are blown down. Moderate damage to banana trees, papaya trees, and most fleshy crops. Large dead limbs, ripe coconuts, many dead palm fronds; some green leaves, and small branches are blown from trees.
1 Minimal Typhoon	74–95 (peak gusts 95–120)	On windward coasts, sea level rise of 4–5 ft (1.2–1.5 m) above normal in open bays and inlets due to storm surge and wind-driven waves; breaking waves inside bays can reach 5–7 ft (1.5–2.1 m) above normal; 2–3 ft (0.6–1.0 m) additional water across reef. Wind-driven waves may inundate low-lying coastal roads where reefs are narrow. Minor pier damage. Some small craft in exposed anchorages break moorings.	Corrugated metal and plywood stripped from poorly constructed or termite-infested structures; and may become airborne. A few wooden, nonreinforced power poles tilted, and some rotten power poles broken. Some damage to poorly constructed, loosely attached signs. Major damage to banana trees, papaya trees, and fleshy crops. Some young trees downed when the ground is saturated. Some palm fronds crimped and bent back through the crown of coconut palms; a few palm fronds torn from the crowns of most types of palm trees; many ripe coconuts blown from

2 Moderate Typhoon	96–110 (peak gusts 121–139)	On windward coasts, sea level rise of 6–8 ft above normal in open bays and inlets due to storm surge and wind-driven waves; breaking waves inside bays can reach 7–9 ft above normal; water is about 3–5 ft above normal across reef flats. Wind-driven waves will inundate low-lying coastal roads below 4 ft on windward locations where reefs are narrow. Some erosion of beach areas, some moderate pier damage, and some large boats torn from moorings.	coconut palms. Less than 10% defoliation of shrubbery and trees; up to 10% defoliation of tangantangan. Some small tree limbs downed, especially from large bushy and frail trees such as mango, African tulip, poinciana, etc. Some rotten wooden power poles snapped and many nonreinforced wooden power poles tilted. Some secondary power lines downed. Damage to wooden and tin roofs and to doors and windows of termite-infested or rotted wooden structures, but no major damage to well-constructed wooden, sheet metal, or concrete buildings. Considerable damage to structures made of light materials. Major damage to poorly constructed signs. Exposed banana trees and papaya trees totally destroyed; 10–20% defoliation of trees and shrubbery; up to 30% defoliation of tangantangan. Light damage to sugarcane and bamboo. Many palm fronds crimped and bent through the crown of coconut palms, and several green fronds ripped from palm trees. Some green coconuts blown from trees. Some trees blown down, especially shallow-rooted ones such as small acacia, mango, and breadfruit, when the ground becomes saturated.
3 Strong Typhoon	111–130 (peak gusts 140–165)	On windward coasts, sea level rise of 9–12 ft above normal in open bays and inlets due to storm surge and wind-driven waves; breaking waves inside bays can reach 11–14 ft above normal; water is about 5–7 ft above normal across reef flats. Wind-driven waves will inundate low-lying coastal roads below 7 ft of elevation on windward locations where reefs are narrow. Considerable beach erosion. Many large boats and some large ships torn from moorings.	A few nonreinforced hollow-spun concrete power poles broken or tilted and many nonreinforced wooden power poles broken or blown down; many secondary power lines downed. Practically all poorly constructed signs blown down, and some stand-alone steel-framed signs bent over. Some roof, window, and door damage to well-built, wooden, and metal residences and utility buildings. Extensive damage to wooden structures weakened by

continues

TABLE 3.1
(Continued)

Category	Maximum sustained winds in mph	Coastal inundation and wave action	Damage
			termite infestation, wet and dry wood rot, and corroded roof straps (hurricane clips). Nonreinforced cinder block walls blown down. Many mobile homes and buildings made of light materials destroyed. Some glass failure due to flying debris, but only minimal glass failure due to pressure forces associated with extreme gusts. Some unsecured construction cranes blown down. Air is full of light projectiles and debris. Major damage to shrubbery and trees; up to 50% of palm fronds bent or blown off; numerous ripe and many green coconuts blown off coconut palms; crowns blown off of a few palm trees. Moderate damage to sugarcane and bamboo. Some large trees (palm trees, breadfruit, monkeypod, mango, acacia, and Australian pines) blown down when the ground becomes saturated; 30–50% defoliation of most trees and shrubs; up to 70% defoliation of tangantangan. Some very exposed panax, tangantangan, and oleander bent over.
4 Very Strong Typhoon	131–155 (peak gusts 166–197)	On windward coasts, sea level rise of 13–18 ft above normal in open bays and inlets due to storm surge and wind-driven waves; breaking waves inside bays can reach 15–24 ft above normal; water is about 8–12 ft above normal across reef flats. Wind-driven waves will inundate coastal areas below 12-ft elevation. Large boulders carried inland with waves. Severe beach erosion. Severe damage to port facilities including some loading derricks and gantry cranes. Most ships torn from moorings.	Some reinforced hollow-spun concrete and many reinforced wooden power poles blown down; numerous secondary and a few primary power lines downed. Extensive damage to nonconcrete roofs; complete failure of many roof structures, window frames, and doors, especially unprotected, nonreinforced ones; many well-built wooden and metal structures severely damaged or destroyed. Considerable glass failure due to flying debris and

5 Devastating Typhoon	155–194 (peak gusts 198–246)	On windward coasts, sea level rise of > 25 ft above normal in open bays and inlets due to storm surge and wind-driven waves; breaking waves inside bays can be > 30 ft above normal; water is about 12–18 ft above normal across reef flats. Serious inundation likely for windward coastal areas below 18-ft elevation. Very large boulders carried inland with waves. Extensive beach erosion. Extensive damage to port facilities including most loading derricks and gantry cranes. Virtually all ships, regardless of size, torn from moorings.	explosive pressure forces created by extreme wind gusts. Weakly reinforced cinder block walls blown down. Complete disintegration of mobile homes and other structures of lighter materials. Most small- and medium-sized steel-framed signs bent over or blown down. Some secured construction cranes and gantry cranes blown down. Some fuel storage tanks may rupture. Air is full of large projectiles and debris. Shrubs and trees 50–90% defoliated; up to 100% of tangantangan defoliated. Up to 75% of palm fronds bent, twisted, or blown off; many crowns stripped from palm trees. Numerous green and virtually all ripe coconuts blown from trees. Severe damage to sugarcane and bamboo. Many large trees blown down (palms, breadfruit, monkeypod, mango, acacia, and Australian pine). Considerable bark and some pulp removed from trees; most standing trees are void of all but the largest branches (severely pruned), with remaining branches stubby in appearance; numerous trunks and branches are sandblasted. Patches of panax, tangantangan, and oleander bent over or flattened.	Severe damage to some solid concrete power poles, numerous reinforced hollow-spun concrete power poles, many steel towers, and virtually all wooden poles; all secondary power lines and most primary power lines downed. Total failure of nonconcrete reinforced roofs. Extensive or total destruction to nonconcrete residences and industrial buildings. Some structural damage to concrete structures, especially from large debris such as cars, large appliances, etc. Extensive glass failure due to impact of flying debris and explosive pressure forces during extreme gusts. Many well-constructed storm shutters ripped from structures.

continues

TABLE 3.1
(Continued)

Category	Maximum sustained winds in mph	Coastal inundation and wave action	Damage
			Some fuel storage tanks rupture. Nearly all construction cranes blown down. Air full of very large and heavy projectiles and debris. Shrubs and trees up to 100% defoliated; numerous large trees blown down. Up to 100% of palm fronds bent, twisted, or blown off; numerous crowns blown from palm trees; virtually all coconuts blown from trees. Most bark and considerable pulp removed from trees. Most standing trees are void of all but the largest branches, which are very stubby in appearance and severely sandblasted. Overall damage can be classified as catastrophic.

Source: Adapted from Guard and Lander (1993).

basin in terms of activity. Several oceans produce more hurricanes annually than the North Atlantic. For example, the most active ocean basin in the world—the Western North Pacific—averages seventeen hurricanes (typhoons) per year. It is for this reason that China has the highest frequency of land-falling tropical cyclones, about nine per year. The second most active is the eastern North Pacific, which averages ten hurricanes per year. In contrast, the North Atlantic mean annual number of hurricanes is 6. Table 3.2 summarizes each basin's average number of hurricanes and total storms (Landsea 2005).

Also notable is where these storms do not occur. Tropical cyclone formation is confined to the tropical and subtropical regions where water is warm. In addition, tropical cyclones do not form along the equator. As discussed in chapter 1, these storms require the Coriolis force, which is too weak within 5° of the equator. It is also noteworthy that tropical cyclones do not form in the South Atlantic and eastern South Pacific. Water tends to be colder in these regions, especially the eastern South Pacific. However, even in the warm-water regions, tropical cyclones still do not occur. One factor is that few troughs occur in these regions, possibly due to the unique shape of South America and southern Africa, which decrease in width toward the South Pole. However, the primary

TABLE 3.2
Mean number of total storms (hurricanes and tropical storms), hurricanes, and intense hurricanes per year in all tropical ocean basins

Tropical ocean basin	Mean annual tropical storms and hurricanes	Mean annual hurricanes	Mean intense hurricanes
Northwest Pacific	27	17	8
North Pacific and Northeast Pacific	16	9	4
Eastern Australia and Southwest Pacific	11	5	2
Western Australia and Southeast Indian	7	4	2
North Atlantic	10	6	2
Southwest Indian	13	7	3
North Indian	5	2	Between 0 and 1
South Atlantic	0	0	0
Southeast Pacific	0	0	0
Global	89	50	21

Numbers based on years 1967 to 2003, except for North India. North India data is from 1980 to 2003. The Atlanta data includes subtropical cyclones. From Landsea (2000), updated by Landsea (2005).

reason is probably because wind shear is large in these two basins. Satellite images show several disturbances on the verge of developing each year, only to be torn apart by westerly winds aloft (Sheets and Williams 2001). Nevertheless, on rare occasions, it is possible for tropical cyclones to form in the South Atlantic. A tropical cyclone with winds possibly of 34 knots formed off the coast of Congo in April 1991 and drifted westward for five days into the central South Atlantic before dissipating. In March 2004 a hurricane formed in the South Atlantic Ocean and made landfall in Brazil. These are the only known occurrences, but quite possibly other tropical cyclones have developed but went unobserved before satellites were launched.

Also, systems with structure similar to tropical cyclones are occasionally observed over arctic and polar regions. This maritime cyclone, known as a *polar low*, forms poleward of zones of strong temperature contrasts, ranges in sizes from 120–700 miles, and has winds near or above gale force. A spectrum of polar low classifications exists. Some form along the frigid continental arctic air near the sea ice boundary over a relatively warmer body of water, others develop in unstable air masses with strong fluxes from the water, and still others are initiated by upper-level forcing from jet stream features. In all cases, water temperatures are in the 40–50°F range, well below the 80°F range observed for tropical cyclones. A polar low can have banding features with an eye similar to a tropical cyclone, so some scientists refer to them as *arctic hurricanes*. However, polar lows form over cold water, have a short life cycle of three to thirty-six hours, have weaker but still significant winds, and are smaller in size. They are observed in the Northern Hemisphere around the Nordic Seas (near Greenland, Iceland, and Norway), the Labrador Sea, the Bering Sea, the Gulf of Alaska, and the Sea of Japan. In the Southern Hemisphere, the eastern Weddell, the Bellinghausen, and the Ross Seas are favored locations. A system with a similar structure to a polar low also occurs in the cold Mediterranean Sea every ten or fifteen years. Since the key ingredient is cold air moving over relatively warmer water in all cases, a key driver of polar lows is heat flux from the ocean. The dynamics of polar lows have some tropical cyclone researchers speculating that too much attention is being paid to water temperature and that heat flux is a more important quantifier for better understanding tropical cyclones. Rasmussen and Turner (2003) provide a comprehensive review of polar lows.

Hurricane Categories around the World

By international agreement, the most general term for all large, cyclonically rotating systems with organized thunderstorms originating over tropical or subtropical water is *tropical cyclones*. This category includes depressions, tropical storms, and hurricanes in addition to two other type of cyclones, *monsoon depressions* and *subtropical cyclones*. Monsoon depressions are weak, cyclonic disturbances that form in the Bay of Bengal with no inner-core structure and track northwestward into the Indian subcontinent, bringing moderate winds and heavy rains. However, should a monsoon depression remain over the ocean long enough to develop an inner-core structure, it can transform itself into a tropical storm or hurricane. Subtropical cyclones are hybrid systems that form over the subtropical waters in the Atlantic and Pacific Oceans and contain a mixture of tropical and polar characteristics. Subtropical cyclones derive some of their energy from horizontal temperature contrasts (whereas tropical storms and hurricanes do not) as well as heat flux from the sea (similar to tropical storms and hurricanes). Subtropical cyclones may have winds up to 74 mph, and the maximum sustained winds tend to be farther from the center (on the order of 60–125 miles) than in a tropical storm or hurricane. Sometimes, the heat flux component begins to dominate, and a subtropical cyclone will transform itself into a tropical storm or hurricane. Even if these transformations do not occur, both types of systems are still a serious threat, and forecasters will issue watches and warnings on them even though they are not entirely tropical in structure. In this chapter, in deference to international readers and to avoid confusion, the term "tropical cyclone" will be used. In the remaining chapters, for U.S. readers, the term "hurricane" will be used generically in the place of "tropical cyclone."

Tropical cyclones are generally a summer phenomenon, but the length of the tropical cyclone season varies in each basin, as does the peak of activity (table 3.3). For example, the Atlantic tropical cyclone season officially starts on June 1 and ends November 30, but most tropical storms and hurricanes form between August 15 and October 15. The Atlantic season peaks around September 10. In contrast, tropical cyclones occur year-round in the Northwest Pacific, with a longer active period lasting from July to November and peaking around September 1. In general, this late-summer

TABLE 3.3
Length of the official hurricane season, when the season is most active, and day when the season typically peaks for all tropical ocean basins

Tropical ocean basin	Hurricane season	Active regime	Peak day(s)
Northwest Pacific (1945–1988)	Year-round	July 1–December 1	September 1
Northeast Pacific (1966–1989)	May 15–November 30	June 1–November 1	August 25
Eastern Australia and Southwest Pacific (1958–1988)	October 15–May 1	December 1–April 1	March 1
Western Australia and Southeast Indian (1958–1988)	October 15–May 1	January 1–April 1	January 15 and February 25
North Atlantic (1886–1989)	June 1–November 30	August 15–October 15	September 10
Southwest Indian (1947–1988)	October 15–May 15	December 1–April 15	January 15 and February 20
North Indian (1891–1989)	April 1–December 30	April 15–June 1, September 15–December 15	May 15 and November 10

"Active" is subjectively defined relative to each basin's hurricane history (dates are shown in parentheses), and is defined as the period when most tropical storms and hurricanes occur. Adapted from Neumann (1993).

peak exists for all basins, although Southern Hemisphere tropical cyclones are six months out of phase. This is because the three favorable conditions—warm water, weak wind shear, and cyclonic disturbances—are optimized in late summer. In particular, water temperature peaks in late summer. This seems paradoxical since the longest day is June 22. However, the days are still longer than nights until fall; therefore, the water is still accumulating heat into late summer. The monsoon troughs are most active in late summer as well, and large-scale circulation patterns favor weak wind shear in late summer. The exception to the late-summer peak pattern is the North Indian basin, where there are peaks in both May and November. This is because strong wind shear occurs over India during the summer and because the Indian monsoon moves inland during the summer.

The classification schemes used in the United States for tropical systems—tropical disturbance, tropical depression, tropical storm, and hurricane—are often called different names in other regions of the world. For example, in the Northwest Pacific Ocean, the equivalent name to the hurricane category is "typhoon," and in India it is "very severe cyclonic storm." Unfortunately, the naming conventions can also be confusing. For example, the Australia region uses the term "tropical cyclone" for storms with maximum sustained winds starting at 39 mph (the U.S. tropical storm category), while countries in the Southwest Indian Ocean region reserve it for maximum sustained winds of at least 74 mph (the U.S. hurricane category); both conflict with the international definition of tropical cyclones, which has no wind speed restriction and only requires a low pressure system in the tropics or subtropics with organized thunderstorms. Furthermore, these categories are further subdivided into additional names by other countries, such as "cyclonic storm" and "severe cyclonic storm," which are actually both tropical storms (by U.S. standards) but of different intensities. These different naming categories are synthesized in table 3.4.

Recall from chapter 2 that intense tropical cyclones are the most destructive. Countries give special recognition to strong tropical cyclones as a result. A tropical cyclone with winds of 111 mph or more are classified as "major hurricanes" in the United States. Countries surrounding the North Indian Ocean call storms with winds above 137 mph "super cyclones," while the Australian region calls them "severe tropical cyclones" for the same wind regime. Similarly, for winds greater than 132 mph, the Southwest Indian Ocean calls them "very intense tropical cyclones."

Finally, in the most active region in the world, the Western North Pacific, extremely intense storms with winds greater than 150 mph are common enough that the special category of "super-typhoon" exists for them. However, it is not known how or when this classification was first conceived. The first use of this term was by Kinney (1955) in reference to large typhoons, but this had no correlation to intensity. The first official use with reference to the 150-mph threshold was in a 1963 Joint Typhoon Warning Center (JTWC) annual typhoon report (JTWC 1970). It is quite probable that the 150-mph delineation was chosen because it is roughly twice the 74-mph threshold for classification as a typhoon.

Recall from chapter 1 that tropical cyclone intensity is defined by maximum sustained winds at a height of 33 feet somewhere near the storm's center, averaged over a time period to remove wind fluctuations. Unfortunately, U.S. forecast centers use one-minute averaging, while the rest of the world uses ten-minute averaging, which adds further confusion to these categories. Statistics

TABLE 3.4.
Different names and wind categories for tropical cyclones used around the world

Maximum sustained surface wind speed	North Atlantic Ocean, Northeast Pacific Ocean, Central North Pacific Ocean	North Indian Ocean	Eastern Australia Ocean, Western Australia Ocean, Southwest Pacific Ocean, Southeast India Ocean	Northwest Pacific Ocean	Southwest India Ocean
<20 mph	Tropical depression	Low	Tropical depression	Tropical depression	Tropical depression
20–31 mph	Tropical depression	Depression	Tropical depression	Tropical depression	Tropical depression
32–38 mph	Tropical depression	Deep depression	Tropical depression	Tropical depression	Tropical depression
39–54 mph	Tropical storm	Cyclonic storm	Tropical cyclone (gale)	Tropical storm	Moderate tropical storm
55–73 mph	Tropical storm	Severe cyclonic storm	Tropical cyclone (storm)	Severe tropical storm	Severe tropical storm
74–102 mph	Hurricane	Very severe cyclonic storm	Tropical cyclone (hurricane)	Typhoon	Tropical cyclone (hurricane)
102–110 mph	Hurricane	Very severe cyclonic storm	Tropical cyclone (hurricane)	Typhoon	Intense tropical cyclone
111–132 mph	Major Hurricane	Very severe cyclonic storm	Tropical cyclone (hurricane)	Typhoon	Intense tropical cyclone
133–137 mph	Major Hurricane	Very severe cyclonic storm	Tropical cyclone (hurricane)	Typhoon	Very intense tropical cyclone
138–148 mph	Major hurricane	Super cyclone	Severe tropical cyclone	Typhoon	Very intense tropical cyclone
>148 mph	Major hurricane	Super cyclone	Severe tropical cyclone	Supertyphoon (valid for winds >150 mph)	Very intense tropical cyclone

Maximum sustained winds are averaged over one minute in the Atlantic and North Pacific and ten minutes in other basins. Conversion factors for ten- and one-minute averaging are provided in the glossary.

show that ten-minute averaging results in winds 13 percent weaker than one-minute averaging. Sometimes, sea-level pressure estimates of the tropical cyclone center from maximum sustained wind speeds are also desired, and unfortunately this will vary by ocean since pressure patterns differ around the world. The lack of international standards can be confusing. For the reader's convenience, wind speed and pressure conversions are contained in the glossary.

One fascinating aspect of Southern Hemisphere tropical cyclones is their direction of rotation. Recall in chapter 1 that tropical cyclones around the United States (as well as the whole Northern Hemisphere) spin counterclockwise due to the Coriolis force. However, in the Southern Hemisphere, the earth's rotational influence is opposite and tropical cyclones there spin clockwise. Visitors to the Southern Hemisphere are often surprised by this observation.

Monitoring, Forecasting, and Warning of Tropical Cyclones Worldwide

The World Weather Watch

The monitoring, forecasting, and warning of tropical cyclones are carried out within the framework of the WMO's World Weather Watch (WWW). The WMO is part of the United Nations and is located in Geneva, Switzerland. The WWW program is based on the international cooperation of approximately 185 participating countries and territories. The main purpose of the WWW is to ensure that each participant's national meteorology service has access to the information it needs to provide effective services. Thus, the WMO's main mission is to facilitate establishing an international network of weather data in order to promote the rapid exchange of this data. This consists of consolidating weather data from 10,000 land-based stations, 8,000 ships and other marine stations, 3,000 aircraft, geostationary and polar-orbiting meteorological satellites, and vertical profile measurements using balloons (called *radiosondes*) needed for analyzing and forecasting the weather. The weather analysis is then conducted by three World Meteorological Centers (the Bureau of Meteorology in Australia, the National Centers for Environmental Prediction in the United

States, and the HydroMeteorology Center in Russia) and thirty-four_Regional Specialized Meteorological Centers (RSMCs). Each RSMC has specific tasks and roles for countries and territories within its region. Some roles of these individual RSMCs include drought monitoring, transport forecasts of pollutants or radiation in the event of an environmental emergency, and tropical cyclone forecasts. These specialized analyses and forecasts are then disseminated back to the national meteorological service in each country or region.

Five of the RSMCs are directly concerned with tropical cyclones. These RSMCs monitor and forecast tropical cyclones; issue information to the international community, including the international media; and provide advisory information and guidance to each country's or territory's national meteorological service in their region. However, official warnings are the responsibility of each national meteorological service. The RSMCs' responsibilities also include assigning names to tropical cyclones, training tropical cyclone forecasters of each national meteorological service, preparing operational performance statistics and annual summaries of tropical cyclone seasons, archiving tropical cyclone data, conducting tropical cyclone research, and leading tropical cyclone public awareness programs. In addition, the RSMCs provide specialized tropical cyclone advisory services to the aviation community and provide information and warnings through the Global Maritime Distress and Safety System to marine interests. The five centers are located in La Réunion Island; Miami, Florida; Nadi, Fiji; New Delhi, India; and Tokyo, Japan. Each center forms part of its country's or territory's national meteorology service (respectively, Météo-France, the U.S. National Weather Service's Tropical Prediction Center, the Fiji Meteorological Service, the India Meteorological Department, and the Japan Meteorological Agency). The centers are called, respectively: RSMC Miami—National Hurricane Center; RSMC Tokyo—Typhoon Center; RSMC—Tropical Cyclones New Delhi; RSMC La Réunion—Tropical Cyclone Centre; and RSMC Nadi—Tropical Cyclone Centre.

The WMO has defined twelve regions with tropical cyclone activity, each designated with a roman numeral (table 3.5). Six of these regions cover a wide area and are the responsibility of the RSMCs—the Tropical Prediction Center covers both the North Atlantic Ocean (Region I) and the eastern North Pacific (Region II). The other areas, which include only a single or a few countries or territories, are the responsibility of Tropical Cyclone Warning

Centers (TWTCs). The TWTCs monitor and forecast tropical cyclones, provide local warnings (unlike the RSMCs), and assign tropical cyclones names. The TWTCs include the U.S. National Weather Service facility in Honolulu, Hawaii; three Australian Bureau of Meteorology facilities in Perth, Darwin, and Brisbane; the U.S. National Weather Service facility in Papua, New Guinea; and the Meteorological Service of New Zealand, Ltd., in Wellington, New Zealand. Tropical cyclone forecasts follow the same procedures as in the United States. Forecasters utilize numerical models for guidance, with subjective alterations based on new observations, satellite imagery, and human intuition.

The U.S. Department of Defense (DOD) also plays a special role in tropical cyclone forecasting worldwide. With a combined staff of air force and navy personnel, the JTWC, located at the Naval Pacific Meteorology and Oceanography Center in Pearl Harbor, Hawaii, conducts forecasts for all the tropical oceans in the Eastern Hemisphere (the North Indian, the Southwest Indian, Southeast Indian/Australian, and the Australian/Southwest Pacific basins) as well as the western Pacific Ocean. The U.S. Naval Western Oceanography Center in Pearl Harbor performs the same task for the eastern Pacific Ocean. The forecasts are intended for DOD decisions regarding ship movements, aircraft sorties, and operational planning, as well as for other U.S. government agency requirements. However, these products are also available to the general public and are a valuable second opinion

TABLE 3.5

Ocean regions, designated by roman numerals I–XII, and their respective Regional Specialized Meteorology Centers (RSMC) and Tropical Cyclone Warning Centers (TCWC)

Region	Ocean Basin	RSMC and TCWC Centers
I–II	Atlantic and Eastern Pacific	National Hurricane Center (RSMC Miami)
III	Central Pacific	Central Pacific Hurricane Center (RSMC Honolulu)
IV	Northwest Pacific	Japan Meteorological Center (RSMC Tokyo)
V	North Indian Ocean	India Meteorological Department (RSMC New Delhi)
VI	Southwest Indian Ocean	Meteo France (RSMC La Reunion)
VII–X	Southwest Pacific and Southeast Indian Ocean	Australian Bureau of Meteorology (TCWC Perth), Australian Bureau of Meteorology (TCWC Darwin), Papua New Guinea (TCWC Port Moresby), Australian Bureau of Meteorology (TCWC Brisbane)
XI–XII	South Pacific	Fiji Meteorological Service (RSMC Nadi), Meteorological Service of New Zealand, Ltd. (TCWC Wellington)

to the WMO-based forecast centers. But, since JTWC uses one-minute averages, its wind information will be higher than described by a non-U.S. tropical cyclone warning center for the same cyclone. JTWC also archives hurricane track data and publishes an annual tropical cyclone summary (see chapter 7).

Climate Variations in Worldwide Tropical Cyclones and Seasonal Forecasting

Just as in the Atlantic Ocean, the most influential modulator of tropical cyclone activity in other ocean basins tends to be El Niño and La Niña events (see chapter 2). In contrast to the Atlantic Ocean, El Niño events increase the number of tropical cyclones in some ocean basins while altering the location of tropical cyclones in others. Because El Niño warms ocean waters in the eastern North Pacific, more tropical cyclones form in this location during such events. Likewise, during an El Niño, tropical cyclones tend to form farther east in the western North Pacific. However, the number of tropical cyclones remains the same (Lander 1993). Australia also sees a similar eastward shift during El Niño years, and storms form earlier in the year (Nicholls 1992). The Southwest Indian Ocean experiences an overall reduction in tropical cyclone activity (Jury, Pathack, and Parker 1999). Generally, opposite results are seen for La Niña years. Thus far, while the North Indian Ocean experiences considerable year-to-year variations in tropical cyclone activity, any correlation to El Niño has thus far been inconclusive (Singh, Khan, and Rahman 2001), and no other climate signals have been detected as an explanation.

The Quasi-Biennial Oscillation (QBO), which plays a pivotal role in Atlantic tropical cyclone activity (see chapter 2), apparently only plays a weak role in other ocean basins. However, it is known that intense typhoons in the western North Pacific are twice as likely during the westerly phase of the QBO (Gray 1993). In addition, clustering of tropical cyclone activity is also observed in the Pacific Ocean and Australia, and some researchers have correlated this to the phase of the Madden-Julian Oscillation (see chapter 1).

Based on the pioneering work of Bill Gray (see chapter 2), other organizations are attempting seasonal forecasting for some of these basins. Australia's Bureau of Meteorology and Research has issued monthly and seasonal tropical cyclone forecasts for years (Nicholls 1992), as has the University of Hong Kong for the western North Pacific (Chan 2000; Chan, Shi, and Liu 2001). In addition, the U.S. Climate Prediction Center recently started issuing seasonal forecasts for the eastern North Pacific.

Worldwide Naming Conventions

Just as in the Atlantic, names are assigned to tropical cyclones in most basins. However, the number of lists varies, as does the rotational cycle of each list. The NHC performs forecasts for the eastern North Pacific (WMO Region II) in addition to its Atlantic responsibilities (WMO Region I) and uses six lists of eastern Pacific names on an annual rotating basis similar to that described in chapter 1. All other regions follow different, and sometimes unique, procedures. The western North Pacific (Region IV) uses five lists, the central North Pacific (Region III) and the Fiji Area (Region XI) use four lists, and all other regions use lists consisting of two or three sets of names. While the Southwest India region follows the NHC's practice of rotating a list each year, most other active basins rotate not based on whether it's a new tropical cyclone season but on when the last name is reached on the list. There are a few exceptions. The northern Indian Ocean (Region V) does not assign names but instead gives an identification number followed by the letter "A" or "B" depending on whether the tropical cyclone originates in the Arabian Sea or the Bay of Bengal. Wellington (Region XII) also does not assign names. The Papua New Guinea region has few tropical cyclones and picks names randomly off its two lists.

Whenever a tropical cyclone has had a major impact in terms of damage or fatalities, any country affected by the storm can request that the name be retired by agreement of the WMO. Retiring officially means that a name cannot be used for at least ten years, although thus far no name has ever been reused after a protracted retirement. As with the United States, all regions have retired names.

All names are determined at international meetings of each region's RSMC or TWTC and reflect the regional culture. For example, eastern Pacific storms tend to have Hispanic names, while central North Pacific storms have Hawaiian names. In regions with multiple countries, names will vary and reflect the different cultures. For example, Southwest Indian tropical cyclone names are based on French, Madagascan, and many African countries.

Before 2000, the U.S. Navy and the U.S. Air Force assigned names for the western North Pacific. Beginning in January 2000, the naming convention in the western North Pacific underwent a dramatic transformation to Asian names. The names used in the western North Pacific have the following origin: Cambodia, China, Korea, Japan, Lao, Macau, Malaysia, Micronesia, the Philippines, Thailand, the United States, or Vietnam. These countries also do not endorse the use of people's name to represent these destructive storms. For instance, Korea, Cambodia, Laos, Malaysia, and Vietnam prefer to name tropical cyclones after natural objects (such as trees, flowers, rivers, or animals); China chose female "pet names" but no pronouns in addition to natural objects; the Philippines chose "stormy" adjectives (such as "swift," "strong," "sharp," "fast"); and Micronesia chose the names of gods and goddesses. Examples include Hagibis (Philippine for "swift"), Nepartak (a famous warrior in Micronesia), and Linfa (a Chinese water lily). The meaning of each western Pacific name, its phonetic pronunciation, and the name pronounced by a native speaker are available on the Hong Kong Observatory website at http://www.weather.gov.hk/informtc/sound/tcname2000e.htm. A list of tropical cyclone names for all ocean basins is available on the tropical cyclone section of the WMO website at http://www.wmo.ch/web/www/TCP/Storm-naming.html. These names are also available on Chris Landsea's Hurricanes, Typhoons, and Tropical Cyclones FAQ website at http://www.aoml.noaa.gov/hrd/tcfaq/tcfaqHED.html as well as the National Hurricane Center's website at http://www.nhc.noaa.gov.

The Philippines independently name tropical cyclones that form in the area bordered by 115°E, 135°E, 5°N, and 25°N. These names are in addition to the designations given by the Japanese Meteorological Agency. This list is used locally since it is felt that familiar names are more easily remembered in the rural areas and that the names emphasize a threat to the Philippines. It also provides more leverage in naming a tropical depression or monsoon depression to indicate that the situation is a serious flash flood threat to the general public. In contrast, by WMO standards, only tropical cyclones with winds greater than 39 mph will have a name. So, when a tropical cyclone in this region has winds greater than 39 mph, it will have two names!

The Bay of Bengal and Arabian Sea storms do not follow the WMO conventions. Tropical cyclone names are called BOB (Bay of Bengal) followed by a four-digit number in which the first two are the year and the last two are the sequential number. Similarly, Arabian Sea tropical cyclones are called ARB with a four-digit number. For example, the second tropical cyclone of 2004 in the Bay of Bengal will be identified as BOB 0402.

Storm Size Variations

Tropical cyclones can vary tremendously in size. Using the radius of gale-force winds as a criterion (see chapter 1), four categories are defined based on a literature review (Liu and Chan 1999; Carr and Elsberry 1997; Merrill 1984). A small tropical cyclone with gale-force winds (35 mph or faster) extending only 40 miles or less from the storm center is known as a *midget*. A small tropical cyclone will have gale-force winds between 40 and 100 miles radially. An average tropical cyclone's gale-force winds will extend between 150 miles and 200 miles outward. Gale-force winds extending out beyond 200 miles would be considered a large tropical cyclone. The smallest on record is the midget tropical cyclone Tracy, with gale-force winds out to 25 miles, that devastated Darwin, Australia, on December 24, 1974. The largest is the western North Pacific tropical cyclone Tip, with gale-force winds extending outward to 683 miles on October 12, 1979. As stressed in chapter 1, storm size is not correlated with intensity, and indeed some of the midget tropical cyclones were catastrophic, such as Tracy as well as the Labor Day Hurricane that hit the Florida Keys in 1935.

It is noteworthy that the largest storm did not occur in the Atlantic. Indeed, the average North Pacific tropical cyclone has a radius of 450 miles, 1.5 times the size of the average Atlantic tropical cyclone of 300 miles. The reason for these size differences is that Atlantic storms, on average, form from tropical waves. In contrast, elsewhere in the world, tropical cyclones form in association with the monsoon trough, a large-scale feature that can stretch across much of the regional ocean. The monsoon trough can also exhibit some bizarre configurations. Every two years in the western North Pacific, the monsoon trough will become a large-scale vortex with a diameter of 1,500 miles, known as a *monsoon gyre*. The monsoon gyre is a very large, nearly circular vortex that rotates counterclockwise for two weeks with a cloud band rimming the southern and eastern periphery (Lander 1994). A series of small tropical cyclones may emerge from the leading edge of the cloud band. It is also possible for the monsoon gyre to become a large tropical cyclone, such as the giant typhoon Gladys in 1991.

4

Chronology

This chapter is divided into three parts. The first part chronologically lists scientific weather advances related to hurricane research and forecasting. The second part chronologically lists significant U.S. land-falling hurricanes since 1900 with commentary about their impact. The last part discusses a few pre-1900 hurricanes that have altered history.

Chronology of Weather Advances Related to the Study and Forecasting of Hurricanes

1675   Gottfried Leibniz invented calculus, a branch of mathematics that expresses how a math expression changes dependent on space or time (Sorbjan 1996). Isaac Newton actually invented it first in 1665 but kept the discovery to himself.

1686   Edmond Halley, known mainly as a great astronomer who predicted the famous comet now named after him, also contributed to meteorology. In 1686, he published the first meteorology map. This map showed that tropical winds have a prevailing direction from the northeast in the Northern Hemisphere and from the southeast in the Southern Hemisphere (Nebeker 1995). These winds are known as *trade winds*. This oceanic surface wind pattern is now known to be very constant in the tropical Atlantic, eastern half of the Pacific, and southern Indian Ocean. The wind consistency gave them the obsolete English use of "trade," meaning steady course; later, it meant a reliable breeze for sailing goods from Europe to North America (Sheets and Williams 2001). Halley also identified pressure differences as the cause of wind.

1687   At the urging (and funding) of Edmond Halley, Isaac Newton published *The Mathematical Principles of Natural Philosophy* (*Philosophiae Naturalis Principia Mathematica* in Latin, often just referred to as *The Principia*), the most important book in the history of the physical sciences. Its laid the foundation of modern mathematical theory for all the physical sciences, including meteorology, by writing them in calculus notation. In addition, Newton demonstrated that the universe operates by certain natural laws. Newton's principles describe acceleration, inertia, fluid dynamics, and planetary motion. It set forth the fundamental three laws of motion and the law of universal gravity. Weather forecasting and analysis ultimately are applications of Newton's second law, force equals mass times acceleration (Atkinson 1981b; Panofsky 1981). Finally, the book revolutionized the method of scientific investigation. A translation is presented in Newton, Cohen, and Whitman (1999).

1735   George Hadley, a lawyer with a dedication to the natural sciences, proposed that as air warms at the sultry equator, it rises high aloft 30,000 feet or higher, moves poleward to the middle latitudes, and sinks. This circulation is today known as the Hadley cell (Allaby 1998). Satellite photos show a band of thunderstorms near the equator known as the InterTropical Convergence Zone (ITCZ), confirming that this branch of the cell exists, although the atmospheric general circulation is much more complicated than what Hadley postulated. Hadley also deduced that the general direction of the winds in the tropics is due to the earth's rotation (Nebeker 1995). The sinking midlatitude air flows back to the equator, and an easterly component is added from the earth's rotation. This helped explain the direction of the trade winds (see "1686" above). The mathematical details were later formulated by Gaspard-Gustave de Coriolis (see "1835" below).

1743   Benjamin Franklin discovered that storms do not move in the same direction as the surface wind direction. The cloud cover from a hurricane in 1743 blocked the eclipse Franklin wished to see one night in Philadelphia. Later, he wrote to his brother in Boston about his disappointment and was surprised to find out that the eclipse had been visible in Boston, and soon afterward a violent storm followed. Franklin noted the difference in time between the onset of the storm in Philadelphia and Boston and the reasons it must have been the same storm traveling from southwest to northeast. Typically when New England is affected by a low-pressure system such as a hurricane, the area is on the back side of the storm system. Due to the counterclockwise rotation of low-pressure systems, this results in a wind blowing from the northeast. Before Franklin's discovery, it had been assumed that these storms move northeast to southwest because the wind is from the northeast. Franklin discovered that this often was not true. This laid the foundation for the discovery that cyclones are rotary in nature (Fleming 1990; Ludlum 1989).

The 1743 hurricane is noteworthy also because it is the first one accurately measured by scientific instruments. John Winthrop, a professor of natural philosophy at Harvard College, measured the pressure and ocean water level during the storm passage (Ludlum 1989).

1752   Leonhard Euler derived the equation of fluid flow, which today is used to predict wind. These equations were written for a fixed point in space (in 1760 Joseph Lagrange derived a similar set of equations, but following the fluid flow). Another consequence of Euler's work was the development of the continuity equation. The continuity equation states that the mass of a system remains constant (Atkinson 1981a; Sorbjan 1996).

1768 Leonhard Euler published a technique for approximating prognostic calculus equations such that future values can be predicted (Chabert 1999). This technique is an application of the field of finite differencing. Finite differencing converts calculus equations into a set of algebraic equations with functions represented on a grid in the x-y-z plane. By representing values on a grid (instead of an infinite amount of points as calculus requires), approximate solutions can be obtained using arithmetical procedures in space. Euler extended this concept by rewriting the prognostic calculus term such that future values on a grid can be related to previous, slightly older values, thus making forecasts possible. Many calculus equations cannot be solved analytically, including the predictive meteorology equations, and this technique allows approximate solutions that make weather prediction possible. Lewis Fry Richardson is the first to test this technique for weather forecasts (see "1922" below). Finite differencing is part of the science of numerical methods in which approximations to calculus equations are solved on computers (Durran 1999).

1806 Admiral Francis Beaufort proposed a scale for wind forces, allotting values from 0 to 12. (See the glossary for more detailed information.)

1819 Professor John Farrar of Harvard University documented the Boston hurricane of September 23, noting that "it appears to have been a moving vortex and not the rushing forward of the atmosphere." Farrar then described the extent of the storm damage, the veering of the wind, and how it turned in opposite directions at Boston and New York, as well as the different times of impacts on these two cities (Ludlum 1989).

1822 The French engineer Claude-Louis Navier, and a few colleagues, added terms to the equation of fluid flow to account for intermolecular viscous forces (internal friction). A Cambridge professor, George Stokes, unfamiliar with French research on this subject, independently derived the mathematical formulation for the viscous terms in 1845. The resulting set of equations is today called the Navier-Stokes equations (Nebeker 1995; Sorbjan 1996).

1831   William Redfield made the first comprehensive analysis of the cyclonic rotation of hurricanes, noting that winds rotate counterclockwise around hurricanes "in the form of a great whirlwind." (In the Southern Hemisphere, the rotation is clockwise; see comments in "1835" below). It should be noted here that several people in the 1600s, mostly mariners, documented that hurricanes had rotary winds (most notably the British explorer William Dampier), but apparently few had knowledge of these reports, so Redfield generally gets the credit (Colon 1980; Sheets and Williams 2001). See chapter 5 for more on Redfield.

1834   Emile Clapeyron derived the ideal gas law, which expresses a relationship between pressure, temperature, and volume (Von Baeyer 1999). It combined previous expressions relating pressure to volume derived by Robert Boyle in 1661, temperature to pressure derived by Jacques Charles in 1787, and volume to temperature derived by Joseph Gay-Lussac in 1802 (Sorbjan 1996). Lord Kelvin also derived a temperature scale based on these results in 1851, showing that an object cannot be colder than – 273.15°C.

1835   Gaspard-Gustave de Coriolis published a paper describing how the earth's rotation causes motion to be deflected to the right (left) in the Northern (Southern) Hemisphere (Persson 1998). Twenty years later, William Ferrel, a self-taught mathematician and schoolteacher, realized that Coriolis's concepts could be applied to weather systems (Sheets and Williams 2001). This apparent force due to the earth's rotation, known as the *Coriolis force*, causes a deflection of the wind to the right of its intended path in the Northern

Hemisphere. Without a rotating earth, air would flow toward low-pressure systems such as hurricanes, but the Coriolis force deflects the airflow to the right. As a result, cyclones (including hurricanes) rotate counterclockwise in the Northern Hemisphere. In the Southern Hemisphere, the Coriolis force is reversed, deflecting wind to the left of its intended path; therefore, all cyclones in the Southern Hemisphere rotate clockwise. The Coriolis force is added to the equation of fluid flow.

1839   Henry Piddington coined the word *cyclone*, from the Greek words *kyklon* and *kyklos*, meaning "moving in a circle, whirling around" and also "coiling of the snake," to describe all rotary storms (see chapter 5 on Piddington).

1843   James Prescott Joules, along with other scientists such as Rudolf Clausius, Willard Gibbs, and Julius Robert von Mayer in the period, developed the field of thermodynamics (Von Baeyer 1999). Thermodynamics is the study of energy and energy conversion. Applications of thermodynamics to meteorology include cloud formation, static instability, and prediction equations for temperature and moisture. The ideal gas law (see "1834" above) is also considered a thermodynamic equation.

1845   The first telegraph line is completed between Washington, D.C., and Boston on April 1, providing the necessary infrastructure for rapid communication of weather observations. Joseph Henry, secretary of the new Smithsonian Institution, envisioned the opportunity to create an observational weather network based on this technology (Fleming 1990). In 1848, Henry issued a press release requesting volunteers to transmit weather data to the Smithsonian. By 1849 there were 150 volunteers, and by 1860 there were 500 volunteers (NOAA 1991; Shea 1987). This network was severely interrupted by the Civil War. The success of this network, and renewed efforts after the war by the Mitchell Astronomy Observatory's director Cleveland Abbe (who obtained cooperation from the U.S. Army, telegraph companies, and the Smithsonian), paved the way for the first government-sponsored synchronous weather observations in 1870 (Sheets and Williams 2001).

1847 Lt. Col. William Reid of the Royal Engineers of England established the first American hurricane warning system while on duty in Barbados. His warning system is primarily based on pressure measurements. See chapter 5 for more on Reid.

1870 President Ulysses S. Grant signed a joint resolution of Congress authorizing the secretary of war to organize a national meteorological service under the auspices of the Army Signal Corps on February 9 (NOAA 1991; Shea 1987). It was thought that a military service would secure the greatest promptness, regularity, and accuracy. Later that year, on November 1, the first synchronous weather observations ever taken in the United States were made at twenty-two stations and telegraphed to Washington, D.C. Fairly primitive weather forecasts also were issued from these observations. This service established a system for receiving daily observations in the Caribbean to monitor hurricane activity in 1873 (Sheets and Williams 2001). The first official weather map emerged from this network on September 28, 1874, showing a hurricane offshore of the Florida-Georgia state line. Originally called The Division of Telegrams and Reports for the Benefit of Commerce, this agency was the forerunner to the U.S. Weather Bureau, later to be called the National Weather Service.

1873 The Army Signal Corps issued its first hurricane warning on August 23 for New England and the Middle Atlantic states.

1875 Father Benito Viñes, director of Belen College at Havana, Cuba, developed the first systematic scheme for hurricane forecasts and warnings. This scheme was based on meticulous daily observations of sea swells, surface winds, and cloud motion aloft, from which he deduced patterns that occurred with hurricanes. Viñes issues the first forecast in 1875 for a hurricane that hit Hispanola and Cuba, then began issuing hurricane warnings routinely for this region. The hurricane system consisted of a pony express between isolated villages and an organization of hundreds of observers around the Cuban coastline. See chapter 5 for more on Father Viñes.

1890   The weather service was made a civilian agency on October 1 after two decades of public dissatisfaction with the Army Signal Corps' weather forecasting and the 1881 indictment of the service's disbursing officer for embezzlement (Larson 1999; Shea 1987). For example, the Signal Corps failed to issue a warning for the 1875 hurricane that destroyed Indianola, Texas, and killed 176 people (Sheets 1990). There is also internal strife in the Signal Corps and a general lack of interest for continuing the labor-intensive observational network since it interfered with military duties. President Benjamin Harrison named the new civilian agency the U.S. Weather Bureau (NOAA 2001).

1898   A concerted effort to establish a comprehensive hurricane warning system for North America was begun in preparation for the Spanish-American War at the urging of Willis Moore, the chief of the U.S. Weather Bureau (Sheets 1990). Fearing for the safety of the American navy, Moore voiced his concerns to President William McKinley, recalling the 1896 hurricane that killed 114 people from Florida to Pennsylvania as an example. Impressed by Moore's concerns, McKinley stated that he feared a hurricane more than the Spanish navy. In response, Congress authorized funds to establish observation stations in the islands of the Caribbean. A Weather Bureau center was established in Kingston, Jamaica, to issue hurricane forecasts for the West Indies. After the war was over, additional observing stations were added, and the headquarters was transferred to Havana, Cuba. However, the Havana office was only authorized to issue hurricane forecasts for the West Indies; hurricane warnings for the continental United States had to be officially issued by the Weather Bureau in Washington, D.C. The reason is not known, but perhaps the Weather Bureau thought the Cuban forecasters lacked the skill (an unfounded notion, since they were schooled in the techniques of Father Viñes) or were afraid that the Cubans would outperform the Weather Bureau. Either way, this turned out to be a tragic mistake. Communications between these two facilities was very slow, and the Washington, D.C., office turned out to be inept with regard to hurricanes. See "1902" and "1935" below for more details.

1900   The first serious study of hurricanes by the U.S. Weather Bureau—*West Indian Hurricanes*, by E. B. Garriot—was published (Garriot 1900). The main contribution was an update of hurricane statistics. Quotes from Redfield, Viñes, and others as well as known hurricane tracks for 1875–1900 were also included.

A major hurricane struck Galveston, Texas, killing at least 6,000 people on the island and 2,000 more in nearby coastal regions with no formal warning from Washington, D.C. (Larson 1999; Sheets 1990). The death toll may actually be higher. This is still the worst natural disaster in U.S. history with regard to number of fatalities.

1902   The Cuban hurricane forecasting office was moved to Washington, D.C., in the hope that it would improve the U.S. warning system (Sheets 1990). However, the move may also have been made to avoid political embarrassment from the Galveston hurricane. Weather Bureau administrators in Washington, D.C., and Havana ignored warnings from Cuban meteorologists that a hurricane was in the Gulf of Mexico and instead stated that the storm was headed into the Atlantic; no warning was ever issued from Washington D.C. (Larson 1999). This rearrangement ultimately was a failure (see "1935").

1905   Advances in wireless telegraphy (radio), developed by Guglielmo Marconi in the previous decade, began to make important contributions to weather information. On December 3, 1905, the first marine weather report was broadcast at sea by the SS *New York* (NOAA 1991), and on August 26, 1909, the SS *Cartago*, near the coast of Yucatan, radioed the first ship report about an existing hurricane. This wireless transmission provided critical information, which the U.S. Weather Bureau utilized in its decision to issue warnings to south Texas for a hurricane that later impacted the Texas/Mexico border. No American lives or ships were lost (Garriot 1909; Tannehill 1956).

1913 O. L. Fassig published a study titled *Hurricanes of the West Indies*, which contains statistics and information on hurricane characteristics and swells (Fassig 1913).

Vilhelm Bjerknes proposed that weather be studied using physical laws governed by the following set of equations: fluid flow (see sections "1752," "1822," and "1835"), continuity (see "1752"), and thermodynamics (see "1834" and "1843"). The result is seven basic variables (pressure, temperature, relative humidity, density, and wind components in the x-y-z plane) with seven equations that in principle could be solved (Lynch 2002). This was a dramatic change in thinking, as most scientists at that time concentrated on empirical methods. These equations form a mathematically complete set and today are used in weather forecast models. While other earlier scientists had similar ideas (notably the Russian meteorologist M. F. Spasskii in 1851 and Cleveland Abbe in 1901), Bjerknes deserves the credit for being most passionate about the idea (Nebeker 1995). Bjerknes's idea inspired Lewis Richardson to formulate a way to solve these equations (see "1922" below). Bjerknes was also the first to employ the technique of graphical calculus to these physical laws.

1922 Lewis Richardson described how solutions to meteorological equations may be approximated for weather prediction using finite differencing (see "1768"). His first forecast attempt using this technique failed due to inadequate formulation of the equations as well as other numerical reasons. The sheer enormity of the tedious calculations also discouraged further work by other meteorologists (Nebeker 1995). However, this groundbreaking research laid the foundation for the first successful numerical weather forecasts as certain numerical problems related to the equations were solved over the next two decades, and when the invention of the computer made the calculations achievable. See chapter 5 for more on Richardson.

Edward H. Bowie published an article on the formation and movement of West Indian hurricanes. He stated that a hurricane moves in the general motion of the air surrounding it and tends to move clockwise around the periphery of a semipermanent feature in the central Atlantic known as the Bermuda high-pressure system (Bowie 1922).

1923   Sir Gilbert Walker discovered a west-to-east circulation pattern in the west Pacific Ocean, today called the Walker circulation. He also noted that this circulation pattern periodically shifts, today known as the Southern Oscillation, and is correlated with an El Niño event (Allaby 1998). El Niño is a significant increase in water temperature over the eastern and central equatorial Pacific occurring at irregular intervals between three and seven years, shifting the Walker circulation. El Niño is the largest modulator of seasonal Atlantic hurricane activity (see "1983" below) and also influences hurricanes in other ocean basins. See chapters 2 and 3 for more discussion on El Niño.

1924   C. L. Mitchell published *West Indian Hurricanes and Other Tropical Cyclones of the North Atlantic Ocean*, which tracks all known tropical storms and hurricanes from 1887 to 1923, with statistics on their frequency, formation points, and motion (Mitchell 1924). This is the most comprehensive study up to that time.

1926   Isaac Cline, the Weather Bureau meteorologist in charge on Galveston Island during the 1900 Galveston Hurricane and in the New Orleans Weather

Bureau during the 1915 hurricane, published *Tropical Cyclones,* which was the most authoritative book in the United States on hurricanes at that time. It contains considerable original work on hurricane rainfall, tides, storm surge, and waves, with the focus on Gulf of Mexico storms. See chapter 5 for more on Cline.

1929   Robert Goddard launched a rocket that included a barometer, a thermometer, and a camera (Purdom and Menzel 1996). When the parachute deployed, the camera recorded the instrument readings. The roots of high-altitude photography and weather monitoring, and ultimately weather satellites, can be traced to this moment (see "1954" below).

1935   Congress appropriated $80,000 to revamp the hurricane warning service. New local hurricane forecast centers were established in Jacksonville, Florida; New Orleans, Louisiana; San Juan, Puerto Rico; and later in Boston, Massachusetts. A twenty-four-hour hurricane teletype system network was set up from Wilmington, North Carolina, to Brownsville, Texas, so that no more hurricanes are "lost" for several days (Sheets 1990; DeMaria 1996). These actions were motivated by discontent from coastal communities, who felt that the Washington, D.C., Weather Bureau hurricane warning service had been inadequate. For example, warnings were issued only six hours before the devastating 1926 hurricane hit Miami; even worse, the warning was issued at 11 p.m., after most residents had gone to sleep. Coastal residents also felt that federal officials lacked sensitivity to hurricane problems since Washington, D.C., was rarely affected by hurricanes. The best example occurred in August 1934, when the Weather Bureau issued a hurricane warning for the Texas coast one Sunday afternoon. Since no additional observations would be available until 7 p.m., and because the forecaster's shift had ended, he went home, planning to return to work at 7 p.m. During his absence, an anxious Galveston Chamber of Commerce wired the Washington Weather Bureau for an update, and the map plotter on duty honestly but imprudently wired back, "Forecaster on golf course—unable to contact" (DeMaria 1996).

Bernhard Haurwitz provided evidence that hurricanes extend from the surface to 20 miles aloft and that the air is anomalously warm at the top of a hurricane (Haurwitz 1935).

Floating automatic weather instruments mounted on buoys began collecting weather data (NOAA 1991).

1937   The first official Weather Bureau radio meteorograph (more commonly called a radiosonde) was made at East Boston, Massachusetts. A radiosonde is an instrument attached to a balloon that measures the vertical profile of temperature, moisture, pressure, and wind to 50,000 feet and sends the data with a radio transmitter. Before the radiosonde, real-time upper-air observations were provided by either kites (1894–1931), pilot balloons manually monitored to deduce cloud height and air motion, or aircraft (1931–1939). Before the radiosonde, balloons could measure weather data, but the measurements had to be retrieved by finding the instruments dropped by a parachute after the balloon burst, often an impossible task. The radiosonde network proved superior and was firmly established by 1939 (NOAA 1991), providing a three-dimensional picture of the atmosphere around the United States. Pan American World Airways, the flagship airline for the United States to Latin America and the Caribbean, also started providing a dense pilot balloon network in the Caribbean (Sheets and Williams 2001), augmenting the network. With the emergence of the radiosonde network and Pan Am data, forecasters quickly realized that mid- and upper-level winds steer hurricanes, providing valuable information for track forecasts (Dunn 1940a).

1940   Gordon Dunn published a paper showing that most Atlantic tropical storms and hurricanes form from tropical waves—not due to cold fronts, which was the popular theory at that time (Dunn 1940b). This was also the first paper documenting and describing tropical waves. See chapter 5 for more on Dunn.

President Roosevelt ordered the Coast Guard to establish ocean weather ships (NOAA 1991).

1941　RAdio Detection And Range (RADAR) with wavelengths capable of detecting rain and snow was first used (Rogers and Smith 1996). Radar detects precipitation by transmitting and receiving reflected radiation from raindrops. In 1942, the U.S. Navy gave the Weather Bureau twenty-five surplus aircraft radar to be modified for ground meteorological use (NOAA 1991). Up to 1957, all Weather Bureau radars were actually military in origin and modified for rain detection.

1942　To increase surveillance of weather systems during World War II, the U.S. Army Air Force began aerial weather reconnaissance operations between North America and Allied Western Europe on August 16. In the summer of 1943, they also began "weather scout" missions in Europe (Nolan and Murphy 2000).

1943　The first intentional (but unauthorized) flight into a hurricane occurred on July 27. British pilots at a flight school in Bryan, Texas, overheard that the AT-6 planes may be moved because of a hurricane off Galveston. Not knowing what a hurricane was, and used to bigger airplanes from their combat experience, the British students started taunting their trainers about flying in bad weather and about the durability of the smaller AT-6 planes. One of the trainers, Army Air Corps Col. Joseph Duckworth (an unassuming man who specialized in teaching military pilots how to fly in bad weather), bet that he could fly into the hurricane and back with an AT-6 plane. Because of this dare, Duckworth flew into the eye of the storm twice that day, once with a navigator and again with a weather officer (Sheets and Williams 2001; Fincher and Read 1999). This same hurricane also seriously damaged refineries in Texas, hurting wartime production of fuel.

1944 Motivated by Duckworth's successful flight and the threat of hurricanes to navy ships and wartime production (such as the refinery incident in Texas), the Joint Chiefs of Staff approved hurricane reconnaissance over the Atlantic using U.S. Army Air Force and Navy planes. The Jacksonville hurricane forecast center was moved to Miami, Florida, to establish the Joint Hurricane Warning Center with the U.S. Air Corps and the U.S. Navy; Grady Norton was appointed as its chief. The first official, authorized flights occurred early in the summer into tropical storms, and on July 17 Capt. Allan Wiggins made the first authorized flight into a hurricane. In September, the hurricane reconnaissance flights had their first success tracking an East Coast hurricane (see Great American Hurricane in table 4.1), providing forecasters unprecedented detailed information on a hurricane (Sheets and Williams 2001).

On December 16, the U.S. naval fleet, commanded by Adm. William "Bull" Halsey en route to support the invasion of the Philippine Islands, suffered devastating damage from a typhoon. The meteorologist in charge, Cdr. George F. Kosco, used what very limited information he had to make a forecast upon which Halsey could steer the fleet. Unknown to anyone at the time was that a scout plane had information about a typhoon that would have altered the fleet's path. The fleet inadvertently sailed into Typhoon Cobra with winds up to 130 mph and 70-foot waves. Three destroyers were sunk, 146 carrier-borne aircraft were destroyed, 9 additional ships were severely damaged, and 778 sailors were killed. After this incident, Halsey requested dedicated reconnaissance flights for hurricanes, but it was turned down because such heavy-duty aircraft were needed for the war effort. A similar incident again occurred to Halsey's fleet in June 1945 en route to an invasion of Okinawa, damaging 33 ships, destroying 76 planes, and killing 6 men. Halsey once again requested regular weather reconnaissance flights for the Pacific (Sheets and Williams 2001).

Horace Robert Byers showed in his classic book *General Meteorology* that heat and moisture transmitted from the ocean partially balance adiabatic expansional cooling as the pressure in a hurricane falls. He showed that surface air temperature does not cool much; otherwise, the surface heat source support for cloud updrafts would become disrupted, and the storm would decay. This explains why hurricanes develop over the ocean, because over land they lack the necessary heat and moisture energy sources. This also explains why hurricanes weaken rapidly upon landfall or when they move over cold water (Byers 1944).

World War II radar observations showed that rainfall in hurricanes occurs in well-defined spiral bands rather than being uniformly distributed throughout the storm. The first radar picture of a hurricane eye was taken at the U.S. Naval Air Station, Lakehurst, New Jersey, on September 14 (Wexler 1947).

Herbert Riehl and Maj. Robert Schafer published a paper showing that hurricanes do not develop or intensify if the vertical wind shear is too strong (Riehl and Shafer 1944). See chapter 5 for more on Riehl.

1945   The hurricane eyewall was discovered (Wexler 1945; Depperman 1946).

Motivated by the need to find faster ways to compute World War II ballistic firing tables, the U.S. Army funded the University of Pennsylvania's Moore School of Electrical Engineering to develop the first general purpose electronic digital computer, called the Electronic Numerical Integrator and Computer (ENIAC), for delivery to the army's Ballistic Research Laboratories (Weik 1961; Ralston and Reilly 1993). Developed by Professor John Mauchly and his graduate student J. Presper Eckert, the top-secret ENIAC was a huge machine, covering 1,800 square feet (150 feet long by 12 feet width) with a height of 10 feet high and weighing 30 tons. It required approximately 17,500 vacuum tubes, 70,000 resistors, 10,000 capacitors, 1,500 relays, 6,000 manual switches, and 5 million soldered joints. ENIAC consumed 160 kilowatts of electricity and caused the city of Philadelphia to experience brownouts. ENIAC could perform 5,000 additions, 357 multiplications, or 38 divisions in a second—slow by today's standards, but 1,000 times faster than any machine of its era and much, much faster than doing the same task by hand. ENIAC also included other computer hardware innovations such as a gate (for the logical "AND" statement) and a buffer (for the logical "OR" statement). Not completed in time before the war ended, ENIAC was used first for classified studies of thermonuclear chain reactions to design a hydrogen bomb. Later tasks included cosmic ray studies, random number studies, wind tunnel design, thermal ignition, and weather prediction (see "1950" below). Since the failure of even one vacuum tube could corrupt the calculations, another by-product of ENIAC was progress in vacuum tube development (Platzman 1979).

During ENIAC's development, it becomes obvious that the computer had a number of shortcomings: it could only store 20 ten-decimal numbers, and manual wiring was required to program the machine. Moore school researchers and world-renowned mathematician John von Neumann (who acted as a consultant) proposed a new computer design to overcome these deficiencies (Ralston and Reilly 1993)—the concept of a "stored-program" computer, in which instructions as well as data are stored together in the primary memory, with the memory being configured to be as large, random access, and fast as possible. Previous computer designs, including ENIAC, stored instructions on external devices such as a tape. This new concept proposed transferring an instruction from memory to the processor, decoding the instruction, and executing it with respect to data that was also retrieved from memory. This radically new design was called the Electronic Discrete Variable Automatic Computer (EDVAC) and would make computer programming easier. EDVAC also contained new hardware configurations to maximize internal memory and used binary math rather than decimal math, which is more economical on vacuum tubes. These key concepts ushered in the academic discipline of computer science. Thus, the need to solve complex math problems in fields such as meteorology spawned the modern computer era.

1946   The U.S. Army Air Force began organized reconnaissance flights scouting for typhoons in the Pacific (Nolan and Murphy 2000). The U.S. Navy also formally began weather reconnaissance, with units stationed in Miami, the Philippines, and Guam (Sheets 1990). The motivation for these flights resulted from the massive damage incurred by Halsey's fleet.

John von Neumann organized the Electronic Computer Project at the Institute for Advanced Study (IAS) in Princeton, New Jersey. The goal of this project was to design and build an electronic computer based on the EDVAC concept that exceeded the power and capabilities of existing computers such as ENIAC to advance mathematical science (Nebeker 1995). The computer architecture, designed by Neumann, proved to be the model for more advanced computers built in the next decade. In 1947, Neumann decided that the main application of the computer would be for weather forecasting and started the Meteorology Group at IAS.

The science of weather modification was initiated by two discoveries in 1946. Rainfall development generally requires the formation of ice crystals in a cloud. Under proper (and complex) conditions, the ice crystals will grow heavy enough to overcome the cloud's updrafts and fall as precipitation, first as snow, then melt to rain if the air below the cloud is above freezing. However, ice crystal initiation needs a floating particle with a molecular structure similar to ice, known as *ice nuclei*. Natural ice nuclei include certain kinds of clay, bacteria, and amino acids, as well as ice crystals from another cloud or another section of the same cloud (Stull 2000). Hence, in theory, if one drops nuclei in a cloud with a structure similar to ice, rain could result. In 1946, Vincent Schaefer accidentally discovered that dry ice (solid carbon dioxide) injected into moist air at $-9.5°F$ causes water vapor to sublime into ice crystals. Realizing the importance of this discovery, on November 13 he flew over a supercooled (meaning below freezing) stratus cloud, dropped six pounds of dry ice into the cloud, and triggered snow (Byers 1974; Chisholm 2003). Dr. Bernard Vonnegut (the older brother of novelist Kurt Vonnegut) was also involved in the study of ice nucleation (Chisholm 2003). On November 14, he released silver iodide smoke particles in a refrigerated chest containing supercooled liquid water droplets. Swarms of ice crystals were produced. Vonnegut later found that silver iodide gets better results in nucleating clouds than does dry ice. These successes lead to many attempts at weather modification, including Project Cirrus in 1947 where clouds were seeded with artificial ice nuclei to induce rainfall. The first attempt to seed a hurricane was also done in Project Cirrus in a hurricane off Georgia, which then changed track and hit Savannah, causing much controversy (Sheets and Williams 2001). The storm motion shift was actually caused by a change in steering currents, not the seeding, but demonstrated that scientists need to be diligent about public perception in weather modification experiments. Later, a more carefully constructed cloud-seeding experiment called Project STORMFURY was conducted on hurricanes and contained strict guidelines on storms eligible for seeding (see "1962" below and chapter 2).

1947　　A celebrated paper from the Department of Meteorology at the University of Chicago provided a rudimentary understanding of the jet stream (a term coined by C. G. Rossby; see chapter 5), its relationship to fronts, and its three-dimensional structure (Lewis

2003). This work was motivated by World War II bombers' unexpected encounters with narrow bands of strong westerly winds at 30,000–35,000-foot levels. Other noteworthy works include observation of strong upper-level winds during a European field experiment in 1935 and the obscure research of Japanese meteorologist Wasaburo Ooishi, who used upper-air kite measurements to document strong westerly winds aloft over Japan in the winter of 1924. The jet stream is intricately related to atmospheric flow and evolution.

1948   Erik Palmen showed that hurricanes do not form over water colder than 80°F, demonstrating that warm water is a requirement for hurricane formation and intensification (Palmen 1948).

Jules Charney became director of the IAS's Meteorology Group. Charney brought key leadership to the program, proposing that a hierarchy of weather models of varying complexity be tested. He derived a simplified system of equations for weather forecasts, making weather models feasible on computers (Nebeker 1995).

1950   The IAS group performed feasibility tests of numerical weather forecasting on the ENIAC computer. The simulations were not good but had potential, showing meteorological features (Nebeker 1995; Cressman 1996).

1952   The IAS computer was completed, and researchers tested a model simulation of an unusually severe storm on the new machine. On November 24, 1950, a strong cyclone formed off the East Coast that had been unpredicted by the Weather Bureau. The simulation accurately predicted the development of the storm. More impressively, the computer only took forty-eight minutes to complete the forecast. The success of this and other computerized test forecasts motivated the formation of the Joint Numerical

Weather Prediction Group (JNWPU) in 1954, jointly funded and staffed by the Weather Bureau, the U.S. Air Force, and the U.S. Navy. The JNWPU pioneered operational weather forecasts on computers, becoming operational in 1955 (Cressman 1996; Nebeker 1995). In 1958, the JNWPU divided into three organizations, which still exist today: the National Meteorological Center (for civilian needs), now called the National Centers for Environmental Prediction; the Global Weather Central (for air force needs); and the Fleet Numerical Oceanography Center (for navy needs), now called Fleet Numerical Meteorology and Oceanography Center. All three units, particularly the NCEP, interact with the National Hurricane Center and operational forecast centers worldwide for day-to-day operational situations. The IAS also spawned a new unit, now called the Geophysical Fluid Dynamics Laboratory (GFDL), that later developed the most skillful Atlantic hurricane track model in the 1990s.

1954 In October, the first color pictures taken from a high-altitude Viking rocket (launched by the Navy Research Laboratory) revealed a tropical depression or tropical storm (D. Roth, personal communication, 2004) over the southwestern United States that had moved inland from the Gulf of Mexico (Vaughan 1982; Vaughan and Johnson 1994; Amato 1997). The photo of the storm's circulation pattern explains why rain occurs in that region. The discovery of this storm, which had not been detected by the conventional weather-observing network, demonstrated the usefulness of high-altitude photography in detecting or enhancing knowledge of weather systems. It convinced the U.S. government that routine meteorological imagery from higher altitudes was needed. The success of the high-altitude rocket experiments, which began in 1944, encouraged scientists that, with a more powerful engine and the addition of upper stages, the Viking rocket could be made a vehicle capable of launching an earth satellite. This led to the

Vanguard satellite program in 1955. Satellites represent the single greatest advancement in observing weather over the tropics.

1955 The Miami Weather Bureau office was officially designated the National Hurricane Center (NHC) (DeMaria 1996; Sheets 1990). Gordon Dunn was the first official NHC director, although many, including Dunn, recognize the well-respected Miami forecaster Grady Norton as the "honorary" first director. Dunn, who suffered from migraines and high blood pressure, performed forecast duties against his doctor's advice during Hurricane Hazel, died of a stroke at his home after the forecast shift. See chapter 5 for more on Norton.

A donated military radar was installed at Hatteras, North Carolina, and all three storms that made landfall in 1955 passed within its range, demonstrating the usefulness of radar for tracking hurricanes (DeMaria 1996). This was further motivation that the Weather Bureau should complete its own radar system, which had been in development since the end of World War II (Rogers and Smith 1996) and already had cooperative networks in Texas, Oklahoma, Louisiana, and Arkansas due to local concerns about tornadoes (Zipser 1989). In 1959, the Weather Bureau commissioned its own first radar at Miami's NHC (NOAA 1991). By the early 1960s, radars thoroughly cover the U.S. coastline from Brownsville, Texas, to Eastport, Maine, and are known as the WSR-57 network.

1956 The National Hurricane Research Laboratory (NHRL) was established through the efforts of Robert Simpson (see chapter 5) and other scientists (Sheets and Williams 2001; Sheets 1990; Zipser 1989). Six significant hurricanes hit the northeastern United States in 1954 and 1955 (Carol, Edna, and Hazel in 1954; Connie, Diane, and Ione in 1955), causing massive damage. This heightened awareness in Washington, D.C., of hurricanes, and Congress authorized funding for the NHRL. Its goals are to examine the structure of all stages of hurricanes and to determine important parameters for hurricane forecasting using instrumented planes and other data sources. Originally located in West Palm Beach, Florida, the NHRL was later moved to Miami in the same building as the NHC. Today the lab is called the Hurricane Research Division (HRD) and is located at Virginia Key, Florida, just east of Miami. HRD research flights have provided our most complete understanding of the three-dimensional structure of hurricanes.

Herbert Riehl (see chapter 5) and colleagues developed the first track forecasting scheme based on empirical statistics. This technique assembled past meteorological observations and developed an equation that predicts what a current storm may do based on similar past situations. The advantage of empirical equations is that they are easy to develop and simple to run on computers. The disadvantage is that the atmosphere responds in a much more complicated fashion than suggested by simple statistics, and therefore statistical schemes often do not work well, especially in unusual situations or when a hurricane is rapidly changing track or intensity. Nevertheless, many schemes have been devised over the years using statistics.

Julian Adem published a paper showing that a vortex could slowly propagate to the northwest even in the absence of a steering current. This is known as the beta drift. Adem demonstrated that hurricane motion is more complicated than a vortex simply following a steering current and that complex interactions between the hurricane and the environment also determine a hurricane's track.

The first meteorological satellites were launched in 1956. Pioneer I and Pioneer II contained television cameras, but the equipment failed (Purdom and Menzel 1996).

1959   On May 1, weather forecasting elements of the U.S. Navy and Air Force in the western North Pacific are combined into a single tropical cyclone warning center, the Joint Typhoon Warning Center (JTWC) in Guam (now located in Hawaii). Over the years, the role of JTWC grew to include the entire Indian and Pacific Oceans.

Attempts at satellite weather monitoring continued. Vanguard II was launched in February 1959, but an uneven separation from the launch vehicle and improper balancing resulted in a wobbling axis, making the data unusable (Schnapf 1982; Vaughan and Johnson 1994). Explorer VI, launched on August 7, captured the first crude images of the earth's cloud cover. Explorer VII, launched on October 13, contained a radiometer designed by Verner Suomi and Robert Parent of the University of Wisconsin. This radiometer contained black and white ping-pong balls on the end of transmission antennas that measure solar and infrared radiation, providing the first radiation budget of the earth—and the first successful measurement of the earth's atmosphere by a satellite (Purdom and Menzel 1996; Vaughan and Johnson 1994).

1960　The meteorological satellite era officially began on April 1 when the first successful experimental weather satellite with a camera, the Television InfraRed Observation Satellite-1 (TIROS-1), was launched by NASA (Vaughan and Johnson 1994). TIROS-1's primary objective was to demonstrate the feasibility of observing the earth's cloud cover by means of slow-scan television cameras in a polar-orbiting, spin-stabilized satellite (Schnapf 1982). The potential of satellites for hurricane monitoring became obvious in April when TIROS-1 photographs a previously unreported hurricane 800 miles east of Brisbane, Australia.

TIROS-1 is an example of a *polar-orbiting satellite*. These spacecraft follow a north-south orbit around the earth's poles, providing pictures centered on different longitudes each hour as the earth rotates underneath them. Today's polar orbiting satellites are important research tools because they provide high-resolution pictures. They are also used to infer (but not directly measure) cloud-top heights, moisture profiles, temperature profiles, sea surface temperature, wind, and many other useful quantities.

Nine more experimental TIROSs followed during the next five years. Other notable achievements include technology called Automatic Picture Transmission (APT), which allowed the direct transmission of real-time pictures; the change to a "wheel-mode" operation so that a sequence of overlapping pictures could be taken and a mosaic of the whole globe could be constructed; and the introduction of sun-synchronous orbits (which means that a satellite passes over any given location at the same solar time each day). The initiation of the TIROS program, its transition from the Department of Defense to the NASA Goddard Space Flight Center, and TIROS's legacy are an interesting story, and readers are referred to Schnapf (1982), Vaughan and Johnson (1994), and Purdom and Menzel (1996) for details.

1961   Swirling vortices in the low cloud decks were observed by high-altitude U-2 aircraft photographic reconnaissance inside the eyes of hurricanes (Fletcher, Smith, and Bundgaard 1961). These small regions of rotation embedded on the inner edge of the eyewall are known today as *mesovortices* (Kossin, McNoldy, and Schubert 2002; Kossin and Schubert 2004) and can be associated with locally strong, damaging winds. It is postulated that mesovortices may be responsible for localized extreme damage in Hurricane Andrew (1992) and perhaps Hurricane Celia (1970). They may also be responsible for nearly crashing a WP-3D NOAA research aircraft during Hurricane Hugo (1989).

1962   Project STORMFURY, a study to assess whether man could modify hurricanes by introducing artificial ice nuclei into the eyewall region, began at the NHRL. See chapters 2 and 6 for more on Project STORMFURY.

1963   Edward Lorenz published a landmark paper showing that there are inherit limits to the predictability of weather (Lorenz 1963; Nebeker 1995). This is because weather forecasts are sensitive to the accuracy of weather observations fed into computers—even a slight change in the decimal point of an observation will yield completely different long-term forecasts. Since measurements of temperature, wind, and moisture will always contain some small errors, long-term weather forecasts will always contain large errors (although short-term forecasts of less than two days generally will be accurate). This discovery started a new science called chaos theory, the study of erratic behavior in predictive equations in the sense that very small changes in the initial state lead to large and apparently unpredictable changes in the late state. This chaotic property eventually makes long-term weather forecasts completely inaccurate. In theory, the predictability limit is ten days, but in practice it is about four to five days. This is because any additional errors introduced into the computer models accelerate forecast inaccuracies. These additional errors are due to (1) data gaps in certain parts of the world, especially over the oceans; (2) incomplete understanding of certain atmospheric features such as clouds, radiation, and rain; and (3) the approximate solution of calculus equations using numerical methods. As a result, forecast errors grow faster in time, especially after two days. Chaos theory also applies to hurricanes, and some research has shown that twenty-four-hour track forecast errors may never be less than 75 miles, on average (Abbey, Leslie, and Holland 1995; Leslie, Abbey, and Holland 1998).

When someone sues the federal government, a legal doctrine known as sovereign immunity becomes important. Under this doctrine, which is based on the ancient English notion that "the King can do no wrong," the government cannot be sued without its consent. The U.S. government does not consent to be sued for exercising discretion or for misrepresentation, and as a result typically the courts dismiss lawsuits involving weather forecasts by the government without even deciding on the merits of the case. An example is *Bartie v. United States*, when Whitney Bartie and hundreds of others sued the federal government, claiming that the New Orleans Weather Bureau was negligent in giving adequate warnings on the intensity, track, and landfall of Hurricane Audrey (1957). The Weather Bureau issued warnings that Audrey would make landfall late on the evening of June 26, 1957, as a category 2 storm, but instead the storm's motion increased and it also rapidly intensified to a category 4 storm. By sunrise of June 26, high water had already inundated the southwest Louisiana coast, and many surprised residents (most several miles inland and thinking they were safe) were unable to leave. The storm surge killed 390 people, including Bartie's wife and five children. The court, however, held that their suit was barred by the discretionary exception because weather forecasting is a subjective process requiring discretion, as are the means of communicating with the public. The court also cited that their suit was barred by the misrepresentation exception, since a bad forecast is generally viewed in this manner. The court observed that the New Orleans weather service advisories did not convey the necessary urgency of the hurricane threat but dismissed the suit based on sovereign immunity (Klein and Pielke 2002).

In truth, the Weather Bureau did give adequate warning of a landfall, but communication problems and complacency on the part of long-time residents contributed to the disaster. First, newly elected city officials in Lake Charles edited text from the warning and advisories before dissemination (Zipser 1989). Second, the warnings directed people to evacuate from "low or exposed areas," but many residents several miles inland at 7–8 feet elevation didn't consider themselves to be at low elevation. Third, the message lacked a compelling sense of urgency or emergency. Finally, some long-time residents stayed because they had experienced many hurricanes with no serious consequences and thought Audrey posed no threat (Sheets and Williams 2001).

1964	The American Meteorological Society (AMS) wrote to the Taiwanese ambassador to the United States deploring the bad treatment of Kenneth Cheng, head of the Taiwan Weather Service, who had been indicted for an incorrect typhoon forecast. The AMS pointed out that if forecasters are indicted for an incorrect forecast, soon there would be no forecasters (NOAA 1991).

Recognizing the importance of meteorological satellites, the U.S. government formed the Environmental Science Services Administration (ESSA), a new laboratory to transition the TIROS program into operational weather satellites (Purdom and Menzel 1996). This laboratory was renamed the National Oceanographic and Atmospheric Administration (NOAA) in 1970 and has evolved into a department of NOAA today called the National Environmental Satellite and Data Information Service (NESDIS).

In parallel with the TIROS transition, NASA began a new experimental satellite program known as Nimbus, with the first of seven successful satellites launched on August 24. The most notable achievement of Nimbus 1 is its infrared radiometer, which produced the first high-quality photographs taken at night. It also shows an exceptionally clear photo of Hurricane Cleo on its first day, immediately contributing valuable information to hurricane forecasters (Haas and Shapiro 1982).

Six more Nimbus launches followed. Other notable achievements include improved camera and stabilization technology, improved radiometers for heat balance studies, and the ability to derive temperature and moisture profiles from onboard spectrometers. Nimbus provided other scientific firsts, laying the foundation for satellite data use in interdisciplinary earth sciences. The roots of NASA's LANDSAT program can be traced to Nimbus. Readers are referred to Haas and Shapiro (1982) and Purdom and Menzel (1996) for details.

1965   George Cressman, the director of the Weather Bureau, issued a plan to improve the hurricane warning service by concentrating the responsibilities at the NHC in Miami. This plan included increasing public awareness, more reliable service, mitigating the cost of hurricane destruction, and minimizing excessive preparation costs resulting from overwarning. The plan also called for hurricane specialists, who would issue official track and intensity forecasts for the Atlantic and eastern Pacific oceans. This essentially describes the mission and forecast staff at NHC today. Cressman's plan also initiated the first of many administrative changes during the next fifteen years involving forecast offices in New Orleans; San Juan; Washington, D.C.; Boston; San Francisco; and Honolulu. Details are contained in Sheets (1990).

1966   ESSA launched its first operational satellite, ESSA-1, on February 3, and a second satellite, ESSA-2, twenty-five days later. These and later ESSA satellites are part of the TIROS Operational System (TOS). Odd-number ESSA satellites provide photos and global weather data to major forecast centers, while even-number ESSA satellites provide direct real-time readouts of the APT pictures to APT ground stations throughout the world. The original ESSA program evolved into the Improved TIROS Operational System (ITOS) in 1970, which combined the abilities of the odd- and even-number ESSA satellites (Purdom and Menzel 1996). It was not until the ESSA series of satellites that routine hurricane surveillance was assured, and the ESSA satellites also motivated forecasters to develop satellite techniques for hurricane forecasting.

On December 6, the first experimental *geostationary* satellite—the Applications Technology Satellite–1 (ATS-1)—was launched. Geostationary spacecraft orbit at the same speed as the earth's rotation, enabling the satellite to remain over the same location and thus providing continuous coverage. In contrast, *polar-orbiting* satellites lack temporal continuity since the spacecraft only passes over a particular location once or twice a day; but, since polar-orbiting satellites are closer to the earth, they capture better resolution than geostationary. What made ATS-1 even more special, however, was a new camera based on spin-scan technology developed by satellite pioneer Verner Suomi at the University of Wisconsin. It worked by constructing a narrow east-to-west image one strip at a time whenever the camera pointed at earth, with each successive spin recording the next narrow strip below the previous one until a complete picture of the earth was captured (Sheets and Williams 2001). This camera was able to provide full-disk visible images of the earth every twenty minutes, astounding meteorologists with the full-disk view and the ability to see clouds in motion when the pictures are put in motion (Purdom and Menzel 1996). By the early 1970s, ATS imagery was used in operational forecast centers.

The merging of defense and civilian satellite requirements proved impractical, and the Department of Defense initiated its own the satellite system, the Defense Meteorological Satellite Program (DMSP), on September 16 (Purdom and Menzel 1996).

The Weather Bureau was renamed the National Weather Service and was administered by ESSA (NOAA 1991).

1968   Robert Simpson (see chapter 5) replaced the retiring Gordon Dunn as director of the NHC. Simpson placed a renewed effort on research and development at the NHC, including the development of satellite applications and statistical equations for predicting hurricane tracks.

Congress created the National Flood Insurance Program (NFIP) in response to the rising cost of taxpayer-funded disaster relief for flood victims and the increasing amount of damage caused by floods (from rain events and hurricane storm surge). See chapter 2 for details.

1968–1969   Numerical solutions of hurricanes began (Yamasaki 1968; Ooyama 1969). Vic Ooyama (see chapter 5) makes the first two-dimensional computer simulation of a symmetrical hurricane. Ooyama also hypothesized that hurricane genesis occurs as a cooperative interaction between cloud elements and a tropical disturbance—a theory known as conditional instability of the second kind (Ooyama 1969). The technical details of this theory have since either evolved considerably or been discredited but did open new ideas on hurricane genesis.

1970   The first successful operational hurricane computer model, called the SANders' BARotropic (SANBAR) model, was implemented. Over the years other numerical hurricane models became available, including the Moveable Fine Mesh (MFM) model, from 1976 to 1987; the Quasi-Lagrangian Model (QLM), from 1988 to 1993; the Beta and Advection Model (BAM), from 1989 to the present; the VICBAR model, from 1989 to the present; the Limited-area sine transform BARotropic (LBAR) model, from 1996 to the present; and the Geophysical Fluid Dynamics Laboratory (GFDL) model, from 1995 to the present (DeMaria 1996; Vigh et al. 2003; Weber 2001).

ESSA was renamed the National Oceanographic and Atmospheric Administration, or NOAA (NOAA 1991).

1971   Richard Anthes (see chapter 5) made the first three-dimensional computer simulation of a hurricane (Anthes 1982).

Roger Lhermitte proposed the installation of Doppler radar on airplanes (Lhermitte 1971). The development of this concept became an important research platform on HRD aircraft.

U.S. Air Force weather reconnaissance missions began to be cut. In addition, the U.S. Navy ended flights in the western Pacific. This indicates that with the advent of satellites, the military considered weather reconnaissance not of a military nature and less of a priority (Nolan and Murphy 2000).

1972 The National Data Buoy Center was formed (Sheets 1990). Its mission was to establish and maintain a buoy network for the coastal United States, Atlantic Ocean, and Gulf of Mexico (see http://www.ndbc.noaa.gov for details). Buoys provide key maritime data that would otherwise be unavailable in these data-sparse regions.

Charles Neumann (see chapter 5) developed the CLIPER (CLImatology and PERsistence) statistical equation, which is still used today for predicting storm tracks. CLIPER is considered a benchmark for the evaluation of prediction schemes and models; if the scheme has smaller errors than CLIPER, it is considered skillful (DeMaria 1996).

Vern Dvorak (see chapter 1) developed a methodology using visible polar satellite data for estimating hurricane intensity based on the storm's cloud shape and banding, since the amount of cloud organization is directly related to tropical storm and hurricane intensity. This methodology, later refined to include infrared and geostationary pictures as well as satellite-measured eye temperature, is known as the *Dvorak technique*. It is still used worldwide today when no direct measurements of hurricane intensity are available (Dvorak 1984; Velden, Olander, and Zehr 1998).

The third satellite in the ITOS series, NOAA-2 (named after its parent organization), was launched October 15. It marked the end of the video camera era and the beginning of the current era of calibrated multichannel sensing data in different regimes of the electromagnetic spectrum other than visible and infrared using high-resolution radiometers (Purdom and Menzel 1996). The use of other spectrums allows the development of many new satellite weather tools, such as accurate sea-surface temperature mapping

Chronology   137

(important for hurricanes), temperature profiles, moisture profiles, improved cloud detection, land types, and many other applications.

1973   After Hurricane Camille devastated the Gulf Coast in 1969, Congress pushed for the Atlantic Air Force Hurricane Hunters to be closer to the Gulf Coast. In 1973, the squadron moved to its current post, Keesler Air Force Base, Mississippi. Previous posts include Bermuda (1947–1953), England (1953–1960), Bermuda (1962–1963), Georgia (1963–1966), and Puerto Rico (1966–1973) (see http://www.hurricanehunters.org for details).

Neil Frank (see chapter 5) replaced the retiring Robert Simpson as NHC director. Frank placed a renewed emphasis on hurricane preparedness and was particularly skillful at using the media to motivate people in hurricane-threatened areas to respond appropriately (DeMaria 1996).

1974   The GARP (Global Atmospheric Research Program) Atlantic Tropical Experiment (GATE) occurred. This was the first full-phase tropical weather field program and involved sixty-six countries, thirty-nine ships, and thirteen aircraft (Kuettner and Parker 1976). This experiment, which occurred in the eastern tropical Atlantic Ocean, provided some of the most complete and well-sampled observations of the tropical atmosphere using a combination of surface, rawinsonde, satellite, radar, and the first use of Omega dropwindsonde (Hock and Franklin 1999) measurements. This allowed better understanding of cloud structure, thunderstorms, radiative processes, and turbulence in the tropics. It also produced some of the most detailed data of tropical waves propagating off of Africa, generating new knowledge about the genesis of hurricanes.

The transition of the experimental ATS satellite system began. A prototype operational geostationary satellite, the Synchronous Meteorological Satellite-1

(SMS-1) was launched in May, followed by SMS-2 the following year (Purdom and Menzel 1996).

1975 The increased use of satellites and data buoys reduced the need for general weather reconnaissance. The U.S. Navy stopped Atlantic hurricane flights. The U.S. Air Force also ended all weather reconnaissance except for hurricane flights, which continued for the Atlantic and West Pacific. Instead, nonhurricane weather reconnaissance was transferred to the U.S. Air Force Reserve, and a new squadron known as the Storm Trackers was formed at Keesler Air Force Base (Nolan and Murphy 2000; see also http://www.hurricanehunters.org).

NOAA assumed a fully operational geostationary satellite system based on the SMS prototypes, known as the Geostationary Operational Environmental Satellite (GOES) program. GOES-1 was launched on October 16. Today's GOES provides nearly continuous pictures of the Atlantic and eastern Pacific from 60°N to 60°S. GOES provides multispectral imagery, transmits weather facsimile to low-cost receiving stations, and relays data for a wide range of weather applications (Purdom and Menzel 1996).

1976 NOAA purchased two WP-3D aircraft for hurricane research. More details are contained in chapter 7 (see "Hurricane Research Division" and "NOAA Aircraft Operations Center").

1977 Meteosat, the first European meteorological satellite to be placed in geostationary orbit, was launched on November 23. Meteosat provided the first continuous pictures of tropical waves as they formed over Africa and propagated westward. The Meteosat satellite (currently in its seventh generation) still is the most important tool for monitoring tropical waves, with images and measurements taken every fifteen minutes (Schmetz et al. 2002).

The Meteosat also included the first water vapor channel (Morel, Desbois, and Szewach 1978), which shows regions of high- and low-moisture content. Unprecedented moisture structure can now be seen, even if no clouds are present. Animation of this imagery reveals circulation patterns that can affect a hurricane's track and wind shear patterns. Today, water vapor imagery is a prominent tool in analyzing the hurricane environment.

The first Geostationary Meteorological Satellite (GMS) satellite was launched on July 14, providing continuous satellite information of the western Pacific region.

The success of weather satellites caused elimination of the last U.S. weather observation ships (NOAA 1991), leaving only commercial ships to provide maritime weather observations on a voluntary basis.

1978   The experimental Seasat satellite was launched and contained a new instrument called a scatterometer. A scatterometer indirectly measures wind speed by emitting microwave radiation toward the ocean and computes the amount of radiation scattered by short, centimeter-scale (capillary) ocean waves. Since capillary waves are in equilibrium with the local wind field, the wind speed can be deduced. By measuring the scattered radiation with multiple beams from different angles, complex algorithms also give wind direction. Based on the success of this technique, other scatterometers have been launched, including the European Remote Sensing (ERS) satellite system, the NASA SCATerometer (NSCAT) on the short-lived ADEOS satellite, and the SeaWinds instrument on the QuikSCAT satellite. Scatterometers are important tools for hurricane forecasters monitoring tropical disturbances as well as for measuring winds on the periphery of hurricanes (Sharp, Bourassa, and O'Brien 2002; Leidner, Isaksen, and Hoffman 2003).

1979   Brian Jarvinen and Charles Neumann developed the statistical scheme SHIFOR (Statistical Hurricane Intensity FORecast) for predicting intensity change. SHIFOR is analogous to CLIPER in that it's used to assess the skill of other intensity prediction schemes (DeMaria 1996).

1981 The first prototype flight using Doppler radar data in Hurricane Gert was performed; no data was taken (Lee, Marks, and Walther 2003).

1982 The Synoptic Flow Experiment conducted by HRD began. In this experiment, HRD research flights released numerous Omega dropwindsondes within 600 miles of a hurricane's center, obtaining critical temperature, moisture, and wind data required for improved track forecasts. Eighteen hurricanes were sampled from 1982 to 1993, and the experiment successfully showed that extensive measurements of the hurricane environment can yield improved track forecasts by computer models (Burpee et al. 1996). As a result, NOAA later purchased a high-altitude jet, the Gulfstream IV, capable of taking meteorological measurements over a large area (see "1997").

The first HRD flight using Doppler radar data was performed in Tropical Storm Debbie (Lee, Marks, and Walther 2003). A single Doppler radar enables scientists to infer the primary circulation pattern of a hurricane. Doppler radar uses the *Doppler effect* to measure motions of objects toward or away from the instrument. The Doppler effect, first discovered by Austrian scientist Christian Doppler, is a shift in wavelength of radiation emitted or reflected from an object moving toward or away from the observer. Essentially, incoming waves are "squeezed," and outgoing waves are "stretched." A single Doppler radar translates this change in frequency into whether air particles, cloud droplets, and raindrops are moving toward or away from the radar and computes their incoming or outgoing speed relative to the radar. (The same principle also applies to sound waves, which explains why the sound of an oncoming train increases in pitch, then suddenly becomes a low pitch after the train passes.)

| 1983 | A workstation running Man computer Interactive Data Access System (McIDAS) software, developed by the University of Wisconsin, was installed at the NHC (Sheets 1990). This system processes, displays, enhances, and animates satellite data and can also overlay the imagery with meteorological measurements. McIDAS is still actively used today by many university institutions and governmental agencies, including the NHC. Also included with McIDAS are cloud-drift winds, which are computed from animation of infrared cloud imagery or water vapor imagery. This provides the NHC with abundant wind information in normally data-void regions and is particularly helpful in determining the environmental fields impacting storm motion and assessing wind shear events that can weaken a hurricane. |

William Gray (see chapter 5) discovered in 1983 that fewer Atlantic tropical storms and hurricanes occur in El Niño years (Gray 1984a). A year later Gray began forecasting months in advance the number of tropical storms and hurricanes that would occur (Gray 1984b) and has since assembled a forecasting team, issuing skillful forecasts several times each year (Owens and Landsea 2003). Gray considers El Niño as well as other atmospheric phenomena such as the Quasi-Biennial Oscillation, wind shear, and Caribbean surface pressure in making his seasonal hurricane prediction. See chapter 2 for more details.

Project STORMFURY ended (Willoughby et al. 1985).

| 1984 | The first Airborne eXpendable Current Profiler (AXCP) was dropped into Hurricanes Norbert and Josephine, providing the first high-resolution ocean current and sea temperature profiles in hurricanes (Sanford et al. 1987). |

| 1987 | The U.S. Air Force, convinced that satellite technology made aircraft reconnaissance obsolete, proposed ending the expensive flights in all oceans. However, |

meteorologists and emergency management officials persuaded Congress that the flights provided the most accurate measurements of a storm's location, central pressure, wind speeds, and other critical data that satellites cannot measure (or can only estimate), and Congress mandated the U.S. Air Force to continue flights in the Atlantic Ocean. Congress did permit the termination of flights in the western North Pacific, leaving only the Atlantic basin with routine flight monitoring of hurricanes (although occasionally flights will be conducted into hurricanes in the eastern Pacific and near Hawaii). A few years later, in 1991, the U.S. Air Force then ended its active participation in Atlantic hurricane reconnaissance due to budgetary constraints, and the Hurricane Hunters were merged with the Storm Spotters in the U.S. Air Force Reserve (Nolan and Murphy 2000; Sheets and Williams 2001; see also http://www.hurricanehunters.org).

The Improved Weather Reconnaissance System (IWRS) became operational onboard the Hurricane Hunter's Air Force WC-130 (Sheets 1990). This system automatically calculates wind, temperature, humidity, pressure, and heights continuously during a flight. When combined with a direct satellite uplink, IWRS system provides a continuous flow of high-density data to the NHC. Prior to IWRS and its satellite communication system, critical weather data (i.e., sea-level pressure) was calculated manually using tables or handheld calculators, then transmitted over high-frequency voice radio circuits across thousands of miles to military radio operators, who then passed the data by phone to Miami.

In June, the first Special Sensor Microwave/Imager (SSM/I) polar-orbiting satellite was launched as part of the Defense Meteorological Satellite Program. The suite of passive microwave channels at 19, 22, 33, and 85 GHz permits the sensor to "see through" most upper-level cloud decks such as cirrus clouds, which can cover a whole hurricane. Upper-level clouds obscure visible and infrared images used by most satellites, but not SSM/I. This unique ability allows a variety of new remote sensing applications to hurricanes (Hawkins et al. 2001). For example, SSM/I's frequencies are insensitive to nonraining clouds but sensitive to raining portions of hurricanes (rainbands), allowing meteorologists to identify significant spiral bands, locate the storm center more accurately, estimate rain rates, and develop new intensity-estimation algorithms. In addition, SSM/I can estimate wind speeds (but not direction) by sensing changes in ocean roughness and sea foam (where rain does not interfere with the signal).

1988 Robert Sheets became the NHC director, replacing the retiring Neil Frank (Sheets 1990). Sheets's term is marked by major technological improvements and a continued emphasis on hurricane preparedness issues related to coastal development and growth.

HRD research flight winds were made available operationally to the NHC through the Aircraft Satellite Data Link (ASDL). Prior to ADSL, data was transmitted by radio or voice links, severely limiting the type of data that could be transmitted from research aircraft to the NHC (Sheets 1990).

For the first time, in Hurricane Gilbert, two planes simultaneously took Doppler radar measurements. Since the simultaneous use of Doppler radars provides detailed wind information about a storm's vortex structure (while a single Doppler radar only provides information about incoming and outgoing airflow), this is a milestone data set (Lee, Marks, and Walther 2003).

The Office of Naval Research and NOAA sponsored a field program to study the upper-ocean response before, during, and after hurricane passage by deploying seventy-six AXCPs and fifty-one Airborne eXpendable BathyThermographs (AXBTs) (Shay et al. 1992).

1990   Mark DeMaria (see chapter 5) and John Kaplan developed a statistical scheme that incorporates some environmental factors known to affect hurricane intensity, such as wind shear (DeMaria 1996). This scheme is called SHIPS (Statistical Hurricane Intensity Prediction Scheme).

The Tropical Cyclone Motion (TCM-90) Experiment was conducted over the western North Pacific Ocean during August and September 1990 (Elsberry 1990). The objective of this experiment was to test several hypotheses regarding hurricane motion, with a focus on the Beta effect. Seven typhoons were analyzed every twelve hours throughout the lifetime of each storm. Two more field programs follow: the Tropical Cyclone Motion minifield experiments in July and August 1992 (TCM-92) and in July and August 1993 (TCM-93). The primary objective of these two follow-up experiments was to obtain simultaneous aircraft measurements and satellite imagery of typhoons and nearby cloud complexes. Their objectives were to test hypotheses regarding tropical cyclogenesis in the cloud complexes and determine how nearby cloud clusters might cause nearby typhoons to deviate from their track (Harr and Elsberry 1996). These flight missions support theories that cyclonic rotation in tropical disturbances begins in the midlevels of the atmosphere.

1991   The Tropical EXperiment in MEXico (TEXMEX) was conducted over the eastern North Pacific Ocean from July 1 to August 8 (Bister and Emanuel 1997). TEXMEX's objectives were to test several hypotheses regarding tropical cyclogenesis. Six cloud clusters with the potential for formation into a tropical storm were investigated, revealing new insights into the rotational development of tropical disturbances.

NASA started a comprehensive program to understand the earth's environmental system (today called the Earth Science Enterprise). Included in these plans is a coordinated series of Earth Observing System

(EOS) polar-orbiting satellites for long-term global observations of the land surface, biosphere, solid earth, atmosphere, and ocean (see "1999").

1993　Radar pictures and Omega dropwindsonde data from HRD research flights were included in the operational database for the NHC and NCEP's use, thus providing more real-time information about a hurricane's cloud pattern and vertical structure (DeMaria 1996).

Installation of the WSR-88 (NEXRAD) Doppler radar network began to replace the WSR-57 radar network. This network is useful for detecting tornadoes in severe thunderstorms and for measuring the wind field of hurricanes near coastal Doppler radars. The NEXRAD network can also detect precipitation, rainfall rate, accumulated rainfall, and the location of the rainfall.

1994　Techniques for extracting the primary circulation of hurricanes from a single airborne Doppler radar were developed (Lee, Marks, and Carbone 1994; Roux and Marks 1996). This allows real-time estimates of a hurricane's vortex field, since dual-Doppler analysis is a time-consuming procedure not suitable for operational use.

1995　High-resolution GOES-8 pictures of hurricanes are taken in one-minute sequences (called "rapid scan"), revealing unsurpassed imagery of a hurricane's small-scale structure (Purdom 1995).

Bob Sheets retired as NHC's director and was replaced by Dr. Bob Burpee. During Burpee's tenure, he was credited with bringing closer collaborations between the NHC's forecasters and researchers.

The Geophysical Fluid Dynamics Laboratory model became operational at the NCEP (Kurihara, Tuleya, and Bender 1998; Soden, Velden, and Tuleya 2001). This is currently the most sophisticated model and generally has the smallest track errors compared to other model guidance.

1997 Building on the success of the SSM/I program, NASA, in a joint venture with Japan, launched the Tropical Rainfall Measuring Mission (TRMM) satellite on November 27 with a unique orbit that oscillates around the equator (Kummerow et al. 2000). This satellite, essentially a "flying rain gauge," contains several sensors for detecting moisture and rainfall. A spaceborne radar is used to measure tropical and subtropical rainfall, and another instrument, the Microwave Imager (MI), measures the vertical distribution of precipitation. The resolution of the radar is two to three times better than the SSM/I satellites, and the MI can penetrate deeper into clouds. Its primary purpose is to calibrate rainfall estimate routines used on other satellites and to understand the global moisture cycle. However, TRMM has proven immensely useful to hurricane forecasting due to these improved sensors—the higher resolution is providing better information on hurricanes in all stages of development, especially since the MI can penetrate deeper into the hurricane. More accurate placement of the storm center is also achieved using TRMM as compared to SSM/I, and new rainfall estimation techniques are being developed (Hawkins et al. 2001).

Starting with Hurricane Claudette in 1997 and many subsequent hurricanes in 1998, routine high-altitude reconnaissance observations were taken in the environment surrounding hurricanes by NOAA's new Gulfstream IV jet, providing valuable data about the steering currents that influence hurricane motion (Aberson 2003).

The Global Positioning System (GPS) dropwindsonde system was first deployed into a hurricane (Guillermo in the eastern Pacific on August 3) during an HRD research flight. This new device measures wind speeds with unprecedented accuracy and vertical resolution. In addition, it works in heavy rain and has the ability to take observations below 1,500 feet—neither of which the older Omega dropwindsondes could do (Franklin, Black, and Valde 2003). Ten of the GPS dropwindsondes were deployed into the eyewall, providing the first high-resolution wind, temperature, and moisture profiles of a hurricane eyewall (Hock and Franklin 1999). The Hurricane Hunters started using GPS dropwindsondes one year later. Between twenty and thirty dropwindsondes are now routinely deployed in hurricanes from both the Gulfstream jet and reconnaissance planes to improve computer-guided track forecasts (Aberson 2002; Aberson 2003).

1998   The first weather robotic aircraft, the Aerosonde, was successfully flown across the Atlantic Ocean on August 21–22 (McGeer and Vagners 1999). The Aerosonde was developed in Australia under the direction of Dr. Greg Holland (see chapter 5) of the Bureau of Meteorology. The Atlantic flight was conducted by the University of Washington and a U.S. engineering company, The Insitu Group. The unmanned 29-pound plane departed from St. John's, Newfoundland, Canada, and completed the 2,000-mile trip to the west coast of Benbecula in the Outer Hebrides of Scotland in just over twenty-six hours. The trip is completed using less than 2 gallons of fuel, continuously measuring meteorological observations the entire time. These drones theoretically can be programmed to fly a fixed path for more than twenty-four hours, continuously ingesting weather observations that improve weather forecasts. Their unique aerodynamic structure and relative weightlessness also allow them to theoretically withstand hurricane winds and strong updrafts in thunderstorms, thereby providing timely and continuous crucial weather data at a cheaper cost than airplane reconnaissance. More information is available at http://www.aerosonde.com.

NASA and NOAA collaborated in the Third Convection And Moisture EXperiment (CAMEX-3) in an effort to collect comprehensive data on Atlantic hurricanes using multiple aircraft, including high-altitude measurements using the Gulfstream IV and NASA's ER-2 aircraft, and a variety of remote sensing technologies (scanning radar altimeter, Doppler radar). On August 20, Hurricane Bonnie became the first storm to be sampled in this project, which included up to six planes at once (the Gulfstream IV, ER-2, NASA DC8, two WP-3D research flights conducted by HRD, and one WC-130 flown by the Hurricane Hunters). Details are provided at http://www.aoml.noaa.gov/hrd.

The largest coordinated measurements to date of land-falling hurricanes were performed for Hurricanes Bonnie, Charley, Earl, and Georges. These include portable instrumented towers, portable wind profiler, and a portable Doppler radar. Texas Tech University, Clemson University, the University of Oklahoma, and the University of Alabama at Huntsville are the principal investigators. Details are provided at http://www.aoml.noaa.gov/hrd.

The NHC's director Bob Burpee retired and was replaced by Jerry Jarrell, who is talented at interacting with the media.

NOAA's first polar-orbiting satellite containing an Advanced Microwave Sounding Unit (AMSU) was launched in May. This passive microwave radiometer can capture the three-dimensional depiction of a hurricane's warm core structure. These temperature profile retrievals also allow the estimation of the storm's surface pressure (Brueske and Velden 2003) and axisymmetric wind structure (Zhu, Zhang, and Weng 2002).

1999   The satellite flagship of NASA's EOS program, Terra, was launched on December 18 (Jones et al. 2004). It contained state-of-the-art radiometers that provide high-resolution measurements of clouds, temperature, moisture, and radiation.

Data was transmitted operationally for the first time to the NHC from the Stepped Frequency Microwave

Radiometer (SFMR), a remote sensing instrument on the NOAA WP-3D research planes. Details are provided at http://www.aoml.noaa.gov/hrd.

2000   Britain sponsored JET2000, in which four flights during August 25–30 deploy dropwindsondes in data-sparse West Africa to study the tropical waves that form over this region (Thorncroft et al. 2003).

2001   NASA sponsored CAMEX-4 from August 16 to September 24 using NASA-funded aircraft (a DC-8 and ER-2) and NASA satellites (TRMM, Terra, and QuikSCAT) to study Tropical Storm Chantal, Tropical Storm Gabrielle, Hurricane Erin, and Hurricane Humberto as they approached landfall in the Caribbean, Gulf of Mexico, and U.S. East Coast. The Hurricane Research Division also participated in this field program. The NASA aircraft investigated upper-altitude regions of these storms not normally sampled by NOAA, providing validation against satellite accuracy. In addition, the NASA aircraft can sample cloud microphysics data and humidity data with more sophisticated instruments than the NOAA aircraft. The goal is to use this knowledge to improve hurricane landfall predictions and to better understand the three-dimensional structure of hurricanes. Details are provided at http://www.aoml.noaa.gov/hrd.

An ocean prediction system was included in the GFDL so that the atmospheric and ocean components are coupled (Bender and Ginis 2000). In the past, water temperature was held constant, which is not realistic because the ocean can experience significant mixing under hurricane-force winds, resulting in colder waters under the storm.

2002   NASA launched its second EOS satellite, Aqua, in May (Jones et al. 2004). Its mission, as implied by the name, is more focused on water properties and the hydrological cycle of the earth's system (whereas Terra is geared toward land properties). In addition to radiometers, state-of-the-art microwave and infrared sounders were included for measuring temperature and moisture profiles. These sounders have 100 times more channels than previous satellites. The sounders will lead to large improvements in defining the temperature and moisture fields surrounding the hurricane environment and above cloud tops. Aqua also has a next-generation microwave imager for determining storm position, structure, and rain estimation.

The China LAndfalling Typhoon EXperiment (CLATEX) was implemented during July and August (Lianshou, Mingyu, and Xiangde 2004). The objective was to better understand the impact of land topography on hurricane intensity change, track, and rainfall.

The Coupled Boundary Layer Air Sea Transfer (CBLAST) experiment was conducted. Its goal was to understand the air-sea interaction in the hurricane environment to help improve hurricane intensity predictions, with a particular focus on the role of waves, sea spray, and turbulence. The experiment resulted in the deployment of numerous instruments that measure ocean waves, ocean currents, and ocean temperature during the passage of a hurricane. Details are provided at http://www.aoml.noaa.gov/hrd.

2003   The first detailed study of the GPS dropwindsondes was published (Franklin, Black, and Valde 2003), providing new insight into low-level winds. It shows that, on average, surface winds are 90 percent of winds at the 700-mb reconnaissance flight level. In the past, the NHC used a conversion factor of 70–90 percent, suggesting that many hurricane surface wind estimates based on flight data had been too low. The study also shows that winds peak at a height of 1600 feet and that people should not stay near the top of tall buildings in a hurricane where winds can be 17–25 percent stronger than surface winds (roughly one Saffir-Simpson Hurricane Scale category larger).

The success of the U.S. Gulfstream IV jet research program motivated Taiwan to initiate a three-year research reconnaissance program called the Dropsonde Observations for Typhoon Surveillance near the Taiwan Region (DOTSTAR). The purpose was to assess the impact of GPS dropsondes in weather models at the NCEP, the Fleet Numerical Meteorology and Oceanography Center (FNMOC), and Taiwan's Central Weather Bureau (Huang et al. 2004).

The CBLAST experiment continued. An Imaging Wind and Rain Airborne Profiler (IWRAP) was installed on the WP-3D planes. When used in combination with SFMR, continuous measurements of the turbulent wind profile right above the surface were available for the first time (Fernandez et al. 2004; Black 2004). In addition, near-surface flights obtain heat flux, moisture flux, and sea spray spectra measurements with unprecedented detail, providing valuable data for understanding hurricane intensity changes (Black 2004).

Chronology of Significant Land-falling U.S. Hurricanes since 1900

Land-falling tropical storms, for the most part, have been excluded from table 4.1. This does not imply that tropical storms should not be taken seriously. Tropical storms can still cause damaging storm surges, especially if the storm is large and moving slowly. Moreover, they are notorious for heavy deluges that can cause serious flooding. For example, Tropical Storm Claudette (1979) dumped 42 inches of rain in one day on Alvin, Texas (Hebert 1980). Tropical Storm Alberto (1994) dumped 10–27 inches of rain in Georgia, Alabama, and western Florida, resulting in large-scale floods that killed thirty people (Avila and Rappaport 1996).

More recently, Tropical Storm Allison's (2001) heavy rains of 20–35 inches produced catastrophic flooding over portions of the upper Texas coastal area (especially Houston) and significant flooding along the remainder of its track through Louisiana, Mississippi, and Alabama (Beven et al. 2003). Forty-one deaths are attributed to Allison's heavy rains, flooding, tornadoes, and high surf.

TABLE 4.1
Chronological order of some significant U.S. land-falling hurricanes since 1900

Dates of Hurricane	Areas Most Affected	Saffir-Simpson Category at Landfall	Deaths (U.S. Only)	Comments and Damage
1900, August 27–September 15 Galveston Hurricane	Texas	Category 4	At least 6,000 on island, and 2000 more inland	Principal damage and most fatalities caused by an 8–15-foot storm surge that inundated Galveston Island. Flying debris, including slate shingles from roofs, caused some deaths. Damage was widespread along Texas coast. Of the island's approximately 12,000 inhabitants who stayed (roughly another 12,000 evacuated), 6,000 were killed, and the others were bruised and battered. Another 2,000 were killed inland. This is still the worst U.S. natural disaster in terms of fatalities. In the postanalysis, it is discovered that most survivors were located behind concentrated elongated areas of debris that acted as a break against the full force of the storm surge, and the notion of a seawall is conceived to help prevent future disasters. A 6-mile-wide seawall was completed in 1905, and the elevation of the entire city was raised 8 feet to a height of 17 feet (Weems 1997; Larson 1999; Hughes 1990).
1906, September 19–29	Florida panhandle to Mississippi	Category 3	134	Destructive winds and unprecedented tides accompany the storm upon landfall on September 27. At Pensacola, FL, the tide is 10 feet above normal, and this is the city's most violent storm in 170 years. At Mobile, AL, property damage is severe with a 10-foot storm surge (NOAA 1993; Tannehill 1956).
1909, July 13–22	Texas and Louisiana	Category 3	41	The hurricane passes directly over Velasco, TX, on July 20. The town is in the calm of the eye for forty-five minutes, followed by devastating winds that destroy half the town. A storm surge of 10 feet occurs at Galveston, with higher surges farther west peaking in Velasco (NOAA 1993; Bunnemeyer 1909; Tannehill 1956). No lives are lost in the seawall region of Galveston, and there is only minimum damage in this protected area while devastation elsewhere is heavy, showing that the seawall can provide satisfactory protection. Southwest Louisiana also experiences a rainfall with amounts of almost 9 inches reported. South Texas receives heavy storm surge and winds that drown hundreds of cattle and destroy the cotton crop.
1909, September 14–21	Louisiana and Mississippi	Category 4	353	The hurricane center passes 50–75 miles west of New Orleans on September 20. A wide region of the Louisiana coast is inundated with a 15-foot storm surge (NOAA 1993; Tannehill

1956). New Orleans and the whole southern state experiences extensive damage to property, vessels, crops, and the railroad system. Half of the coal fleet, anchored in the Mississippi River for shelter, is lost, but this is considered a partial victory since otherwise all would have been lost without the Weather Bureau's warnings to seek protection. Most deaths occur outside the New Orleans levee system in rural fishing communities, where timely warnings are not possible. A few deaths in New Orleans result from falling chimneys or contact with live power lines.

Date	Location	Category	Deaths	Description
1915, August 5–21	Texas and Louisiana	Category 4	275	This hurricane makes landfall on the western end of Galveston Island at 1 A.M. on August 17. Despite the seawall and recently raised island, a 12-foot storm surge inundates the Galveston business district, with water rising up to 6 feet in parts. Nevertheless, the value of the seawall and raised elevation is evidenced by the fact that only 8 are killed in the city (Hughes 1990). Along adjacent coastal areas, Galveston Bay, and the parts of Galveston Island without the seawall, destruction is massive. Vessels of all sizes are sunk or damaged. The causeway bridge connecting the island is badly damaged. Fires also destroy buildings. Houston experiences severe infrastructure damage, and crops in half the state are wiped out. Strong winds and heavy rain cause destruction from Texas northeastward to New York, with eastern Missouri in particular experiencing wind damage. Several marine vessels in the open Gulf of Mexico and in Texas are sunk, causing a large number of fatalities. The biggest tragedy is the sinking of the steamship *Marowijne* in the Yucatan Channel with its crew of 96 people. Sixty-nine people are killed on dredge boats and a tugboat in Galveston Bay. When combined with other lost vessels, 8 in Galveston city, 42 on Galveston Island, and other scattered fatalities, 275 are killed (Frankenfield 1915; NOAA 1982). Louisiana also experiences gale-force winds and high water.
1915, September 22–October 1	Louisiana	Category 4	275	This dangerous hurricane makes landfall in southeast Louisiana on September 29 and passes 12 miles west of New Orleans. The 15–20 foot storm surge tops the Mississippi River levees south of New Orleans. The surge also overflows levees in western New Orleans, flooding a large section of the city. Because the city is below sea level, it takes 3–4 days for the drainage system to remove this water. Ninety percent of buildings are destroyed

(continues)

TABLE 4.1.
(Continued)

Dates of Hurricane	Areas Most Affected	Saffir-Simpson Category at Landfall	Deaths (U.S. Only)	Comments and Damage
				over a large area of Louisiana south of New Orleans and along Lake Pontchartrain. The area's infrastructure and businesses (shipping, coal, crops, and railroads) are severely impacted. Most heed warnings, but 275 casualties result from those who stay in low-lying areas (Cline 1915; NOAA 1993; Tannehill 1956).
1916, June 29–July 10	Mississippi to northern Florida	Category 3	7	This hurricane arrives unexpectedly on July 5 with a 12-foot storm surge, leaving shoppers and businessmen stranded in downtown Mobile hotels. Eight feet of water inundates the lobby of the Grand Hotel at Point Clear. This hurricane is very destructive along the coast from Mobile to Pensacola (NOAA 1982; Tannehill 1956).
1919, September 2–15	Florida, Louisiana and Texas	Category 4 in Florida Keys and category 4 in Texas	287 on land, 500 on ships at sea	This hurricane is severe in both Florida and Texas. The slow-moving storm reaches an intensity of 927 mb 65 miles west of Key West on September 10. Ten vessels are lost at sea, accounting for 500 casualties. The hurricane moves slowly westward, causing minor storm surge along the Gulf Coast, such as a 6-foot storm surge on Louisiana's central coast. The hurricane goes inland south of Corpus Christi on September 14, inundating the region with 16 feet of water and killing 287 people. Additionally, much of the Texas coast is devastated by high water (NOAA 1993; Tannehill 1956).
1926, September 11–22 Great Florida Storm; also call the Miami Hurricane	South Florida, Florida panhandle, and Alabama	Category 4 in southeast Florida; category 3 at Alabama-Florida border	Officially 243, probably 372 (Mykle 2002; Pfost 2003)	This hurricane is very destructive in the Miami area and from Pensacola into southern Alabama. The hurricane makes landfall in Miami on September 18 with an 8–12 foot storm surge. Many go outside during the eye's calm interval and are killed when the strong winds return (Mitchell 1926). The storm bends Miami's newest skyscraper, the Myer-Kiser Bank Building, 15 degrees, forcing the building to be torn down. The 12-foot surge drives a steamer into the middle of Miami. Miami's economy, which had been experiencing a development boom, suffers due to the destruction and lost investment. The hurricane's winds

			push water out of Lake Okeechobee and cause extensive flooding in the town of Moore Haven on the lake's western shores, killing about 150 (Reardon 1986). The hurricane makes landfall between Pensacola and Mobile at Perdido Beach on September 20, hammering the region with a 5-foot storm surge. It takes an unusual inland path, going west along the coast. The eye passes over Biloxi, Gulfport, and Pass Christian, finally passing northwest of New Orleans and dissipating in Texas. The back side of the hurricane actually flushed water out into the sea along this path. For example, water levels in Mobile River decreased by 9 feet, stranding boats seeking shelter and causing mud slides along the riverbanks (Mitchell 1926).
1928, September 6–20 Okeechobee Hurricane	Southern Florida	Category 4	Officially 1,836, probably at least 2,500; could be as high as 3,000 (Mykle 2002; Pfost 2003).	After making landfall near Palm Beach on September 16, the storm travels up the Florida peninsula over Lake Okeechobee. This hurricane officially kills 1,836 people and injures 1,849 on the shores of Lake Okeechobee, the second largest freshwater lake in the United States. More likely, at least 2,500 are killed; water covers the region for weeks after the storm, making recovery efforts difficult, and bodies are still being discovered after funding ends for the posthurricane recovery, (Mykle 2002; Pfost 2003). Moreover, many are poor migrant agricultural laborers with no easy way to trace their origin; their bodies are buried in mass graves or burned. Unfortunately, racial attitudes probably also contribute to a lack of concern for the deceased individuals. This hurricane is the second deadliest natural event in U.S. history. Most of the deaths are by drowning, but some are due to snakebites as people climb into trees to escape the floodwaters, only to find hordes of venomous water moccasins also seeking shelter. Although the lake is actually inland, the hurricane forces a storm surge that breaks the eastern earthen dike on the southern end of Lake Okeechobee (Reardon 1986). This calamity occurs within a few miles of a large city and a world-famous resort, yet so isolated is the location that no one knows about the tragedy for three days afterward. As a result of flooding from this storm and the 1926 hurricane, a levee (called the Herbert Hoover Dike in honor of the president who supported its construction) that compares in size to the Great Wall of China is built to prevent further disasters (Will 1990).

(continues)

TABLE 4.1
(Continued)

Dates of Hurricane	Areas Most Affected	Saffir-Simpson Category at Landfall	Deaths (U.S. Only)	Comments and Damage
1932, August 11–14	North Texas coast	Category 4	40	Landfall occurs slightly east of Freeport and directly over eastern Columbia on August 13. The hurricane is small in diameter; winds damage rice and some cotton near the coast, but accompanying rains are beneficial to the drier interior areas (NOAA 1982; Tannehill 1956).
1933, August 17–26 Chesapeake-Potomac Hurricane	North Carolina, Virginia, Maryland	Category 3	47	The storm makes landfall August 23 at Nag's Head on the Outer Banks and then impacts Norfolk, VA; Washington, D.C.; Baltimore, MD; and Atlantic City, NJ. It is the only storm to strike the Chesapeake-Potomac area this century. Heavy damage occurs in northeastern North Carolina and resorts on Maryland, Delaware, and New Jersey coasts. Downtown Norfolk is flooded. Crop damage is heavy in Maryland and Virginia (NOAA 1982; Tannehill 1956).
1933, August 28–September 5	South Texas coast	Category 3	40	The hurricane goes inland just north of Brownsville on September 1. Heavy property damage takes place from Corpus Christi to northeastern Mexico. The citrus crop is almost completely destroyed (NOAA 1982; Tannehill 1956).
1933, August 31–September 7	Florida	Category 3	2	The hurricane makes landfall on September 3 at Jupiter Inlet. Much property destruction occurs between Jupiter Inlet and Fort Pierce. The citrus loss is nearly complete along the coast (NOAA 1982; Tannehill 1956).
1933, September 8–21	North Carolina	Category 3	21	The hurricane strikes just west of Cape Hatteras on September 16. However, it then recurves to the north and northeast off the Atlantic coast, thus avoiding a repeat of the Chesapeake-Potomac Hurricane. There is damage at Norfolk and Cape Henry, but less than the recent August storm. However, a high storm surge inundates Pamlico and Albermarle Sounds. South of the Virginia Capes to New Bern, NC, damage is very severe. Beaufort, NC, is hit particularly hard (NOAA 1982; Tannehill 1956).
1934, June 4–21	Louisiana	Category 3	6	This hurricane hits the central Louisiana coast on June 16, moving at an unusual rate of 27 mph. Property and crop damage is moderate (NOAA 1982; Tannehill 1956).

Date	Location	Category	Deaths	Description
1934, July 21–25	Florida and Texas	Category 2	11	This storm has an unusual track. It forms off the North Carolina coast, then travels southwest and hits Florida as a weak disturbance. Upon entering the Gulf of Mexico, it intensifies into a moderate hurricane and makes landfall just north of Corpus Christi on July 25. Heavy rains severely damage the Texas cotton crop (NOAA 1982; Tannehill 1956).
1935, August 29–September 10 Labor Day Storm	Florida Keys and Northwest Florida	Category 5 in Florida Keys; category 2 in Northwest Florida	409	The September 3 barometer reading of 26.35 inches (892 mb) on Long Key is lowest on record in the Western Hemisphere until Hurricane Gilbert (1988). It is still the strongest U.S. land-falling hurricane (Gilbert never hit the U.S.). Peak winds are estimated at 150–200 mph on some keys with an 18-foot storm surge. The Labor Day storm is an example of a midget hurricane in which hurricane-force winds are confined to a diameter less than 35 miles, and a striking example that hurricane intensity has little to do with size. It clears the entire landscape of every tree and every building on Matecumbe Key. The dead include residents and at least 160 World War I unemployed veterans building a road as part of the Work Projects Administration. A train is sent to evacuate them, but when it returns north the storm surge had washed out the track. Shortly thereafter, all cars but the locomotive and the tender are washed off the track, killing many (McDonald 1935; NOAA 1982; Tannehill 1956; Sheets and Williams 2001). Ernest Hemingway visits the site and writes about the incident in a scathing article titled "Who Murdered the Vets?" for *New Masses* magazine (Hemingway 1935).
1935, October 30–November 8 Yankee Storm	Southern Florida	Category 2	5	This storm forms near Bermuda and moves north to south along the East Coast, hitting the Miami area. Because of its unusual track, locals called it the "Yankee Storm" to counter hecklers in the Northern states who started calling these storms "Florida hurricanes" because of the recent frequency of Florida landfalls. The storm hits Miami on November 4, causing considerable damage. It enters the Gulf of Mexico, then turns around and hits southwest Florida on November 8 (NOAA 1982; Tannehill 1956).
1938, September 10–22 New England Hurricane	Long Island, N.Y.; southern New England	Category 3	564	This hurricane is usually called the New England Hurricane but is also called the "Long Island Express" because of its rapid northward motion (up to 56 mph) toward the Long Island coast, making landfall on September 21. This storm surprises New England residents, who are unaccustomed to hurricanes (the last hurricane had hit New England seventy years earlier). Because of the hurricane's fast motion as it undergoes an extratropical transition,

(continues)

TABLE 4.1
(Continued)

Dates of Hurricane	Areas Most Affected	Saffir-Simpson Category at Landfall	Deaths (U.S. Only)	Comments and Damage
				and because the Weather Bureau had thought it would recurve offshore and not hit the United States, there is little preparation time available. Even worse, the storm hits during some of the higest tides of the year. Ten-foot waves atop a 14–18-foot storm surge kill 564 people and cause immense property damage. Damage to the marine industry and fishing fleet is catastrophic. Barrier islands are swept so bare that rescue workers used phone company charts to determine where houses once stood. Sustained hurricane winds penetrate far inland, causing extensive damage to roofs, trees, and crops. In Connecticut, downed power lines result in catastrophic fires to sections of New London and Mystic. Rainfall of 7 inches or more results in severe flooding across sections of Massachusetts and Connecticut (Fisher 2003). This storm holds the record for the most property damage in the United States until 1954 (McCarthy 1969; Allen 1976; Minsinger 1988; Pierce 1939).
1940, August 5–15	Georgia and the Carolinas	Category 2	50	This is the first severe hurricane to directly strike the Charleston-Savannah area in thirty years, making landfall on August 11. Severe damage to crops and property occurs. Heavy flooding in the southeastern states penetrates as far inland as Tennessee from hurricane-induced rains. Thirty deaths result from the floods (NOAA 1982; Tannehill 1956).
1941, September 16–25	Texas	Category 3	4	This major hurricane hits Matagorda on September 23. Local property damage is high, and the rice crop is ruined. The storm then turns northeast, passing near Houston where considerable damage occurs. Most low, exposed places are evacuated in response to good warnings, resulting in low casualties (NOAA 1982; Tannehill 1956).
1943, July 25–29 The Surprise Hurricane	Upper Texas coast	Category 2	19	Because of a dare from his students, U.S. Army Air Corps Col. Joseph P. Duckworth flies the first intentional flight into this hurricane (Sheets and Williams 2001). This hurricane is also noteworthy because it hit with little warning on July 27, killing nineteen people, injuring

				hundreds, and causing significant property damage along Galveston Bay. Eleven of these lost lives are due to the sinking of the dredge *Galveston* at the entrance to Galveston Bay. Because of World War II censorship, a forecaster's major source of information—ship reports—is unavailable so that German U-boats would not get weather information. Therefore, the Weather Bureau thought this was a tropical storm and did not know it was a category 2 hurricane until it makes landfall. Even after landfall, the forecast advisories are delayed several hours since they required security clearance. Finally, barometer readings are also restricted from the general public as the hurricane approached. All of this contributes to a confused and uninformed public. It could have been much worse if Galveston had no seawall or if the storm had hit the island itself (it made landfall on the Bolivar peninsula to the east). It is also a small hurricane in diameter, limiting the areal extent of the damage (Tannehill 1956). After this, never again are forecast advisories censored from the general public (Fincher and Read 1999). This hurricane also seriously damages refineries in Texas, hurting wartime production of fuel (Sheets and Williams 2001).
1944, September 9–16 The Great American Hurricane	North Carolina to New England	Category 3	46 in U.S., 298 offshore	This storm receives its name because it travels up the Atlantic coast from North Carolina to the Northeast, making landfall in Rhode Island on September 14. (It's also called the Great Atlantic Hurricane of 1944.) Heavy damage occurs, but it is one-third as great as the 1938 hurricane. This storm is the first success for the Hurricane Hunters; U.S. Navy and Army Air Force planes track the storm all the way from Puerto Rico, thus enabling better warnings (Sheets and Williams 2001). U.S. fatalities are reduced (46 killed) due to better preparation. New England is less impacted than the 1938 hurricane because the stronger portion of the storm remains offshore of New England, so winds and surge are less than in 1938. However, a higher storm surge occurs farther north in New Jersey, where there is great damage (Tannehill 1956). The U.S. Navy and the U.S. Coast Guard lose five ships offshore, killing 298 servicemen.
1944, October 12–23	Southwest Florida	Category 3	18	Damage from the storm surge at landfall on October 19 between Sarasota and the Everglades is heavy. Crop damage is excessive. Warnings and evacuation prevent heavier casualties (Tannehill 1956; NOAA 1982).

(continues)

TABLE 4.1
(Continued)

Dates of Hurricane	Areas Most Affected	Saffir-Simpson Category at Landfall	Deaths (U.S. Only)	Comments and Damage
1945, September 11–20 Kappler's Hurricane	Southeast Florida	Category 3	4	This hurricane makes landfall on September 15, causing very heavy damage in Dade County in a 40-mile swath, particularly to the area south of Miami. Evacuation of exposed locations prevents heavy loss of life (Sumner 1946). This hurricane is named after U.S. Army Air Corps 2nd Lieutenant Bernard Kappler, a Hurricane Hunter who discovered the storm during a weather reconnaissance mission (Tannehill 1955). His flight and subsequent flights encountered extreme turbulence, which the new Hurricane Hunters had never seen before.
1947, September 4–21	Southeast Florida, Louisiana, and Mississippi	Category 4 in Florida; category 3 in Louisiana and Mississippi	51	This severe hurricane crosses over Florida and hits Louisiana and Mississippi. The center of this very large and intense storm hits Pompano, FL, on September 17. Most of the Florida east coast is subjected to hurricane-force winds. Fortunately, this region is sparsely populated. Lake Okeechobee's levee holds, but rains of 6–8 inches fall on the already-saturated soil in the croplands. The sugarcane crop is devastated, and a large number of cattle are lost. After passing over the Everglades, the storm retains much of its energy and impacts the lower west coast of Florida. Heavy damage occurs from Everglades City to Sarasota, with the greatest damage in the Fort Myers-Punta Gorda area from winds and a surge as high as 5 feet. After leaving Florida, the huge hurricane takes a northwesterly course with increased forward motion over the Gulf of Mexico and sweeps onto the Mississippi and Louisiana coasts on September 19. Tides rise to 12 feet at Biloxi Bay, St. Louis, and Gulfport, MS. The eye of the storm passes directly over New Orleans and later Baton Rouge, LA. Mississippi and Louisiana suffer heavy damage, mostly from the surge. A few levees in Louisiana are breached. A total of 51 lives are lost--17 in Florida, 12 in Louisiana, and 22 in Mississippi. (NOAA 1993; Sumner 1947).

1947, October 9–16	Southern Florida; Georgia and South Carolina	Category 1 in south Florida; category 2 in Georgia and South Carolina	1	This northeast-moving hurricane passes over the extreme southern portion of the Florida peninsula on October 12. It moves offshore east of Florida, then turns west. The Sarasota region is skirted by the storm but has hurricane-force winds. Because of the previous month's hurricane, there is little left to destroy. However, heavy rains of 5–13 inches from Lake Okeechobee southward climax a very wet season, causing flooding. The hurricane makes landfall just south of Savannah, GA. Hurricane-force winds are confined to a 40-mile region. A 12-foot surge, winds, and falling trees cause much structural damage along the Georgia–South Carolina border (NOAA 1982; Sumner 1947).
1948, September 18–25	Southern Florida	Category 3	3	This hurricane skirts just east of Key West on September 21, then makes landfall just east of Everglades City with a significant storm surge, crosses Lake Okeechobee, and enters the Atlantic at Jenson Beach near Stuart. Eight to eleven inches of rain cause flooding throughout the region. High water drowns cattle and destroys the citrus crop (NOAA 1982; Sumner 1948).
1948, October 4–9	Southern Florida	Category 2	0	The center passes over the Florida Keys on October 5 at noon, and seven hours later the eye passes over Miami. The damage is not as great as could be expected, since much of area had been hit by the September storm two weeks earlier (NOAA 1982; Sumner 1948).
1949, August 23–31	Florida to the Carolinas	Category 3	2	This hurricane makes landfall in West Palm Beach on August 26, causing significant property damage, then passes over Lake Okeechobee. The water rises 12 feet on the lake's east side, but the levees built since 1928 prevent overflow and casualties. The center then passes northwestward through the heart of the citrus belt, devastating the crops (Zoch 1949).
1949, September 27–October 6	Texas	Category 2	2	This hurricane makes landfall in Freeport, TX, on October 3, then curves northeast over Houston. The Houston Ship Channel has an 11-foot storm surge. Heavy rain values up to 14.5 inches are observed. Four-fifths of the damage is to crops, with oil rigs and roads also being damaged (Zoch 1949).
1950, September 1–9 Hurricane Easy	Northwest Florida Peninsula	Category 3	2	Also called the "Cedar Key Hurricane," this storm's path exhibits an unusual double loop near the fishing village of Cedar Key. During the first loop it made landfall just

(continues)

TABLE 4.1
(Continued)

Dates of Hurricane	Areas Most Affected	Saffir-Simpson Category at Landfall	Deaths (U.S. Only)	Comments and Damage
				south of Cedar Key on September 5, where it made another loop; in making this loop the calm center moved over Cedar Key from the southeast and then moved south. This path resulted in Cedar Key being exposed to the same side of a hurricane twice, with 1.5 hours of calm in between. The coast from Sarasota northward suffers extensive wind, storm surge, and flood damage, and Cedar Key's fishing fleet is destroyed. Rain amounts are also heavy, between 10 and 25 inches in most places. Nearly 39 inches of rain fall on Yankeetown, FL, in twenty-four hours. The coastal area inland from Yankeetown to Tampa is flooded for several weeks (NOAA 1982; Norton 1951).
1950, October 13–19 Hurricane King	Southeast Florida	Category 3	4	This small, violent storm passes directly over Miami on October 17, then up the entire inland Florida peninsula. Property and crop damage is heavy along the entire path (NOAA 1982; Norton 1951).
1954, August 25–31 Hurricane Carol	North Carolina to New England	Category 2 in North Carolina; category 3 in New York, Connecticut, and Rhode Island	60	This hurricane passes near Cape Hatteras on August 30 with minimal damage. The storm moves rapidly north-northeast and crashes into the New England states. Sixty are killed, and there is record (up to this time period) property and crop damage. A significant storm surge inundates many low-lying regions in New England. Providence, RI, is under 8–10 feet of water for several hours (Davis 1954).
1954, September 2–14 Hurricane Edna	New Jersey to New England	Category 3	21	Edna skirts east of Cape Hatteras with minimal effect. On September 11, it passes over Cape Cod and then into eastern Maine. Heavy damage occurs again in New England (NOAA 1982; Davis 1954).
1954, October 5–18 Hurricane Hazel	South Carolina to New York	Category 4 in South and North Carolina	95	Hazel makes landfall at Holden Beach on October 15, inflicting heavy damage on the North and South Carolina coasts due to a high storm surge superimposed on the highest ocean tide of the year. Structural damage is immense, and downed trees litter the state. Rainfall is heavy along and to the west of the storm track in North Carolina, with amounts up

to 11 inches. Hazel is the most destructive storm in North Carolina's history (Barnes 1998; Davis 1954). Hazel retains its intensity inland as it joins with another storm system during an extratropical transition to devastate inland communities from Virginia to Ontario, Canada. Washington, D.C., experiences its strongest winds on record. A consequence of Carol, Edna, Hazel, as well as other previous East Coast hurricanes, is that Congress increases funding for hurricane research, and the National Hurricane Research Laboratory is established (Sheets 1990).

In addition to the U.S. casualties of 95, Hazel also kills 83 people in southern Ontario (Palmen 1958; Knox 1955).

Date / Hurricane	Location	Category	Deaths	Description
1955, August 3–14, Hurricane Connie	North Carolina	Category 2	25	Connie makes landfall at Morehead City on August 12, causing a 5–8 foot storm surge and considerable beach erosion. Crops are devastated. Heavy rains of 6–12 inches falls between North Carolina and New England. This rainfall saturates the ground and fills the streams, thus setting the stage for the destruction caused by Diane (NOAA 1982; Dunn, Davis, and Moore 1955).
1955, August 7–21, Hurricane Diane	North Carolina to New England	Category 1	184	The weak and poorly defined Diane makes landfall near Wilmington on August 17. Surge and damage is minimal on the North Carolina coast. As Diane moves inland, a catastrophic situation emerges. Due to high watersheds and saturated soil from Connie, Diane's heavy rainfall (up to 12 inches in some spots) causes severe floods for the entire Northeast. Relatively small river basins rise quickly from mountain streams, and flooding occurs within hours, killing many. The worst-hit areas are Pennsylvania, Massachusetts, Rhode Island, Connecticut, and southeastern New York, but flooding is widespread in the northeastern United States. Damage exceeds any prior storm on record, earning the storm the nickname the "Billion-dollar Hurricane"—the first storm to reach this damage threshold. Ironically, later estimates show the dollar figure to be less, but damage is truly extensive (NOAA 1982; Dunn, Davis, and Moore 1955).
1955, September 10–23, Hurricane Ione	North Carolina	Category 3	7	Ione hits North Carolina on September 19. It is the third hurricane in eastern North Carolina within five weeks and the fourth in eleven months. It then recurves back out to sea. Most damage is to crops (Dunn, Davis, and Moore 1955).

(continues)

TABLE 4.1
(Continued)

Dates of Hurricane	Areas Most Affected	Saffir-Simpson Category at Landfall	Deaths (U.S. Only)	Comments and Damage
1956, September 21–30 Hurricane Flossy	Louisiana to northern Florida	Category 2 in Louisiana; category 1 in Florida	15	Flossy crosses the Mississippi delta just north of Burrwood near Pilottown on September 24. The center then passes just south of Pensacola, FL, and makes landfall near Fort Walton. The storm becomes extratropical shortly afterward and travels over Virginia out into the Atlantic. Damage occurs from the mouth of the Mississippi River to New Orleans and eastward to western Florida, with a storm surge up to 8 feet near Louisiana and Florida landfall points. Flossy completely submerges Grand Isle and causes extensive coastal erosion in Louisiana marshlands. Flossy also dumps 16 inches of rain along its track, resulting in flooding. Hundreds lose homes, cattle drown, and citrus, sugar, and pecan crops are heavily damaged. Eastern sections of New Orleans seawall are overtopped with inland flooding. Heavy rain also occurs in Alabama. Gulf Coast evacuations are a success, as deaths are mainly due to plane and car accidents. An extratropical Flossy also causes a surge in Norfolk, VA, with 2.5 feet of water in its streets. Beach erosion occurs as far north as Delaware (NOAA 1982; Dunn, Davis, and Moore 1956).
1957, June 25–28 Hurricane Audrey	Texas and Louisiana	Category 4	526	Hurricane Audrey makes landfall near the Texas-Louisiana border on June 27 with devastating effects and is the most destructive hurricane to strike southwestern Louisiana in history. Its central pressure deepens considerably in the last five hours before landfall. Its forward motion also increases, making landfall earlier than expected. There are many deaths as the result of a storm surge in excess of 12 feet, which inundates the flat coast of Louisiana as far as 25 miles inland in some places. Many homes are destroyed, and offshore oil installations are heavily damaged. Texas also experiences a high storm surge, with 450 people injured. Loss of life is the greatest since the New England 1938 hurricane. Audrey later undergoes extratropical transition, generating heavy floods south of the Great Lakes, particularly in Illinois and Indiana. Strong winds also occur from western Pennsylvania through New York (Moore and Staff 1958).

1959, September 20–October 2 Hurricane Gracie	South Carolina to Virginia	Category 3	22	This storm generates much controversy regarding emergency communication issues and adequate hurricane preparedness, including an unsuccessful lawsuit filed by one of the survivors (Klein and Pielke 2002; Sheets and Williams 2001). Storm passes inland on the South Carolina coast near Beaufort on September 29. Some flooding occurs, particularly in the Carolinas and Virginia. Crop damage due to wind and rain is heavy in South Carolina and eastern Georgia. Wind damage is severe near the eye's landfall and is the most severe in the history of Beaufort, SC. However, landfall at low tide minimizes storm surge damage. Twelve deaths are associated with tornadoes in Virginia; car accidents, falling trees, and live wires cause ten deaths (NOAA 1982; Dunn et al. 1959).
1960, August 29–September 13 Hurricane Donna	Florida to New England	Category 4 in Southwest Florida and Keys; category 2 in North Carolina; category 1 in New York and Connecticut	50	Donna devastates southwest Florida from the combination of strong winds and storm surges up to 13 feet, particularly northeast of the track. It makes landfall as a hurricane in the Florida Keys, southwest Florida, North Carolina, New York, and Connecticut — the first time that happens since records were first taken. Vast expanses of mangrove stands in the western areas of Everglades National Park incur losses from one-half to total destruction. The great white heron population is decimated by 35–40%. Peak maximum sustained winds of 135 mph affect the central Florida Keys, with a broad swath of category 3 winds from Maderia Bay to Cape Sable, FL, northward to Fort Myers Beach. This is the costliest hurricane to strike the Florida region up to this time period. Serious damage takes place along northeast sections of coastal North Carolina. Considerable damage to coastal properties from the combined effects of wind and storm surge occurs all along the Middle Atlantic and New England coasts. Serious flooding happens farther inland, particularly from the Catskills to Long Island (Dunn 1961; Dunion et al. 2003).
1961, September 3–15 Hurricane Carla	Central Texas coast to Louisiana	Category 4	46	Carla is the largest and most intense Gulf Coast hurricane in decades, impacting Corpus Christi to the Louisiana border with hurricane-force winds. The eye passes over Port O'Connor on September 11, totally devastating the town. However, severe damage occurs

(continues)

TABLE 4.1
(Continued)

Dates of Hurricane	Areas Most Affected	Saffir-Simpson Category at Landfall	Deaths (U.S. Only)	Comments and Damage
				along a wide expanse of the Texas coast from unusually prolonged winds, high surges up to 16 feet, and flooding due to the hurricane's slow movement. In addition, Galveston is seriously impacted by a tornado. The region's rice crop is ruined, and entire herds of cattle drown (Dunn and Staff 1962; Hogan 1961).
1964, August 20–September 5 Hurricane Cleo	Southern Florida, Georgia, Eastern Virginia	Category 2	0	This is the first hurricane in the Miami area since 1950. Moderate wind damage is confined to a 20–30-mile strip between Miami and Melbourne. Record rainfall amounts up to 14 inches in twenty-four hours fall in Hampton Roads southward in Virginia, resulting in widespread flooding. Tornadoes are reported in southeast Florida and the Carolinas (Dunn and Staff 1965).
1964, August 28–September 16 Hurricane Dora	Northeastern Florida, Southern Georgia	Category 2	5	First storm of full hurricane force on record to move inland from the east over extreme northeastern Florida, causing extensive wind damage, inland flooding, and crop damage in the Jacksonville area; Georgia also suffers damage. Storm slows considerably before landfall, perhaps allowing time for residents who are unaccustomed to hurricanes to evacuate. The persistent onshore winds produce an estimated storm surge of 12 feet, which severely erodes beaches, inundates several coastal communities, and washes out beach roads (Dunn and Staff 1965).
1964, September 28–October 5 Hurricane Hilda	Louisiana, North Carolina	Category 3 in Louisiana; remnants in North Carolina	38	Damage from wind, rain, and tornadoes is extensive in southeast Louisiana. On the morning of October 3, several tornadoes occur in southeastern Louisiana in prehurricane squall lines. One tornado at Larose, LA, kills 22 and injures 200 people. Three other tornadoes cause much damage but no deaths in the New Orleans metropolitan area. Most nontornado fatalities occur in Erath, LA, when a large water tower falls on City Hall, where Civil Defense activities are being directed. Almost a complete evacuation of the entire Louisiana coast accounts for the low death toll, aside from the tornado and water tower

		incidents. Offshore oil facilities also are damaged, and a storm surge on Lake Pontchartrain causes damage to the New Orleans lakefront. Remnants of Hilda cause heavy rains and flooding in North Carolina, already saturated from recent rains (Dunn and Staff 1965).
1965, August 27–September 12 Hurricane Betsy	Southern Florida, Louisiana Category 3 in Florida and Louisiana 75	Betsy moves south through the Bahamas, then west over the Florida Keys. Damage from winds, high tides and wave action is confined to an area from Ft. Lauderdale, FL, southward. Flooding over the upper Keys is extensive. Betsy turns to the northwest upon entering the Gulf of Mexico, with its eye arriving at Grand Isle, LA, the evening of September 9. High water causes great devastation on the central Gulf Coast from the point where the center makes landfall to Mobile, AL. In addition, Betsy pushes a storm surge up the Mississippi River. Since it also impedes the river's flow, unprecedented flooding occurs as water topples or breaches New Orleans' weak levees composed mostly of trees and mud (Adler 2001). The city's worst flooding in decades occurs, with water reaching the eaves of houses in the eastern part of the city. Ships, tugs, and barges are torn from their moorings. Hundreds of barges between New Orleans and Baton Rouge are either sunk or driven aground. In Louisiana, 300,000 people evacuate to safe shelters. However, 58 people lose their lives because of widespread wind gusts in excess of 100 mph and floods. In particular, because the New Orleans area is 5 feet below sea level, an extensive pumping system removes rainwater, but the system becomes overloaded and also experiences power failures. Four deaths take place in Florida, and other lives are lost in the adjacent waters of the Gulf and the Atlantic. The total of 75 deaths in Betsy is the greatest loss of life along the Gulf Coast since Audrey in 1957. Extensive crop damage occurs in Florida, Alabama, Mississippi, and Louisiana. Louisiana oil and marine industries suffer heavy economic losses (Sugg 1966). Since Betsy, the Army Corps of Engineers has raised the levees to 16 feet surrounding the city, added material that improves the strength of the levees, and improved the pumping system. However, many think this city, which essentially lies in a depressed bowl below sea level, is still the most vulnerable to a catastrophic hurricane that would topple the levees, especially since the land is subsiding about a foot every 30 years (Bazile 2004) and because its surrounding marsh has experienced severe erosion (Brouwer 2003). Moreover, the levees are not designed to protect the city from a category 4 or 5 hurricane.

(continues)

TABLE 4.1
(Continued)

Dates of Hurricane	Areas Most Affected	Saffir-Simpson Category at Landfall	Deaths (U.S. Only)	Comments and Damage
1967, September 5–22 Hurricane Beulah	Southern Texas	Category 3	15	Beulah makes landfall between Brownsville, TX, and the mouth of the Rio Grande about daybreak on September 20. A storm surge of at least 18 feet occurs, creating thirty-one new cuts on Padre Island. This slow-moving, erratic storm dumps torrential rains of 10–30 inches in a 40,000-square-mile area between San Antonio, the Rio Grande, and the south Texas coast. Rivers swell, with levels 10–20 feet above flood stage. The ensuing floods cause considerable damage and kill 10 people. In addition, a record number of tornadoes — 155 — form, which kill 5 people (Sugg and Pelissier 1968).
1969, August 14–22 Hurricane Camille	Mississippi, Louisiana, Alabama, Virginia, West Virginia	Category 5	297	Camille, with sustained winds of at least 180 mph, produces a storm surge of 25 feet in Pass Christian, MS, at landfall on August 17. Floods from the storm surge penetrate 8 miles inland in the Waveland-Bay St. Louis region, and in river estuaries the surge extends 20–30 miles upstream. Total devastation on the immediate Mississippi coastline and severe damage farther inland result. Camille kills 172 people in Mississippi, including 41 who are missing and assumed swept out to sea (Bergeron 1999; Sullivan 1987; Pielke, Simonpietri, and Oxelson 1999; Wilemon 1999). The bodies of three unidentified females are buried in a permanent memorial and renamed "Faith," "Hope," and "Charity"—a testimony to this sad event and to the perseverance of this community, which has fully recovered and today has a booming tourism and casino economy. Hurricane Camille spawns several myths as well. Legend states that 24 residents of the Richilieu Apartments stayed, threw a "hurricane party," and then are killed by the storm surge. The truth is that only 8 stayed, believing the building to be safe because they were told it was steel-reinforced and a designated bomb shelter; 4 of these residents perished. No hurricane party ever occurred (Bergeron 2000; Bergeron 2003). The origin of the myth is unknown, but it has been perpetuated by one resident; it has since been documented that the "hurricane party" did not happen.

Camille's storm surge also pushes water over both levees near the mouth of the Mississippi River in Louisiana, "removing almost all traces of civilization" as one U.S. Department of Commerce report states (ESSA 1969). The marsh region northeast of New Orleans, close to the storm's center, also suffers extreme damage (Roberts 1969). Alluvial City, Shell Beach, Hopedale, and Ycloskey suffer a 7–12 foot storm surge and 120–150 mph winds. Some camps in this region, which had been standing for 30–50 years, are destroyed. Many summer homes along the north shore of Lake Pontchartrain and the Rigolets are also wiped out. Nine in Louisiana die.

Camille slices or erodes many offshore islands; Pelican Island off Mobile Bay is completely washed away, and other islands are gouged with new cuts, including slashing Ship Island into two pieces. The Chandeleur Islands decreases from 7800 acres to 5016 acres (a 35% decrease), and it takes almost 20 years for the breaches to seal and for the island chain to begin growing (Penland and Couathors 1999a).

Camille dumps 12–30 inches of rain in 6 hours in the Blue Ridge Mountains, triggering devastating flash floods and mud slides that killed 114 people in Virginia and 2 in West Virginia (Simpson, Sugg, and Staff 1970).

| 1970, July 23–August 5 Hurricane Celia | Texas | Category 3 | 11 | Celia intensifies rapidly during the fifteen hours before it crosses the coast north of Corpus Christi, TX. As it moves over land, spectacular damage occurs from high-energy winds of short duration (also called downbursts or microbursts), which exceed the prevailing sustained hurricane-force winds by a factor of 2 or 3. It is also possible that these localized extreme winds are from mesovortices. The extreme winds rake across the residential and business areas in less than half an hour. It is estimated that winds reach as high as 160 mph for several seconds. During those disastrous seconds, incredible damage occurs at the airport and an adjacent mobile home park which is completely demolished. These winds cause most of the damage, as only a tiny fraction can be attributed to the storm surge and flooding. Storm surges of 9 feet are observed in Port Aransas, however. Fortunately, only 11 die in the Corpus Christi area due to the state of preparedness by its disaster prevention agencies. This is the costliest hurricane to strike the Texas coast up to this time period (Simpson and Pelissier 1971). |

(continues)

TABLE 4.1
(Continued)

Dates of Hurricane	Areas Most Affected	Saffir-Simpson Category at Landfall	Deaths (U.S. Only)	Comments and Damage
1972, June 14–23 Hurricane Agnes	Florida to New York	Category 1 in Florida and New York	117	Agnes is one of the largest June hurricanes in history, with a circulation over 1,000 miles in diameter. It makes landfall in Cape San Blas in the Florida panhandle already in a weakening state due to an unfavorable environment. A storm surge of 6–7 feet, local windstorms on Lake Okeechobee, and tornadoes kill 9 people. As it moves northward, Agnes interacts with another storm system, becomes an extratropical system, and actually regains winds of tropical-storm strength over land in North Carolina. The remnants of Agnes move offshore, and the storm intensifies a little more and makes landfall on the western tip of Long Island, NY. At the same time, the other storm system moves northeast as well. The combination of heavy rains from this complex weather system and saturated soil from recent rains produce devastating floods from North Carolina to New York, with many record river crests and 108 deaths. This storm is the costliest natural disaster in U.S. history up to that time period, even exceeding Betsy and Camille (Simpson and Hebert 1973).
1974, August 29–September 10 Hurricane Carmen	Southern Louisiana	Category 1	1	Carmen moves inland just east of Vermilion Bay. Damage occurs primarily to the sugarcane crop, offshore installations, and the shrimping industry (Hope 1975).
1975, September 13–24 Hurricane Eloise	Florida panhandle and eastern Alabama	Category 3	21	Eloise makes landfall between Fort Walton Beach and Panama City, FL, early on September 23, producing a storm surge of 12–16 feet. It's the first major hurricane to impact the Panama City area in fifty years. The storm surge and wind cause extensive damage to structures along the 25-mile beach strip from Fort Walton Beach to Panama City, FL. High winds destroy property and crops over eastern Alabama. The moisture from Eloise combines with a stalled front, causing 10 inches of rain in mountainous regions of New York and 14 inches of rain near Harrisburg, PA. The resulting floods drown 17 and damages property and crops (Hebert 1976).

1979, August 25–September 7 Hurricane David	Eastern Florida through Middle Atlantic States	Category 2 in Florida, Georgia, and South Carolina	5	This storm makes landfall just north of Palm Beach, parallels the eastern coast of Florida, and enters the Atlantic Ocean just north of Cape Canaveral. David is the first hurricane to strike the Cape Canaveral area since 1926. It moves inland again near Savannah, GA. Damage is not great in any one area, but the cumulative total caused by winds, storm surge, floods, and tornadoes is large because of the wide area affected. Numerous tornadoes occur along inland path. Wind and rain are responsible for widespread power outages all along the eastern seaboard (Hebert 1980).
1979, August 29–September 14 Hurricane Frederic	Southern Alabama, southern Mississippi, and Florida panhandle	Category 3	5	The eye passes over Dauphin Island, AL, on September 13 with sustained winds of 120 mph. Frederic is the first hurricane to strike Mobile directly since 1932. This storm causes extensive damage due to a 15-foot storm surge near the mouth of Mobile Bay. Major wind damage occurs in southern Alabama, southern Mississippi, and the extreme western Florida panhandle, causing it to be one of the most costly Gulf Coast hurricanes in history. Heavy rains in Ohio and New York result from the interaction of a weakening Frederic and a cold front that produce record floods and major damage in this region (Hebert 1980).
1980, August 1–11 Hurricane Allen	South Texas	Category 3	15	Originally a very powerful category 5 hurricane, Allen slows down and weakens to a category 3 hurricane before making landfall just north of Brownsville, hitting a largely unpopulated area. Allen deluges south Texas with 15–20 inches of rain 100 miles inland, causing severe local flooding. A storm surge up to 12 feet along Padre Island causes numerous cuts. Two oil rigs are destroyed, and 13 people perish in a helicopter crash during an oil rig evacuation (Lawrence and Pelissier 1981).
1982, November 19–24 Hurricane Iwa	Kauai and Oahu, Hawaii	Category 1	2	Iwa's center moves just north of Kauai on November 24. Hurricane Iwa's intensity had peaked, but its rapidly accelerating forward motion of 25–30 mph make winds and swell conditions within the dangerous right semicircle extend over Kauai. Iwa is the most damaging Hawaiian storm of record up to this time. The south shore of Kauai is particularly hard hit by wave action with very severe damage around Poipu. The Waianae coast of Oahu also has stretches of severe surf damage. In fact, all islands reported some surf damage along their southwest-facing shores. Wind damage is widespread on Kauai.

(continues)

TABLE 4.1
(Continued)

Dates of Hurricane	Areas Most Affected	Saffir-Simpson Category at Landfall	Deaths (U.S. Only)	Comments and Damage
				Pockets on Oahu also receive heavy wind damage, such as the Wahiawa area of central Oahu and portions of the windward coast (NWSTM 1982).
1983, August 15–21 Hurricane Alicia	North Texas	Category 3	21	An unusual hurricane, Alicia forms in an environment that appears unfavorable for development. Galveston officials decide against complete evacuation of coastal areas. When Alicia intensifies significantly the last 18 hours before landfall, it is too late to evacuate these areas. Only the presence of Galveston's seawall prevents a large loss of life. Alicia makes landfall 25 miles southwest of Galveston, then passes directly over Houston, blowing out skyscraper windows and causing extensive damage. This is the first major hurricane to impact the Houston area since Hurricane Carla (1963), and the first hurricane in forty years to pass directly over Houston. Alicia also produces twenty-six tornadoes. Even though Alicia is only a small- to medium-sized storm, it leaves a legacy as the costliest hurricane in Texas history up to this time period (Case and Gerrish 1984; Kareem 1985; Hohertz and Coauthors 1983; Sallee 1993).
1985, August 28–September 4 Hurricane Elena	Louisiana to west Florida	Category 3 near Apalachicola, Florida; category 2 in Biloxi, Mississippi	4	Elena is best remembered for its bizarre track that confounded forecasters. On August 29, Elena moves northwest, which would have brought landfall somewhere between Louisiana and Florida. But, on August 30, the steering currents change and Elena moves east toward the west coast of Florida. Within a couple hundred miles of Cedar Keys, FL, Elena stalls for two days before turning back to the west-northwest, finally making landfall near Biloxi, MS. Nearly a million people are evacuated, with a large section of the middle Gulf Coast being asked to evacuate twice within a three-day period. Elena's maximum intensity is reached when its center is 75 miles from Apalachicola, FL, devastating its oyster industry. Extensive beach erosion and damage from the mouth of the Mississippi River to Fort Myers, FL, occurs during this storm's lifetime (Case 1986).

Date	Location	Category	Description
1985, September 16–October 2 Hurricane Gloria	North Carolina to New York	Category 3 in North Carolina; category 2 in Long Island	8	Gloria reaches a minimum pressure of 919 mb about 900 miles southeast of Cape Hatteras, which is the lowest pressure ever measured by reconnaissance aircraft over the East Coast Atlantic region. Gloria receives much attention from the media because of the threat it poses to the Northeast. Fortunately, Gloria weakens as it moves northward, somewhat minimizing casualties and damage. It crosses over the Outer Banks of North Carolina on September 27, passes just offshore of the mid-Atlantic states, and makes landfall over western Long Island, NY, ten hours later. The strongest winds of Gloria generally remained offshore throughout this period, and the hurricane hits New York during low tide, thus minimizing the 3–6-foot storm surge. Downed trees cause extensive power outages for hundreds of thousands in the Long Island and New England areas. Beach erosion is severe from North Carolina to New York, and North Carolina experiences coastal flooding (Case 1986).
1985, October 26–November 1 Hurricane Juan	North Texas to Florida	Category 1 in Louisiana; tropical storm in Florida	12	Juan is a weak but persistent storm that creates much havoc on the northern Gulf Coast, with three landfalls resulting from multiple loops in its track. Even before Juan became a tropical storm, the tropical disturbance creates a broad zone of gale-force winds, making evacuation of offshore rigs very difficult as Juan is developing. Nine deaths are caused by toppled oil rigs or sunken boats rescuing oil workers. Juan makes its first loop just off the Louisiana coast on October 28 and then makes landfall near Morgan City on October 29. The following day, Juan makes a second loop around Lafayette and emerges over Vermilion Bay, where it becomes better organized. It moves over the mouth of the Mississippi River near Burrwood, LA, and then makes its final landfall just west of Pensacola, FL. Because of Juan's slow, erratic movement, much damage is caused by its persistent onshore flow (prolonging and enhancing the storm surge with water as high as 8 feet) and excessive rain (8–17 inches) from the upper Texas coast to the extreme western Florida panhandle over a five-day period. This includes damage to offshore oil rigs; crop devastation; drowned livestock when high water overflows levees in south Louisiana; and flooding in Louisiana's marsh and Lake Pontchartrain (Case 1986).

(continues)

174 Chronology

TABLE 4.1
(Continued)

Dates of Hurricane	Areas Most Affected	Saffir-Simpson Category at Landfall	Deaths (U.S. Only)	Comments and Damage
1988, September 8–19 Hurricane Gilbert	South Texas	Did not hit U.S.; hit Mexico as category 3	3	Gilbert is noteworthy because it produced the lowest surface pressure ever recorded in the Western Hemisphere of 26.23 inches (888 mb) while over open ocean. This storm causes much damage and a number of fatalities in Mexico, Jamaica, Haiti, and Central America. U.S. damage is confined to twenty-nine tornadoes in south Texas from the remnants of Gilbert, beach erosion along Padre Island, and minor wind damage from the coast through the lower Rio Grande Valley (Lawrence and Gross 1989; Willoughby, Masters, and Landsea 1989).
1989, September 10–22 Hurricane Hugo	South Carolina and North Carolina	Category 4	21	Hugo is the most powerful hurricane to strike the United States since Camille in 1969, with sustained winds of 130 mph or more and extreme storm surges of 20 feet in some places. It is also the costliest hurricane until Andrew in 1992. Hugo makes landfall near Charleston at Sullivans Island on September 22, devastating waterfront properties and beaching (or destroying) an incredible number of watercrafts. Hugo's fairly swift motion carries the hurricane well inland before it can significantly weaken. Almost two-thirds of South Carolina suffers wind damage, as does much of central North Carolina along a wide north-south band through the Charlotte area. Falling trees and limbs do much of the damage; otherwise, roof damage is common. Almost 9,000 square miles of forest (seven times the area of Rhode Island) are flattened, wiping out the timber industry. Long-term power outages are widespread (Case and Mayfield 1990; Fox 1992; Barnes 1998). A NOAA WP-3D hurricane research aircraft nearly crashes when it is damaged by an eyewall mesovortex during a low-level penetration a week before landfall (Black and Marks 1991; Marks and Black 1990).
1991, August 16–29 Hurricane Bob	New England	Category 2	17	Bob makes landfall on Rhode Island, then passes between Boston and Sciuate, MA. Although Bob is a weakening hurricane, it still causes widespread damage. Dozens of cottages and boats on Buzzards Bay and Narragansett Bay are damaged or destroyed by a

1992, August 16–28 Hurricane Andrew	South Florida and Louisiana	Category 5 in Florida; category 3 in Louisiana	49	7-foot storm surge. Heavy rains of 4–7 inches in northwest Rhode Island cause considerable urban flooding. Other portions of New England also experience damage from storm surge, downed trees, flooding from heavy rain, and tornadoes. Power outages are widespread. Apple crops in southern Maine suffer extensive damage. Three people survive for ten days on a life raft after their boat is sunk off Cape Hatteras (Pasch and Avila 1992). Andrew is a small but extremely destructive hurricane that devastates south Florida and is the most expensive natural disaster in U.S. history. Originally classified as a category 4 hurricane at landfall, later analysis shows that Andrew is a category 5, only the third U.S. land-falling hurricane of this status in the twentieth century (Landsea et al. 2004; NOWCAST 2002). Injuries are numerous, and 250,000 people are left homeless. A record 17-foot storm surge is recorded near the Burger King headquarters south of Miami, causing much localized damage However, most inland damage is caused by winds in Andrew's eyewall that level homes and businesses. Exceptional devastation occurs from the Kendall district southward through Homestead and Florida City to near Key Largo, with government officials saying that this area looks like a "war zone." The damage is comparable to a 10-mile tornado. In addition, smaller-scale trails of devastating damage are evident (Wakimoto and Black 1994). Some have speculated that this damage is caused by eyewall mesovortices. The National Guard set up tent cities and enforced a curfew to stop rampant looting. So expensive is the damage that eight insurance companies cannot cover the claims and are driven out of business. Fifteen people are killed as a direct result of the storm, mostly from flying debris and collapsing roofs or walls or falling furniture. Twenty-six others are killed indirectly during the clean-up process. Natural coral reefs and marine life are damaged. Andrew inflicts catastrophic damage to the mangrove forests in Everglades National Park and Biscayne National Park (Smith et al. 1994). A weaker but still potent Andrew makes landfall in a sparsely populated area of central Louisiana on August 26. Damage is still significant to offshore platforms, barrier islands, boats, barges, freighters, and coastal properties due to winds and an 8-foot storm surge. One strong tornado causes damage in Laplace and Reserve, LA. An upwelling of bottom sediment kills 187 million fish in the Atchafalaya Basin, tremendously impacting the fishing

(continues)

TABLE 4.1
(Continued)

Dates of Hurricane	Areas Most Affected	Saffir-Simpson Category at Landfall	Deaths (U.S. Only)	Comments and Damage
				industry. Louisiana's sugar and soybean crops are devastated. More than 200 oil and gas platforms are damaged, and a few are destroyed. Six fishermen are killed when their boat sinks offshore, and two are killed by the strong tornado (Mayfield, Avila, and Rappaport 1994).
1992, September 5–13 Hurricane Iniki	Kauai, Hawaii	Category 4	6	A rare land-falling major Hawaii hurricane, Iniki pummels the south side of the Kauai island around Poipu. The northern shores of the island are equally ravaged. Wind damage is extremely heavy throughout Kauai. Many homes and buildings are flattened or lose their roofs; 14,350 homes are severely damaged, making a third of the population temporarily homeless. Nearly all of the island's seventy hotels suffer severe damage. Seven thousand of the island's 8,202 hotel, condo, and bed and breakfast rooms are shut down after the hurricane. Electricity and telephone service are not restored to many areas for several months. The hurricane cripples the island's tourist industry, which accounts for 45% of the island's economy. The high winds and surf also annihilate the island's entire crop of macadamia nuts and sugarcane, the island's other two large industries. Newly planted coffee trees and guava crops are destroyed. It takes years for Kauai's tourism industry to rebuild and recover. Other islands are also impacted. The areas most affected on Oahu are the leeward coast from Barbers Point through Makaha and Kaena Point, with lesser damage along the south shore from Ewa Beach to Hawaii Kai. Some damage also occurs on the islands of Maui County and the Big Island of Hawaii, where swell and heavy surf from southwesterly direction pound exposed shorelines and anchorages (NWSTM 1992).
1993, August 22–September 6 Hurricane Emily	North Carolina	Category 3	3	Emily's western eyewall passes over the North Carolina Outer Banks. Damage is confined mainly to Hatteras Island, where a 10-foot storm surge impacts the area and severely damages coastal dwellings (Pasch and Rappaport 1995).

Date / Hurricane	Category	Deaths	Description
1995, July 31–August 6, Hurricane Erin	Category 1	6	East Florida and Florida panhandle. Erin comes ashore near Vero Beach on August 1, causing widespread wind damage and beach erosion, mostly in Brevard County. Some flooding occurs in the Melbourne and Palm Bay areas. A 234-foot gambling and cruise ship sinks east of Cape Canaveral, killing its three crew members. After reemerging over the Gulf of Mexico, Erin makes landfall on August 3 near Pensacola, causing significant structural damage in Navarre Beach and Pensacola Beach. Farther inland, Erin knocks down trees and power lines. About half of the cotton crop and 25% of the pecan crop is damaged. This is Pensacola's first direct hit by a hurricane since 1926. Remarkably, it would be hit by Opal two months later (Lawrence et al. 1998).
1995, September 27–October 5, Hurricane Opal	Category 3	9	Florida, Alabama, Georgia, and the Carolinas. Opal makes landfall on October 4 near Pensacola Beach as a minimal category 3 hurricane, causing storm surge damage and erosion from southeastern Mobile Bay and Gulf Shores, AL, eastward through the Florida panhandle to Cedar Key, FL. The structural damage is most extensive in the Fort Walton Beach area. Strong winds spread damage well inland. Opal downs numerous trees and knocks out power to 2 million people in Florida, Alabama, Georgia, and the Carolinas. The combination of Opal and a front result in heavy rains, with up to 10 inches in Florida, Alabama, and Georgia; up to 5 inches in the Carolinas; and less farther north. The remnants of this system bring beneficial rains to the Northeast, which was experiencing a drought. Opal generates much concern the day before landfall. During the night of October 4, Opal unexpectedly intensified to a strong category 4 with sustained winds of 150 mph. Even worse, Opal begins accelerating as it approaches Florida. Since most people are sleeping, it's difficult to alert the public about this new, unanticipated danger. Many people near Pensacola Beach awake the next morning with a near-category 5 hurricane just offshore. The last-minute evacuation procedures clog the roads and Interstate 10. Fortunately, Opal begins weakening, and coastal evacuations are completed in time (Lawrence et al. 1998).
1996, July 5–July 14, Hurricane Bertha	Category 2	12	Carolinas. Bertha makes landfalls on July 12 between Wrightsville Beach and Topsail Beach in the southern North Carolina coastal area, east of Wilmington. Many homes and businesses on

(continues)

TABLE 4.1
(Continued)

Dates of Hurricane	Areas Most Affected	Saffir-Simpson Category at Landfall	Deaths (U.S. Only)	Comments and Damage
				barrier islands sustain heavy damage, mostly due to the 6–8 foot storm surge. High winds cause a large loss in the corn, soybean, and tobacco crops. This same area will be impacted by the even more intense Hurricane Fran two months later (Pasch and Avila 1999).
1996, August 23–September 8 Hurricane Fran	Carolinas, Virginia, West Virginia, Maryland, and Pennsylvania	Category 3	34	Fran makes landfall on September 5 over Cape Fear, NC, but hurricane-force winds extend over a wide area, causing extensive damage to trees, roofs, and power lines. A storm surge as high as 12 feet destroys or seriously damages numerous beachfront houses on the North Carolina coast. Most of the deaths are caused by flash flooding from 6–12 inches of rain in the Carolinas, Virginia, West Virginia, Maryland, Washington D.C., and Pennsylvania (Pasch and Avila 1999).
1997, July 16–26 Hurricane Danny	Louisiana, Mississippi, Alabama	Category 1	9	Danny forms right off the Louisiana coast and within forty-eight hours makes landfall on July 18 at Grand Isle, LA. It moves across the Mississippi River marsh, drifts slowly eastward, and stalls just south of Mobile Bay. On July 19, it moves into the lower reaches of Mobile Bay and stalls again, where it acquires its peak intensity. It finally drifts ashore in southern Baldwin County that evening and slowly meanders northward. The very slow movement produces extraordinary rains of 20–25 inches in Mobile Bay. When combined with a 4–5-foot storm surge, serious flooding results throughout the bay system. Danny's remnants also cause significant inland flooding in Charlotte, NC, with rain totals of 8–12 inches (Rappaport 1999).
1998, August 19–30 Hurricane Bonnie	North Carolina and Virginia	Category 2	3	Bonnie passes just east of Cape Fear and makes landfall near Wilmington, NC, on August 27. A storm surge of 5–8 feet occurs on the eastern beaches of Brunswick County. Winds cause tree, roof, and structural damage, with Hampton Roads, VA, hardest hit. Widespread power outages affect eastern North Carolina and Virginia (Pasch, Avila, and Guiney 2001).

Date/Storm	Location	Category	Deaths	Description
1998, September 21–30 Hurricane Georges	South Florida, Louisiana, and Mississippi	Category 2 in Florida; category 2 in Mississippi	1	After leaving a trail of destruction with more than 600 dead in the Caribbean region, Georges (pronounced "Zhorzh") strikes the Florida Keys, leaving more than 900 homes with minor damage, 500 with major damage, and more than 150 completely destroyed, including 75 houseboats on so-called "Houseboat Row." Georges then heads toward Mississippi, with the eye passing within 5 miles of the Chandeleur Islands. Georges ravages this island chain, causing worse damage than Camille, scouring away land and creating new inlets. The Chandeleur Islands decreases in size from 5,013 acres to 3,003 acres, a 40% reduction (Penland and Coauthors 1999b). Georges makes its final landfall on September 28 just east of Biloxi, MS, causing a 6–8 foot storm surge east of the eyewall all the way to Mobile Bay, AL, where parts of downtown Mobile are inundated. Beaches in this area are badly eroded. Storm surges above 7 feet also occur in Lake Pontchartrain, LA, destroying or damaging a number of fishing camps there. Rainfall totals of 15 inches in southeast Mississippi and Alabama, along with an incredible 26 inches in the Florida panhandle, cause widespread flooding. Pascagoula, MS, is particularly hard hit due to the combination of an 11-foot storm surge and flooding on the Pascagoula River. However, the Biloxi, Tchoutacabouffa, Mobile, Peridido, Escambia, Blackwater, Yellow, and Choctawhatchee river basins also experience near-record or record levels. Approximately 200 residents in the Florida panhandle are rescued by the Coast Guard due to flooding in that region, and a portion of Interstate 10 near the Alabama-Florida state line is destroyed or washed over (Pasch, Avila, and Guiney 2001; Turnipseed et al. 1998).
1999, August 18–23 Hurricane Bret	South Texas	Category 3	0	Bret is a potentially devastating storm but makes landfall over sparsely populated Kenedy County on August 23. Furthermore, Bret is a small storm, with hurricane-force winds confined to a radius of 35 miles from the center in the north semicircle and only 12–24 miles in the south semicircle. As a result, Corpus Christi, 40 miles to the north, and Brownsville, 50 miles to the south, are spared. Most wind damage is to ranch lands, and an estimated 10-foot storm surge causes some cuts in Padre Island. Rain amounts of 12 inches cause notable river flooding in the Rio Grande Valley, and heavy rains are also observed in nearby Mexico. Bret is the first hurricane to strike Texas since Hurricane Jerry (1989) and is the strongest Texas hurricane since Hurricane Alicia (1983) (Lawrence et al. 2001).

(continues)

TABLE 4.1
(Continued)

Dates of Hurricane	Areas Most Affected	Saffir-Simpson Category at Landfall	Deaths (U.S. Only)	Comments and Damage
1999, August 24–September 5 Hurricane Dennis	North Carolina, Virginia	Tropical Storm	0	Dennis threatens the southeastern United States as a category 2 storm but turns sharply away from the coast as it nears landfall. Its legacy is its persistent nature—Dennis loops back, then stalls for a week before making landfall. The result is days of rain, gusty winds, and high surf. Its rain is particularly important in North Carolina, as it saturates the ground and sets the stage for the severe inland flooding from Hurricane Floyd (Lawrence et al. 2001).
1999, September 7–17 Hurricane Floyd	North Carolina to New York	Category 2	56	Floyd is one of the most destructive and deadly hurricanes of the decade. Because of its size and intensity, hurricane warnings are issued from south Florida to Massachusetts. The last hurricane to require warnings for as large a stretch of coastline was Hurricane Donna (1960). As a result, the largest evacuation in U.S. history occurs as coastal residents from south Florida to North Carolina leave their homes. Floyd eventually hits Cape Fear, NC, on September 16. Floyd's center passes over extreme eastern North Carolina later that morning and then across Norfolk, VA. Storm surge values up to 10 feet are reported along the North Carolina coast. About fifteen tornadoes occur, causing structural damage. However, Floyd's legacy will be extraordinary flooding, dumping 15–20 inches of rain over portions of eastern North Carolina and Virginia; 12–14 inches over portions of Maryland, Delaware, and New Jersey; 4–7 inches over eastern Pennsylvania and southeastern New York; and up to 11 inches over portions of New England. Since the soil is already saturated from Hurricane Dennis and other recent rain events, an inland flooding disaster ensues that drowns 50 people (Lawrence et al. 2001). Six others die from other factors. Floyd is the deadliest U.S. hurricane since Agnes (1972). In Bound Brook, NJ, serious flooding from the Raritan River results in the rescue of 800 people, many stranded on the second floors of homes or on porch roofs.

			However, North Carolina is hardest hit, with towns such as Franklin, Rocky Mount, and Greenville under 10 feet of water and 30,000 homes flooded. In addition, more than 100,000 hogs and 1 million chickens are killed in the flooding. When combined with damaged water treatment plants, sewage plants, chemical plants, farming operations, and water wells, water pollution and environmental problems result. This also wipes out the oncoming shrimp harvest. On October 16, the fringes of Hurricane Irene dump more rain in North Carolina. Rain from all three hurricanes (Dennis, Floyd, and Irene) after the nation's largest lagoonal estuary, the Pamlico Sounds. The salinity in Pamlico Sound declines by two-thirds, and nitrogen levels increase by 50%, causing serious damage to the area's marine life.
2002, September 14–27 Tropical Storm Isidore 2002, September 21–October 4 Hurricane Lili	Louisiana	Isidore, tropical storm; Lili, category 1	5 from Isidore, 2 from Lili	Isidore and Lili hit Louisiana with back-to-back-blows, making landfall a week apart. It is difficult to discuss Lili without describing Isidore (a tropical storm at landfall), but a large storm and therefore still devastating), so both storms will be treated as a single event here. Lili is the first hurricane to make landfall in the United States since Irene hit Florida in 1999. Isidore makes landfall as a large tropical storm just west of Grand Isle, LA, on September 26, bringing a 6–8-foot storm surge to the Louisiana marshes, inland bays, and Mississippi coast and causing considerable structural water damage. The storm surge is the worst since Hurricane Juan (1985). It also unleashes torrential rains of 15–18 inches in southern Louisiana. The area's water pumps are unable to keep up, resulting in some homes flooding. The combination of surge and rain damage homes and property along the northshore of Lake Pontchartrain. The lake breaches nearby levees, flooding homes in Slidell, LA. Beaches in Mississippi and Alabama suffer erosion (Pasch et al. 2004; Krupa 2002; Krupa, Staley, and Bartels 2002; Brown and Cannizaro 2002). Already suffering from high water and heavy rain from Tropical Storm Isidore, Hurricane Lili makes landfall along the south-central coast of Louisiana near Intracoastal City on October 3. A tide gauge at Crewboat Channel near Calumet measures a storm surge of 12.3 feet, and another at Vermillion Bay measures 11.7 feet. Water levels are already 2–4 feet above normal from Isidore before Lili's landfall, enhancing the storm surge. Lili

(continues)

TABLE 4.1
(Continued)

Dates of Hurricane	Areas Most Affected	Saffir-Simpson Category at Landfall	Deaths (U.S. Only)	Comments and Damage
				brings less rain, with the highest amount of 8.5 inches just north of landfall, but contains stronger winds. Still, it could have been much worse. Lili had rapidly intensified to a category 4 hurricane the day before landfall, and forecasters predicted it would make landfall as a major hurricane. Fortunately, Lili's eyewall collapsed, and it rapidly weakened during the thirteen hours before landfall. A combination of storm surge and heavy rain caused levees to fail at Montegut and Franklin, Louisiana. Tide gauge water levels were already 2–4 feet above normal prior to Lili's arrival. Because the soil is saturated from Isidore, Lili's strong winds topple trees onto houses and into roadways. Winds strip shingles from roofs and blow out windows. The wind and driving rain from both storms flatten sugarcane fields throughout southern Louisiana (Pasch et al. 2004). Lili's maximum sustained winds increased by 40 mph within twenty-four hours, then weakened by 45 mph the following twelve hours before landfall. This is a truly remarkable — and fortunate — collapse. No hurricane on record has ever weakened at a greater rate immediately after rapid intensification in the Gulf of Mexico (Frederick 2003).
2003, September 6–19 Hurricane Isabel	North Carolina to New Jersey	Category 2	16	Isabel makes landfall near Drum Inlet, NC, on September 18. The storm brings hurricane conditions to portions of eastern North Carolina and southeastern Virginia, with a 6–8 foot storm surge at landfall. Widespread wind and storm surge damage results in this region. Wind damage also occurs northward to New York. Downed trees are particularly numerous in Virginia due to a three-year drought, which weakened tree roots, along with a wet summer that loosened the soil (BAMS 2004). Normally, the storm surge is still significant but less away from the landfall area. However, because the storm's inland track parallels Chesapeake Bay, southeast winds continue to pile water into the estuary system. Surge values of 6–8 feet are observed in the upper reaches of Chesapeake Bay near Annapolis and Baltimore, MD, and in most

Date / Storm	Location	Category	#	Description
				adjacent rivers. The Potomac River in Washington, D.C., also has an 8-foot storm surge. Even higher surges occur at the heads of rivers, such as 8.5 feet at the Richmond City locks along the James River. Water levels exceed previous record levels established in the Chesapeake-Potomac Hurricane (1933) in Washington D.C., Baltimore, Annapolis, and communities near the Blackwater National Wildlife Refuge. Because this storm surge occurs far from Isabel's landfall, many residents are also caught by surprise. Structural and boat damage is extensive. Losses to recreational boats alone total $150 million (Anon. 2003).
2004, August 9–14 Hurricane Charley	Southwest Florida, South Carolina, North Carolina	Category 4 in Florida; category 1 in the Carolinas	10	Charley came ashore August 13 with 145-mph winds at Charlotte Harbor and Punta Gorda, causing severe damage in southwest and central Florida. Ten thousand homes are destroyed in Charlotte Harbor with another 16,000 severely damaged. Charley tracks northeastward across central Florida, severely affecting Orlando and Daytona Beach. It moves offshore, making another landfall as a weak hurricane at Cape Romain, SC, followed by landfall in Myrtle Beach, NC. Charley's hurricane-force winds are confined to a narrow region, but the devastation is immense. Damage totals $15 billion, making Charley the second costliest hurricane in history (only Andrew in 1992 is worse). North Carolina and South Carolina also suffer between $20–30 million in damage. Charley is a controversial storm because it was forecast to hit Tampa Bay as a category 3 but hit 100 miles south in Port Charlotte as a category 4, surprising many residents. Port Charlotte was in the hurricane warning region, but many residents confidently believed it was going to Tampa and did not expect the slight track turn or the rapid intensification. Charley deepened explosively just before landfall, going from a Category 2 to a Category 4 hurricane in five hours. Charley serves as a lesson that all residents in a hurricane warning region should prepare for landfall even if not directly in the track forecast. Track forecast errors within twenty-four hours can still be 50–100 miles wrong, and rapid intensification is often not predicted. Incredibly, south Florida will be hit by two more hurricanes this year: Frances and Jeanne.

(continues)

TABLE 4.1
(Continued)

Dates of Hurricane	Areas Most Affected	Saffir-Simpson Category at Landfall	Deaths (U.S. Only)	Comments and Damage
2004, August 25–September 8 Hurricane Frances	Southeast Florida, Georgia	Category 2	6	Frances hit Labor Day weekend with 105-mph winds on the Atlantic coast near Stuart, FL. A very large, slow-moving storm, Frances damaged much of the state's central east coast, inundating the shoreline with a storm surge of 6–8 feet, stripping roofs, smashing boats, and damaging space and military facilities in Cape Canaveral. Frances also produces copious amounts of rain in excess of 10 inches over large portions of the central and northern Florida peninsula and southeastern Georgia, with some areas receiving almost 16 inches. Totals of 5–10 inches were common elsewhere along Frances's inland track. Frances spawns 101 tornadoes — 23 in Florida, 7 in Georgia, 45 in South Carolina, 11 in North Carolina, and 15 in Virginia. Many of the tornadoes occur in an outbreak over South and North Carolina on 7 September.
2004, September 2–24 Hurricane Ivan	Alabama, west Florida panhandle, southwest Louisiana	Category 3 in Alabama; tropical depression in Louisiana	26	Ivan rakes the eastern Alabama and western Florida panhandle on September 16 after landing on the Alabama coast with 130-mph winds and a 10–15 foot storm surge. Ivan also produces incredible waves as high as a record 91 feet offshore and 10–15 feet inshore. The combination of wind, surge, and high surf levels Perdido Key, FL, with extreme damage along the eastern Alabama coast and western Florida panhandle. Extensive beach erosion undermines homes, apartments, and condominiums, and the coastal highway system is ruined. Ivan damages or destroys thousands of coastal homes. The severe wave action combined with the storm surge dismantles a quarter of a mile of the Interstate 10 bridge over Pensacola Bay, sliding the roadway off its concrete pilings and pushing much of it into the water. Millions of acres of forest are wiped out. Gulf of Mexico oil production is

severely disrupted, resulting in higher oil and natural gas prices. Underwater mudslides damage thirteen pipelines, with many more pipelines damaged by the anchors of mobile drilling rigs pushed by Ivan. Twenty-four drilling platforms are significantly damaged, and seven completely destroyed. Ivan spawns 111 tornadoes inland, with the majority in Virginia, Georgia, and Florida. Ivan is the most destructive hurricane in this region in 100 years. The total damage cost is 14.2 billion dollars, the third costliest hurricane ever behind Andrew (1992) and Charley (2005).

Ivan will also be remembered for its quirky path and rebirth after going inland. Its remnants track northeastward into the mid-Atlantic, then its low-level circulation turns southwestward across Florida into the Gulf of Mexico, where it reintensifies into a weak and short-lived tropical storm. It makes landfall as a tropical depression in Cameron Parish, LA, on September 23.

| 2004, September 13–28 Hurricane Jeanne | Southeast Florida | Category 3 | 4 | After doing a loop in the Atlantic, Jeanne makes landfall near Stuart, FL, with 120-mph winds on September 25, within a startling 2 miles of Frances's landfall three weeks earlier. Many homes already damaged by Frances were leveled. Being a stronger hurricane, Jeanne also inflicts new damage with its stronger winds and higher storm surge. Jeanne undercuts foundations and guts the first floors of condominiums, leaving bedrooms and lobbies filled with sand and debris; breaches seawalls; washes away walkovers and decks; and seriously erodes the beachfront, turning the dunes into sheer sand cliffs, some 10 feet tall.

Widespread rainfall up to 8 inches occurs in its path along east, central, and northern Florida, with isolated amounts up to 13 inches. Rainfall amounts of 4–7 inches also are observed in central Georgia, the western Carolinas, and Virginia.

The combined effect of Charley, Frances, and Jeanne decimates the area's citrus crops of oranges, grapefruits, and tangerines, resulting in the smallest crop in thirteen years.

For the first time in recorded history, four hurricanes hit Florida in a single season. The four hurricanes that strike Florida this season match the most to hit a single state since |

(continues)

TABLE 4.1
(Continued)

Dates of Hurricane	Areas Most Affected	Saffir-Simpson Category at Landfall	Deaths (U.S. Only)	Comments and Damage
				1886, when Texas was battered four times. The havoc caused by Hurricanes Charley, Frances, Ivan, and Jeanne prompts the largest relief effort ever undertaken by the Federal Emergency Management Agency. Property damage in the U.S. is estimated at near $45 billion — the costliest hurricane season on record.
2005, August 23–30, Hurricane Katrina	south Florida, Louisiana, Mississippi, Alabama	Category 1 in Florida, Category 4 in Louisiana	estimated between 1000 and 2000 at press time	see epilogue at back of book

Also included are (1) the area or areas affected; (2) the Saffir-Simpson category at landfall (should be treated with caution); (3) the number of U.S. fatalities associated with the storm (earlier storms are generally estimates, and some numbers include direct and indirect deaths, while others are direct deaths only); and general comments on the storm and the damage it caused.

A final example is Tropical Storm Gabrielle (2001), which made landfall on Florida's west coast near Venice. This storm strengthened with sustained winds to 65 mph right before landfall, causing storm surges of 3–6 feet. Rainfall totals were generally in the 4- to 7-inch range along the storm track over west-central Florida, resulting in major floods on the Manatee River, Little Manatee River, Myakka River, Peace River, and Horse Creek. The combination of storm surge from northeasterly flow ahead of the storm, high astronomical tides, and 6–12 inches of rain in northeast Florida caused near-record flooding of the lower St. Johns River.

In addition, many weak land-falling hurricanes that caused relatively little damage or few fatalities have been excluded from table 4.1. However, be assured that any land-falling storm is a major event to the community it impacts. Detailed tables of all U.S. land-falling hurricanes, as well as yearly storm tracks, are provided in Neumann et al. (1993), the National Hurricane Center website at http://www.nhc.noaa.gov, and the Unisys weather website at http://weather.unisys.com.

Finally, the United States is blessed with solid emergency preparedness, stout infrastructure, and first-rate communication outlets. Other countries along the Atlantic Ocean are not as fortunate. Many hurricanes in table 4.1 also devastated nations along the Caribbean Sea and south of the United States. History is littered with hurricanes that killed thousands of people in floods, mud slides, poorly constructed buildings, and storm surges and left 100,000 or more homeless. Hurricane Georges (1998) killed almost 600 and left 350,000 homeless in the Dominican Republic and Haiti (Pasch, Avila, and Guiney 2001). Hurricane Mitch (1998) killed 9,000 in Honduras and Nicaragua with severe destruction of those countries' infrastructure (Pasch, Avila, and Guiney 2001). Hurricane Donna (1960) produced 114 fatalities from the Leeward Islands to the Bahamas, including 107 in Puerto Rico in floods (Dunn 1961).

Chronology of Some Hurricanes That Impacted History

1274   A fleet from the Mongolian Empire consisting of 40,000 men and 1,000 ships attempted to invade Japan when a powerful storm sank 300 of the ships and killed 13,000. It is not known if this was a typhoon. Undaunted, Mongolia again attempted to invade Japan in 1281 with a massive fleet of 150,000 men. Once again, a strong storm (probably a typhoon) annihilated the fleet, sinking 4,000 ships and drowning 100,000 men. Japan believed it had been protected from certain conquest with godly protection. This great fortune for Japan gave rise to the term "kamikaze," or "divine wind" (Jennings 1970; Sheets and Williams 2001).

1565   France established a settlement east of the present site of Jacksonville, Florida, in 1564, naming it Fort Caroline. Spain, which had amassed a huge colonial empire in Mexico, Central America, parts of South America, and much of the Caribbean Sea considered the small outpost a threat to its shipping routes. Spain quickly established a fort at St. Augustine 40 miles to the south. Religious differences between Protestant France and Catholic Spain only enhanced tensions. In September 1565, France sailed from Fort Caroline to attack St. Augustine. At the same time, Spain, unaware of the pending assault, marched northward and attacked the outnumbered Fort Caroline with an easy victory. Spain immediately started to march back to St. Augustine when it became clear that an attack was imminent. However, bad weather (possibly a tropical storm or hurricane) shipwrecked the French fleet. Spain captured or killed stranded French soldiers along the coast. The French therefore lost their bid for control of North America to the Spanish (Rappaport and Partagas 1995; Sheets and Williams 2001).

1588   Spain, at war with England over religious differences (England is Protestant, Spain is Catholic) and seapower superiority for trade and expansion in the New World of the Americas, dispatched 130 ships of the largest naval force in history to invade England. The English stopped the invasion, and the armada attempted to return to Spain by sailing north around Scotland so as to avoid the English Channel.

Although it's difficult to know for certain, the armada encountered a bad storm that was quite possibly a hurricane in an extratropical stage. Many of the fleeing Spanish ships were sunk or wrecked on the coasts of Scotland and Ireland. Because of the storm and the battle, only half (65) of the armada made it back to Spain. The Englishman Lord Burghley sarcastically called the Spanish Fleet "The invincible armada" after England's victory. Spain lost control of the sea, and therefore its domination of the New World, to England (Mattingly 1974; Martin and Parker 2002).

1609   The British ship *Sea Venture,* en route to Jamestown, Virginia, was grounded on Bermuda by a hurricane. The stranded people are the island's first inhabitants, who think the island is "paradise" and claim it for England. Ten months later, the passengers and crew arrive in Jamestown in boats built from the island's cedar trees and the wreckage of the flagship (Hughes 1990). This incident inspired Shakespeare to write one of his best plays, *The Tempest.* Today Bermuda is a territory of the United Kingdom.

1640   A hurricane partially destroyed a large Dutch fleet poised to attack Havana, Cuba. This helped the Spaniards secure control of Cuba (Rappaport and Partagas 1995).

1666   England loses a fleet of seventeen ships and 2,000 troops in a hurricane near Guadeloupe. The French captured the remaining survivors, resulting in France's control of Guadeloupe through the twentieth century (Rappaport and Partagas 1995).

1780   A catastrophic hurricane, called the Great Hurricane, struck the Lesser Antilles in the Caribbean Sea on October 10 and was the most deadly hurricane in Western Hemisphere history. Approximately 22,000 deaths occurred, with 9,000 lives lost in Martinique, 4,000–5,000 in St. Eustatius, and 4,326 in Barbados (Rappaport and Partagas 1995). Thousands of deaths also occurred offshore, including great losses to the fleets of Britain and France, who were planning assaults on each other to claim the Antilles Islands. After the Great Hurricane of 1780, the governor of Martinique freed the imprisoned English soldiers, even though the French and English were at war. He declared that in such disasters all men are brothers (Ludlum 1989).

1837　A large major hurricane, dubbed Racer's Storm, traveled along much of the Gulf of Mexico coastline. The storm, with an estimated diameter of 1,600 miles, literally filled the Gulf of Mexico. Racer's Storm crossed the Yucatán Peninsula, headed north along the Texas and Mexico coast, then turned east along the Louisiana coast before finally making landfall along the Alabama-Florida border. It then exited into the Atlantic just below Charleston, South Carolina. It caused destruction all over the Gulf Coast, wiped out the Mexican navy, and destroyed several U.S. ships. Another ship, the *Racer*, survived the storm; went to Havana, Cuba, for repairs; and provided valuable information about the storm to William Reid. The storm was named after this ship (Ludlum 1989). Off North Carolina, the paddlewheel steamer *Home*, then hailed as the nation's fastest ship and on pace to set a new speed record during the trip, encountered the hurricane and sank off the North Carolina coast. Only two life preservers were on the ship. Forty passengers struggled to shore, and the rest (mostly women and children) drowned. This event caused Congress to rewrite maritime laws requiring a life preserver for every person on a ship (Hughes 1990).

1889　In March 1889, war was prevented between the United States and Germany by a typhoon. Late in 1888, a German naval force removed the native chief of the Samoans and set up a German protectorate in his place. The natives rebelled, and twenty-two German soldiers were killed. The Germans retaliated by sending three warships into the island's harbor and shelling a native village, incidentally destroying property of U.S. citizens located there. The Germans also ripped down and burned an American flag. The United States sent three warships of its own to protect its citizens' lives and property. On March 16, the U.S. and German fleets were in the harbor, poised to attack should either side provoke the other. However, before any overt action occurred, a typhoon struck the island (Hughes 1990).

The six warships and six other merchant ships were pummeled against reefs, sunk, or beached. About 150 sailors lost their lives. But during the storm, all three factions—the Germans, Americans, and Samoans—united as allies and helped each other, performing many heroic acts. Afterward, Germany and the United States decided to resolve their differences peacefully, resulting in the Act of Berlin of 1889, which states that Samoa should remain an independent territory with its own ruler advised by the U.S., British, and German consuls. This treaty lasted until tribal warfare broke out in 1899, resulting in the death of several U.S. soldiers. The treaty was annulled, western Samoa became German territory, and the United States gained what is now called American Samoa. Britain withdrew in exchange for Tonga and part of the German Solomons. Nevertheless, this incident made the United States realize that it was a major power in the world and required a larger navy to curb other aggressive countries. In other words, the Samoa incident is indirectly responsible for the founding of the modern U.S. Navy. Robert Lewis Stevenson was a witness to the whole event and published a book on it (Stevenson 1997).

1898   This section now ends on a lighter note. In 1898, Cuba was part of the oppressive Spanish Empire. The USS *Maine* was sunk under mysterious circumstances in Havana, Cuba, and the United States declared war on Spain. A U.S. regiment called the Rough Riders under Lt. Col. Theodore Roosevelt quickly overwhelmed the Spanish garrison and liberated Cuba. Lookout posts were established along the Cuban coastline to watch for a Spanish counterattack. On August 12, the war ended and Cuba was free. The U.S. weather service saw great potential in the Cuban lookout posts and took them over as part of a hurricane early warning system. Outpost employees were dismayed to find that the outdoor privy (an outhouse) had been stolen from one of the stations. This caused such uproar that the local who had stolen it for his own personal use returned it. Very shortly afterward, on October 2, both the observation post and its privy were destroyed by a hurricane, immortalizing the Privy Hurricane of 1898 (Hughes 1990).

5
Biographical Sketches

Many individuals have contributed to hurricane research, forecasting, and preparedness in the past two centuries. Biographical sketches of past and current hurricane scientists who have made important contributions to the field are included below. This list should be considered just a sample, since many people have provided key contributions to tropical meteorology, weather forecasting, disaster mitigation, and hurricanes in the last 200 years.

Much of the information in these biographical sketches was provided by the researchers themselves. Additional resources consulted include Burpee (1988, 1985), Byers (1960), Cabbage (1998), Daintith et al. (1994), DeAngelis (1989), Gillispie (1975), Kessler (2004), Kutzbach (1996), LeMone (1989), Sheets and Williams (2001), Williams (1992, 1998), and Zipser (1989).

Sim Aberson (1964–)

Sim D. Aberson has worked as a meteorologist with the National Oceanic and Atmospheric Administration's Hurricane Research Division (HRD) in Miami, Florida, since his senior year in high school in 1981. He was born in Herkimer, New York, and moved to the Miami area when he was 7 years old. His next door neighbors were retired public school science teachers who first got him interested in hurricanes by showing him how to listen to NOAA Weather Radio and plot the coordinates of hurricanes on a tracking map. That eventually led to his work in independent study on hurricanes and finally to a position with the Community

Laboratory Research Program of Miami-Dade Public Schools working at the HRD.

He attended Pennsylvania State University to earn his bachelor of science (1985) and master of science (1987) degrees and continued working at the HRD during school breaks. Upon graduation, he returned to the HRD as a full-time meteorologist to work on hurricane track forecasting with dynamical models. His work improved dynamical track forecast models based on simplified mathematics of the atmosphere known as barotropic equations. This led to early research into prediction of the forecast errors of models, predictability of tracks, probabilistic forecasting, ensemble forecasting (see chapter 2 for an explanation of this technique), and techniques to target regions where observations will most improve track forecasts. He then attended the University of Maryland at College Park to achieve his doctorate in meteorology in 2003 while continuing research on track forecasts.

Aberson's current research involves collaborations with scientists at the National Taiwanese University and Taiwanese Central Weather Bureau on targeting regions for improved typhoon forecasts and with scientists at the University of Munich and the Canadian Hurricane Centre on improvements to forecasts of tropical cyclones undergoing extratropical transition. He was coauthor of a paper on targeted observations in 1996 that won the Environmental Research Laboratories Outstanding Scientific Paper Award, was part of a team awarded a Department of Commerce Gold Medal, and was named NOAA Research Employee of the year in 2003.

Richard Anthes (1944–)

Richard A. Anthes, president of the University Corporation for Atmospheric Research (UCAR) since September 1988, is a highly regarded atmospheric scientist, author, educator, and administrator who has contributed considerable research to weather modeling, thunderstorms, and hurricanes. Anthes was born March 9, 1944, in St. Louis, Missouri. Growing up in Waynesboro, Virginia, he knew as a very young child that he wanted to be a meteorologist.

While attending the University of Wisconsin-Madison to earn his bachelor's degree, Anthes pursued his interest in meteorology by working as a student trainee for the U.S. Weather Bureau at the (NOAA) during the summers of 1962 through 1967. During this period, he discovered that an area of particular interest to him was hurricanes. His theses for his masters and doctorate degrees, obtained in 1967 and 1970, respectively, from the University of Wisconsin-Madison, reflected this interest. In particular, Anthes' doctorate work resulted in the first three-dimensional computer simulation of a hurricane. During 1968–1971, he was also a research meteorologist at NOAA's National Hurricane Research Laboratory.

In 1971, Anthes started teaching and conducting research at Pennsylvania State University, where he attained a full professorship in 1978. During this period, he also was a visiting research professor at the Naval Postgraduate School in Monterey, California. Anthes' research group studied a variety of weather issues related to hurricanes and thunderstorms. He frequently published papers on hurricanes, including one book titled *Tropical Cyclones: Their Evolution, Structure, and Effects.*

Since hurricanes and thunderstorms contain small-scale features requiring high-resolution and more complicated equations, Anthes modified the weather model he used for his doctorate to incorporate the proper physics for studying them. Research and development of this model has continued ever since, and several universities and government laboratories are continuously improving the model. Today this model, known as the Fifth-Generation NCAR/Penn State Mesoscale Model (MM5), is a state-of-the-art public domain model available from the University Corporation for Atmospheric Research. MM5 has been used by hundreds of graduate students and scientists for conducting modeling research and weather forecasts.

The National Center for Atmospheric Research (NCAR) welcomed Anthes in 1981 when he became the director of NCAR's Atmospheric Analysis and Prediction Division. In 1986 Anthes was selected to become the director of NCAR, and in 1988 he was selected to become the president of the University Corporation for Atmospheric Research (UCAR). UCAR is a nonprofit consortium of sixty-one member universities that awards doctorates in atmospheric and related sciences. UCAR manages NCAR in addition to collaborating with many international meteorological institutions through a variety of programs.

Anthes was elected as an American Meteorological Society (AMS) fellow in 1979. In 1980, he was the winner of the AMS's

Clarence L. Meisinger Award as a young, promising atmospheric scientist who had shown outstanding ability in research and modeling of tropical cyclones and mesoscale meteorology. In 1987, he received the AMS's Jule G. Charney Award for his sustained contributions in theoretical and modeling studies related to tropical and mesoscale meteorology.

Anthes has participated in or chaired more than thirty different national committees (for agencies such as NASA, NOAA, AMS, NSF, the National Research Council, and the National Academy of Sciences), including his present chairmanship on the National Weather Service Modernization Committee. Anthes has published more than ninety peer-reviewed articles and books. One book in particular, *Meteorology* (7th ed., 1996) is widely used at colleges and universities as a general introductory book to the field of meteorology for nonmeteorological majors.

Isaac Cline (1861–1955)

Isaac Cline is known for being the Weather Bureau's meteorologist in charge at Galveston Island during the Galveston Hurricane of 1900, which killed 6,000 people. He was also a distinguished forecaster and prolific author. Isaac Monroe Cline was born in 1861 and entered the U.S. Army Signal Corps in 1882. At that time, he entered the weather service school at Fort Myer, Virginia. Some of his instructors were distinguished physicists, such as William Ferrel, T. C. Mendenhal, and Cleveland Abbe. Upon completing school, Cline was assigned to offices in Little Rock, Arkansas. In his spare time, Cline attended the University of Arkansas medical school so that he could study how climate affected health; he graduated in 1885. Cline next served posts at Fort Concho, Texas (near San Angelo), and Abilene, Texas. In 1889, Cline was assigned as chief meteorologist of the Galveston, Texas, Weather Bureau office. Cline turned this office, which had failed an inspection prior to his arrival, into a showpiece for the Weather Bureau and received rave reviews at the next inspection.

Unfortunately, in July 1891, Cline wrote an article for the *Galveston News* stating that Texas was "exempt" from intense hurricane landfalls and that any landfalls would be from weak hurricanes. He reasoned that since no intense hurricanes had made landfall in recent history on Galveston, strong hurricanes must recurve before reaching Galveston. In fact, two deadly hurricanes had hit Indianola 120 miles to the south in 1875 and 1886, but he unknowingly or deliberately labeled these as weak storms and did not mention the fatalities. He also wrote that "it would be impossible for any cyclone to create a storm wave which could materially injure the city." Cline theorized that "the shallow slope of the seabed off Galveston would wear down incoming seas before they struck the city" (in fact, the opposite is true) and that the storm surge would spread first over the lowlands surrounding Galveston. It is unclear how this article impacted his judgment of the Galveston Hurricane or that of the general public nine years later, but it possibly may have contributed to some facets of the disaster.

On September 5, 1900, a tropical system passed over Cuba and developed into a hurricane in the eastern Gulf of Mexico. The Weather Bureau in Washington, D.C., had mistakenly concluded that this system instead was east of the Florida Keys and would be moving into the Atlantic. Tragically, Cuban meteorologists had telegraphed a statement concluding that a hurricane had formed in the Gulf, but for political and competitive reasons the Weather Bureau banned all Cuban weather telegraphs. After the disaster, Weather Bureau chief William Moore released a statement falsely claiming that his agency had warned Galveston a hurricane was coming, and unfortunately many journalists accepted Moore's statement. In fact, no hurricane warning was ever issued from Washington, D.C., to Galveston.

Cline's role in warning the local public preceding the hurricane is controversial. On September 8, 1900, Cline observed increasing ocean swells, stronger winds, and minor flooding, and he relayed these observations to Washington, D.C., every two hours. By the afternoon, with the water level at several feet and winds increasing rapidly, Cline undoubtedly became concerned. By 2:30, Isaac had his brother Joseph (who also worked for the Weather Bureau) telegraph Washington, D.C., with the statement, "Gulf rising, water covers streets of about half of the city." Both brothers then waded home to Isaac's house. This was the last message out of Galveston. It is not known how much interaction Cline had with the public between telegraphs to Washington, D.C. In his autobiography, Cline claims that he hitched up his horse and cart and traveled up and down the beach, telling homeowners to move to higher ground and vacationers to go home.

However, no survivor remembers seeing Cline on the beach or even receiving a warning. Cline's autobiography also says that he told the Weather Bureau "of the terrible situation, and stated that the city was fast going under water, that great loss of life must result, and stressed the need for relief." However, no Weather Bureau documents state anything about a hurricane. In fact, Cline may have not even known it was a hurricane or thought a hurricane was a serious threat until it was too late. One of his Washington, D.C., statements claims that the inclement weather may be an "offspur" of a Florida storm.

By late afternoon, the full fury of the hurricane hit. That evening Cline's house, already pummeled by the still-increasing storm surge, started to float. When his house rolled, he briefly lost consciousness but recovered and found all his family except his pregnant wife, who had been ill in bed. Cline drifted for several hours on one sinking piece of debris after another. With him were his three daughters, his brother Joseph, and a little girl Joseph had pulled from the water. Finally, they came to rest on wreckage that turned out to be right back in their own neighborhood. One month later the body of Cline's wife was found under that very wreckage.

Despite his personal grief, Cline later wrote a technical paper for *Monthly Weather Review* on the Galveston Hurricane titled "Special Report on the Galveston Hurricane of September 8, 1900."

Cline was promoted to chief of the Forecast Center for the Gulf States, located in New Orleans, and moved from Galveston with his three daughters in 1901. The tragedy motivated Cline to conduct research on hurricanes, particularly on the storm surge. Cline noted that water rising above tide levels was a precursor to hurricane landfall. This research paid off in 1915 when a hurricane was in the Gulf of Mexico. Water levels continued to rise rapidly in Burrwood, Louisiana, while receding in Galveston, so Cline concluded a landfall just west of the Mississippi River was imminent. Cline's office issued warnings by telegraph, telephone, and special messenger, resulting in mass evacuations. A category 4 hurricane made landfall later that afternoon, passing directly over New Orleans (see chapter 4). Many lives were saved, and the New Orleans newspaper *The Times Picayune* wrote the next day, "Never before, perhaps in the history of the Weather Bureau, have such general warnings been disseminated as were sent out by the local bureau." In this respect, Cline issued one of the first National

Weather Service local-impact statements. (It should be noted that relying on the storm surge as the sole determinant for landfall prediction is only effective for small storms moving slowly in a straight path, such as the 1915 New Orleans hurricane). Cline also distinguished himself in 1927 during the great Mississippi Valley flood, and for this he later received a special commendation from President Herbert Hoover.

Cline was the author of a multitude of meteorological papers and books, of which the most noteworthy was his book *Tropical Cyclones,* published in 1926 and the most authoritative book in the United States on hurricanes at that time. It contains considerable original work on hurricane rainfall, tides, and waves, with the focus on Gulf of Mexico storms. Another noteworthy book is *Storms, Floods, and Sunshine,* which contains two parts. The first part is his memoirs, and the second part describes the characteristics of hurricanes.

Cline spent his later years in the New Orleans Weather Bureau. In 1935 he retired and opened a small art shop in the city's French Quarter. Isaac Cline died in 1955 at the age of 95.

Mark DeMaria (1955–)

Mark DeMaria was born September 30, 1955, in Middleburg, Pennsylvania, although he spent most of his early life in Miami, Florida. He developed an interest in hurricanes at a young age after witnessing Hurricanes Cleo, Betsy, and Inez during 1964–1966. He obtained a bachelor of science degree in meteorology from Florida State University in 1977 and a master of science and doctorate degrees in atmospheric science in 1979 and 1983, respectively, from Colorado State University. He developed a hurricane model under the direction of Dr. Wayne Schubert for his doctorate research. After short periods of employment at the National Center for Atmospheric Research (NCAR) and North Carolina State University, DeMaria returned to Miami in 1987 to work at the Hurricane Research Division (HRD). In 1995, he left the research community to become the chief of the technical support branch at the Tropical Prediction Center/National Hurricane Center in Miami.

The primary emphasis of DeMaria's work is on the development of techniques for hurricane track and intensity forecasting. His early research focused on track forecasting, and he won the

Banner Miller Award from the American Meteorological Society (AMS) in 1985 and again in 1987 for studies on barotropic vortex motion. As part of his work at the HRD, DeMaria implemented an experimental hurricane track forecast model (VICBAR) using a nested spectral method developed by Dr. K. V. Ooyama. The VICBAR model was also used in the first study to demonstrate statistically significant track forecast improvements for cases that included experimental dropwindsonde data collected in the vicinity of hurricanes with the NOAA WP-3D aircraft; both VICBAR and dropsondes are now used operationally. In collaboration with John Kaplan of the HRD, DeMaria developed a Statistical Hurricane Intensity Prediction Scheme (SHIPS) for the Atlantic basin. In 1996, SHIPS became part of the routine forecast guidance run in real time at the NHC. After the 1996 hurricane season, SHIPS was reformulated to more accurately evaluate the forecast predictors. Since that time, SHIPS is the only model that has demonstrated statistically significant intensity forecast skill for Atlantic storms. Also with John Kaplan, DeMaria developed an empirical model for predicting the inland decay of hurricane winds. In cooperation with the Federal Emergency Management Agency (FEMA), results from this algorithm were distributed to many emergency management agencies in the southeastern United States for planning purposes. DeMaria (with John Kaplan) received an NOAA Bronze Medal in 1997 and the AMS Banner Miller Award in 2002 for this work on intensity forecasting.

In late 1998, DeMaria returned to Colorado to become the leader of the NESDIS Regional And Mesoscale Meteorology (RAMM) Branch at Colorado State University. He is continuing his research on hurricane forecasting, with emphasis on the use of remote sensing data. His current work is focused on further improving the SHIPS model by incorporating storm inner core and subsurface ocean information obtained from geostationary satellite observations and satellite ocean altimetry data. He is also actively involved in research to optimize the use of future operational polar-orbiting and geostationary satellites for hurricane and severe weather analysis and prediction. DeMaria has published more than forty refereed articles and book chapters on hurricanes and tropical meteorology, and in 1995 he was invited to contribute a review article on the history of hurricane forecasting to a special Diamond Anniversary volume of the American Meteorological Society. In 2003, he was invited (with James Gross from the NHC) to a write a review article on the history of Atlantic hurricane forecast models for a special American Geophysical Union book on hurricanes edited by Robert Simpson.

Gordon Dunn (1905–1994)

Gordon Dunn was a weather enthusiast with an innate ability to synthesize observations and forecast weather and hurricanes accurately. Dunn was born on August 9, 1905, in Brownsville, Vermont. He developed a fascination with the weather as a young boy, and as a teenager he would hitch rides to the Weather Bureau office 18 miles away to study weather charts and discuss weather events with the station's sole employee. Dunn applied this knowledge to practical weather problems on his father's dairy farm.

Dunn accepted a job as a messenger with the Weather Bureau in Providence, Rhode Island, in 1924, where he took surface observations, carried messages to the Western Union office, and drew the weather map for the morning paper. He transferred to the Weather Bureau in Tampa, Florida, with a promotion to junior observer two months later. In 1926 he requested relocation to an area where he could complete his college education and transferred to the Weather Bureau's Central Office in Washington, D.C. In this location, he attended George Washington University to pursue an associate bachelor's degree in political science, which he completed in 1932. However, he also took courses in meteorology at other universities whenever possible, such as Florida State University, the Massachusetts Institute of Technology, and the University of Chicago. In 1931, he was promoted to a meteorologist in the Washington Weather Bureau.

In 1935 Congress appropriates $80,000 to revamp the Weather Bureau's hurricane warning service. Four new hurricane forecast centers were established, including one in Jacksonville, Florida. Grady Norton (see biography in this chapter) was transferred to Jacksonville as the senior forecaster, and Dunn was transferred there as the junior forecaster. Dunn and Norton faced the challenge of forecasting several devastating and deadly hurricanes, including the Labor Day Hurricane (1935) and the New England Hurricane (1938) (see chapter 4). During this period, Dunn noticed areas of rising and falling pressure that progressed westward every three to four days in the Atlantic Ocean with an inverted, wavelike pressure signal; he also noticed that some of these systems became hurricanes. In 1940 Dunn published a paper showing that most Atlantic tropical storms and hurricanes form from these disturbances, called tropical waves (also called easterly waves), not due to cold fronts, which was the popular theory at that time. This was also the first paper documenting and describing tropical waves.

In 1939, Dunn was transferred to Chicago, where he soon became the meteorologist in charge, a position he held until 1955. Dunn developed close ties between the Weather Bureau and the well-respected University of Chicago meteorology program and is credited for improving working relationships between forecasters and professors. During World War II, Dunn was assigned to Calcutta, India, where he assisted the military in weather forecasts in the China-Burma-India region. The military awarded Dunn the Medal of Freedom for his outstanding weather analyses and forecasts.

In 1955 the Miami Weather Bureau office was officially designated the National Hurricane Center. Dunn was named the first official NHC director, although many, including Dunn, recognized the well-respected Grady Norton, who died in 1954 at home after performing forecast duties during Hurricane Hazel, as the honorary first director. Dunn continued to work there until his retirement in 1968. During this period, Dunn originated procedures for improving public awareness of the hurricane threat and improved international cooperation, including training and upgrading of meteorological services throughout the Caribbean.

Dunn received many awards for his efforts as the NHC director. The Department of Commerce awarded him its Gold Medal in 1959. In 1966, the University of Miami awarded Dunn an honorary doctorate of science. The American Meteorological Society named him an honorary member in 1992. Dunn coauthored the 1960 book *Atlantic Hurricanes* with Banner Miller.

Dunn died in south Miami on September 12, 1994.

Russell Elsberry (1941–)

Russell Elsberry grew up in Colorado and attended Colorado State University, where he earned a bachelor of science degree in mechanical engineering in 1963. During his undergraduate studies, he worked as a computational assistant for Herbert Riehl (see biography in this chapter) performing tasks related to meteorology. He decided that studying hurricanes was more exciting than mechanical engineering and pursued a doctorate with Riehl as an advisor. Upon completing his doctorate in 1968, Elsberry became an assistant professor in the Department of Meteorology at the Naval Postgraduate School, where he has remained. Elsberry was promoted to associate professor in 1972 and to professor in 1979 and was awarded distinguished professor status in 1994.

Elsberry has remained very active in hurricane research using observational and numerical modeling approaches, particularly with regard to understanding hurricane motion. His research group has studied how hurricane circulation's interaction with the environment affects the motion. The researchers have also developed a systematic approach for forecasting hurricane motion that is being extended from the western North Pacific to other hurricane basins. Elsberry was technical director of the Office of Naval Research Tropical Cyclone Motion Project, which coordinated a field program called the Tropical Cyclone Motion (TCM-90) Experiment over the western North Pacific in August and September 1990. TCM-90's objectives were to test several hypotheses regarding hurricane motion, with a focus on environmental interactions. This was followed by the Tropical Cyclone Motion (TCM-92 and TCM-93) minifield experiments in July and August 1992 and 1993. The primary objective of TCM-92 and TCM-93 was to obtain simultaneous aircraft measurements into typhoons and nearby cloud complexes. The objectives were also to test hypotheses regarding how these nearby cloud clusters may cause typhoons to deviate from their track and how tropical cyclogenesis may occur from the cloud complexes.

Elsberry was editor of two books that provided an international overview on tropical cyclones: *A Global View of Tropical Cyclones* (published in 1987 by the University of Chicago Press) and *Global Perspectives on Tropical Cyclones* (published in 1995 by the World Meteorological Organization). Elsberry has published numerous papers on hurricanes and is a fellow of the American Meteorological Society. He has served as the Hurricane Landfall Science Coordinator for the U.S. Weather Research program since 1998. The hurricane landfall program has included cooperative field experiment activities during 1998 and 2001 by NASA and the NOAA Hurricane Research Division that resulted in extensive data sets. In addition, the Office of Naval Research Coupled Boundary Layer Air-Sea Transfer field experiments have collected unique atmospheric and oceanic measurements in category 4 and 5 hurricanes. Elsberry has also been a leader in the USWRP-sponsored Joint Hurricane Testbed that has as its objective rapidly and efficiently moving transfer research results to operational practice.

Elsberry has participated actively in all five International Workshops on Tropical Cyclones (IWTC) sponsored by the World Meteorological Organization to bring together forecasters and researchers to review progress over the past four years and plan for the future. He served as director of the Fifth IWTC in Cairns, Australia, during 2002.

Neil Frank (1931–)

Neil Frank is credited with improving public awareness on hurricane preparedness. He attended Southwestern College in Kansas, where he played on the basketball team for four years. Encouraged by his coach to pursue a degree in the sciences, Frank earned a bachelor of science degree in chemistry with intentions of being a science and physical education teacher. After college, Frank joined the U.S. Air Force and served in Okinawa, Japan, where he became fascinated by the typhoons that frequently impacted that region. As a result, he pursued graduate studies in tropical meteorology at Florida State University, where he earned master of science and doctorate degrees in meteorology. In 1961, Frank began work at the NHC as a forecaster and in 1974 became its director. In 1987, Frank left government employment for his current position as chief meteorologist at KHOU-TV in Houston, Texas.

Frank's interest in public awareness began in 1965 while investigating Hurricane Betsy's storm surge in Miami. While surveying a damaged area, a muddy survivor of the storm surge angrily asked Frank why no one at the NHC had informed her that she would have been safe across the street where the surge ended. Based on this experience, Frank emphasized hurricane preparedness during his tenure as NHC director, frequently giving slide show presentations to coastal audiences, presenting background hurricane briefings to the press, and effectively using the news media as a means to disseminate information when a hurricane threatened the United States. Frank was one of the first to employ storm surge models to graphically demonstrate possible hurricane landfall scenarios. He also encouraged upgrades to local community hurricane preparedness plans, because most civil defense plans still focused on Cold War plans for a nuclear attack. He served as chairman of an International Hurricane Committee that coordinated hurricane warning procedures for North American countries.

Frank is married with three grown children. He has published in journals, been quoted in popular magazines, and won first place in the 1989 Texas Associated Press Awards for Best Weathercast.

William M. Gray (1929–)

William M. Gray is probably the best-known hurricane scientist among the general public today, frequently being quoted by media sources throughout the country. Gray was born October 9, 1929, in Detroit, Michigan. Afflicted with a stuttering problem, he struggled in school but excelled at sports. Hours after graduating from high school in Washington, D.C., Gray left for Florida to join the Washington Senators training camp but didn't make the team. Instead, he pitched in college while pursuing a degree in geography at Wilson Teachers College (1948–1950) and George Washington University (1950–1953). He hurt his knee at the university, so instead of pursuing a sports career after earning his degree, he joined the U.S. Air Force, which utilized his scientific background as a weather forecaster from 1954 to 1957 in the Azores Islands and England. Gray then attended the University of Chicago to study meteorology under the mentorship of the famous tropical weather scientist Dr. Herbert Riehl (see biography in this chapter), where he earned a master of science degree in 1959 and a doctorate in 1964. However, much of the PhD work was actually completed as a faculty member at Colorado State University, where Riehl started a new meteorology program in 1961 and invited Gray along. During graduate school, Gray developed a passion for hurricane research.

From 1961 to 1983, Gray developed a reputation as one of the premier hurricane researchers in the world. He formulated the general conditions for genesis to occur and developed detailed data sets of hurricane structure by compositing observations from data-sparse regions and reconnaissance flights. In 1983 Gray discovered that fewer Atlantic hurricanes occur during El Niño years. However, while he was well-respected among his peers, he was relatively unknown to the general public during this period.

In 1984, Gray announced a scheme for forecasting the number of tropical storms and hurricanes in the Atlantic. He also boldly made this forecast available to the general public. The forecasts initially generated skepticism from scientists and nonscientists alike. However, with each year the forecasts showed skill, making Gray a household name among coastal residents. Gray's success is considered one of the most influential works in the field of seasonal predictions, in which a forecaster predicts below-average, average, or above-average weather conditions months in advance; the weather conditions may be temperature, rainfall, hurricane activity, etc. Gray's seasonal hurricane forecasts are based on the relationship between hurricane activity and factors such as El Niño, African rainfall, sea-surface temperature, and Caribbean surface pressure, as described in chapter 2. These forecasts are initially issued in December and updated in April, June, and August of each year. These predictions are widely distributed by the press and are available at http://tropical.atmos.colostate.edu.

Gray has served on or chaired more than twenty important committees and panels and is author or coauthor of eighty-three reviewed publications. He has served as a visiting scientist in many countries such as Japan, China, and England. Among the awards he has received are the Jule G. Charney Award (1993) and the Neil Frank Award of the National Hurricane Conference (1995). He was the ABC television Person of the Week in September 1995 after accurately predicting a very active Atlantic hurricane season. He was an invited lecturer to the Twelfth WMO Congress, Geneva, in June 1995, an honorary award given to senior scientists in recognition of lifetime research achievements. Gray has graduated several dozen students with a master's of science or PhD, many of whom have also enjoyed successful meteorology careers. Gray is a fellow of the American Meteorological Society.

Greg Holland (1948–)

Greg Holland is an Australian meteorologist who has studied many aspects of tropical meteorology and hurricanes with a unique Southern Hemisphere perspective. Holland was born August 15, 1948, in Sydney and grew up on a farm. He obtained a bachelor of science in physics with first-class honors in mathematics and a minor in mathematics from the University of New South Wales in 1972. He joined the Australian Bureau of Meteorology as a cadet while still at the university and also received a diploma of meteorology from the Central Training School in the Bureau of Meteorology in 1972. From 1973 to 1975, Holland was a forecaster for the Darwin Tropical Analysis and Regional Forecasting Centre. He was on forecast duty when Tropical Cyclone Tracy wreaked havoc on the town. From 1975 to 1977, he lectured at the Central Training School in the Bureau of Meteorology in Melbourne, Victoria. From 1978 to 1984, the Bureau of Meteorology sent Holland to Colorado State University, where he obtained a master's of science in 1981 and a PhD in 1983 with William Gray (see biography in this chapter) as his advisor. Holland remained at the Bureau of Meteorology after graduation under a variety of titles. He was also an honorary associate of the faculty at Monash University and an honorary professor at the University of New South Wales, under which he advised several doctoral students.

The concept of flying unmanned robotic aircraft (Aerosonde) into hurricanes drew Holland into the private sector, where he led the development and commercialization of these aircraft, first as a research director and later as CEO of the company. Holland returned to meteorology research in 2004 as director of the Mesoscale and Microscale Meteorology Division at NCAR, his current position.

Holland has participated in numerous hurricane research activities during his career. He has studied mechanisms for their formation, intensification, and motion. He has participated and even been lead investigator in many major field experiments. For twelve years, Holland served as the chairman of the Tropical Meteorology Research Program, and he is active on a variety of other WMO committees related to weather in the tropics. Holland is a fellow of the American Meteorological Society.

His publications include major contributions to six textbooks and forecast manuals, and he is the chief editor of the WMO manual *Global Guide to Tropical Cyclone Forecasting*. He has published more than eighty research papers on hurricanes, monsoons, the sea breeze, severe thunderstorms, and the Aerosonde.

Yoshio Kurihara (1930–)

Yoshio Kurihara has been a leader in theoretical and operational hurricane modeling since 1980. He was born on October 24, 1930, in Japan. Kurihara obtained a bachelor of arts degree at the University of Tokyo in 1953 and a PhD at the University of Tokyo in

1962. He worked at the Japan Meteorological Agency in Tokyo from 1953 to 1959, at the Meteorological Research Institute in Tokyo from 1959 to 1963, at the Geophysical Fluid Dynamics Laboratory (GFDL) from 1963 to 1965, and again at the Meteorological Research Institute from 1965 to 1967. In 1967, Kurihara returned to GFDL, where he remained until his retirement in 1998. Kurihara then returned to Japan, where he is conducting hurricane research at the Frontier Research System for Global Change in Tokyo.

Kurihara was the group leader of the hurricane research project at GFDL that was initiated in 1970. His group developed one of the first three-dimensional models of hurricanes and has conducted a number of numerical simulation studies in order to understand the basic mechanism of hurricane evolution. Many of these early studies dealt with spiral band structure and hurricane landfall. While this model was originally developed for pure research, tests from 1986 to 1992 showed that their model was adept at operational track forecasts. The GFDL was tested semi-operationally in 1993. Unlike several other models, the GFDL model correctly predicted that Hurricane Emily would take a sharp turn from North Carolina's Outer Banks and just graze the area. The GFDL model has officially been used as a hurricane forecast model since 1995 and generally has the smallest track errors compared to other model guidance. A similar model is used by the U.S. Navy in the western Pacific.

Kurihara has received the following awards during his career: the Meteorological Society of Japan Society Award in 1975, the Distinguished Authorship Award from NOAA in 1983, the Banner Miller Award from the American Meteorological Society in 1984 and in 1997, the Environmental Research Laboratories Distinguished Authorship Award from NOAA in 1992, the Department of Commerce Gold Medal Award in 1993, the Meteorological Society of Japan Fujiwara Award in 1994, and the Jule G. Charney Award from the American Meteorological Society in 1996. Kurihara has been a fellow of the American Meteorological Society since 1980.

Chris Landsea (1965–)

Chris Landsea was born February 2, 1965, in Urbana, Illinois, but considers himself a native of south Florida, where he developed an interest in the hurricanes that are always a threat to that region. He first participated in hurricane research as a high school intern at NOAA's Hurricane Research Division (HRD), then attended the University of California in Los Angeles where he earned a bachelor of science degree in atmospheric science in 1987. He then studied under the mentorship of Dr. William Gray (see biography in this chapter) and obtained a master's of science and PhD in atmospheric science in 1991 and 1994, respectively. He then spent seven months as a visiting scientist at the Australian Bureau of Meteorology Research Centre in Melbourne and Macquarie University in Sydney to study the monsoon environment of hurricanes in that region. He is currently employed as a research meteorologist at HRD.

Landsea's hurricane research determined the connection between African rainfall and Atlantic hurricane activity. He noted that the occurrence of major hurricanes (category 3 or better) increases during rainy years in Africa. He also noted that the U.S. East Coast had experienced a downturn in major hurricane landfall during the 1970s to early 1990s at the same time that Africa experienced a drought. Both the major hurricane and African rainfall variations have been explicitly linked to the Atlantic Multidecadal Mode (AMM), a natural fluctuation of the Atlantic Ocean. The AMM switched to a favorable warm phase in the late 1990s, and along with it have come more intense hurricanes and increased African rainfall. Such conditions are hypothesized to last another two to three decades. Landsea has also been a coauthor of Gray's and NOAA's annual hurricane seasonal forecasts. His most recent research has been toward a reanalysis of the Atlantic basin hurricane database, which is utilized in meteorology, emergency risk assessment, insurance rates, building codes, and climate change and variability studies.

Landsea is the corecipient of the American Meteorological Society's Banner I. Miller Award along with William M. Gray, Paul W. Mielke Jr., and Kenneth J. Berry for the paper "Predicting Atlantic Seasonal Hurricane Activity 6–11 Months in Advance" at the May 1993 meeting of the 20th Conference on Hurricanes and Tropical Meteorology. The award was given for the "best contribution to the science of hurricane and tropical weather forecasting published during the years 1990–1992." Landsea is also the recipient of the American Meteorological Society's Max A. Eaton Prize for the Best Student Paper given at the 19th Conference on Hurricanes and Tropical Meteorology in May 1991. He is on the

1997–2000 AMS committee for Hurricanes and Tropical Meteorology and is an associate editor for the AMS journal *Weather and Forecasting*.

Landsea served as the chair of the AMS Committee on Tropical Meteorology and Tropical Cyclones for the years 2000–2002. In 2000, Landsea was a corecipient of a U.S. Department of Commerce Bronze Medal "for issuing the accurate and first official physically based Atlantic seasonal hurricane outlooks for the 1998/1999 seasons, based upon new research." In 2002, Landsea was given the AMS Editor's Award for reviews for *Weather and Forecasting*. He currently is serving on the Editorial Board for the *Bulletin of the American Meteorological Society* as the subject matter editor in tropical meteorology. The media frequently interviews Landsea about hurricanes.

Frank Marks (1951–)

Frank Marks became interested in meteorology at 13 years old through the influence of his neighbor, a science teacher at the local high school who was also an amateur meteorologist and ran a weather club as part of his science class. During the next five years Marks learned how to take weather observations, plot and analyze the observations on a map, analyze the observations, and make forecasts of the local weather by posting them and reading them with the high school morning announcements. He obtained an undergraduate degree in meteorology at Belknap College in New Hampshire in 1973, followed by graduate studies at the Massachusetts Institute of Technology, where he received a master of science in 1975 and a doctorate in 1980. While a graduate student he participated in the Global Atmospheric Research Program Atlantic Tropical Experiment (see chapter 4), spending 100 days in Dakar, Senegal, during the summer of 1974, and six weeks in Borneo for the Winter Monsoon Experiment (WMONEX) during the winter of 1978–1979. These two experiences were invaluable in setting his future path as a research meteorologist, because Marks experienced how scientists design and execute experiments in the field. In addition, these scientists became mentors and one (Dr. Robert Burpee) eventually offered him a job at NOAA's Hurricane Research Division (HRD).

Marks specializes in radar remote sensing (ground-based, airborne, and satellite) of hurricanes and tropical disturbances in order to understand their wind, temperature, moisture, and precipitation structure. He has flown in more than 160 research reconnaissance missions and has participated in field experiments throughout the world. He is the recipient of several honors, including an award for best wireless application for development of satellite-cell based WLAN for NOAA WP-3D aircraft in 2002, the U.S. Department of Commerce Silver Medal for performance as the Research Mission Manager for the NOAA High Altitude Jet procurement in 1997, the U.S. Department of Commerce Gold Medal for HRD's Performance in Hurricane Andrew in 1992 (group award), and a Distinguished Authorship Award from NOAA in 1989. He is the lead author or coauthor on numerous journal publications and is a fellow of the American Meteorological Society. He is currently the director of HRD.

Charles Neumann (1925–)

Charles Neumann participated in some of the earliest typhoon reconnaissance flights and is the primary developer of statistical hurricane track forecasts still being used today. Neumann was born in 1925 in New York City, New York. Participating in a World War II extension of the Naval ROTC program, he spent his freshman year at Holy Cross College in 1943, then transferred to the Massachusetts Institute of Technology where he earned a bachelor of science in meteorology in 1946. Upon graduation, he volunteered as a U.S. Naval flight meteorologist for the first organized typhoon reconnaissance missions in the western North Pacific and served in that capacity in 1946 and 1947. Upon discharge from the navy in 1947, he entered graduate school at the University of Chicago, where he earned a master of science degree in meteorology in 1949. From 1950 to 1952, Neumann was a civilian research meteorologist at the Air Weather Service Scientific Services Directorate in Washington, D.C. In 1952, he resumed his navy career, having been recalled to active duty during the Korean War, and again served as a flight meteorologist—this time on Atlantic hurricanes. After one such flight in September 1953, which terminated at a Caribbean Island, he contracted polio and retired from the navy in 1954.

After two years of partial recovery from that illness, Neumann became involved in operational forecasting, serving as chief forecaster at Homestead Air Force Base (1956–1962) and aviation forecaster in Miami (1962–1966). During 1966–1971, he was a meteorologist with the NOAA Spaceflight Meteorology Group, which was collocated with the National Hurricane Center, and provided forecasts for the Gemini and Apollo programs. In 1971, he joined a newly formed National Hurricane Center Research and Development Unit and became its chief in 1976, a position he held until his NOAA retirement in 1987. Since 1987, Neumann has been semiretired but works part-time as a senior scientist for Science Applications International Corporation (SAIC) providing hurricane risk information to military and insurance interests. He is currently working as a team member in revising the Atlantic hurricane historical database and will apply this data to a revised hurricane risk model for the Atlantic.

Neumann is the developer of numerous global statistical and statistical-dynamical models for the prediction of hurricane motion and intensity. These models include the Atlantic NHC 72, NHC 73, and NHC 90; SHIFOR; CLIPER; Eastern Pacific PE90; and many versions of these models for other ocean basins. Many of these models are still being used today at various worldwide forecast centers, including the National Hurricane Center. Neumann has also developed statistical models for the assessment of hurricane risk and hurricane strike probability.

Neumann has authored more than 100 papers, mainly on hurricanes, and is the recipient of several awards. In 1971 he shared the Department of Commerce Silver Medal with fellow forecaster John Hope for "highly competent skill and ingenuity in developing objective techniques for use in hurricane predictions." He is the recipient of the 1977 American Meteorological Society's Banner I. Miller Award for "best published paper in tropical meteorology," and NOAA awarded NHC's Research and Development organization a citation for the "development of objective hurricane forecast aids in the eastern Pacific Ocean." In 1983, he was awarded the AMS Award for Outstanding Contributions to Applied Meteorology. In 1986 Neumann received the Department of Commerce Gold Medal for "exceptional scientific achievement in statistical/dynamical track prediction modeling." Neumann is a fellow of the American Meteorological Society and has been a member of that organization for more than fifty years.

Grady Norton (1894–1954)

Grady Norton was an extraordinary hurricane forecaster in an era before computers and satellite provided weather guidance, and he was blessed with the ability to communicate with coastal residents during hurricane threats. The son of a farmer, Norton was born in 1894 in Womack Hill, Alabama, and became fascinated with severe weather as a young boy. He entered the Weather Bureau in 1915 and was drafted into the army near the end of World War I. He served ten months with the Signal Corps in 1918–1919 and attended a Signal Corps' meteorology program at Texas A&M University.

Upon completing his military service, Norton returned to the Weather Bureau and worked at several different weather offices in the south-central United States, where he developed a reputation for having a knack at making good forecasts. Norton originally did not specialize in hurricane forecasting, but in 1928 he visited some relatives in south Florida in late summer and happened to drive into West Palm Beach during a mass funeral for the 2,500 victims of the Lake Okeechobee hurricane. Norton overheard a remark that such loss of life would not have occurred had adequate warnings been issued by the weather service. Although the comment was not entirely true, this event made a lasting impression on Norton, and he resolved to dedicate his life to the prevention of such tragedies.

In 1935, Congress appropriated funding for four hurricane forecast centers, one of which was established in Jacksonville, Florida. Norton was transferred there and named the chief hurricane forecaster at the Jacksonville office. A few months later the Labor Day Hurricane hit the Florida Keys, killing 409 people even though Norton had issued warnings more than twelve hours in advance. This renewed Norton's resolve to make accurate forecasts and to better communicate warnings to the public. Fortunately, in the 1940s the number of upper-air observations from radiosondes increased dramatically, and Norton studied the relationship between hurricane tracks and upper-air wind observations provided by the radiosondes. He noted that hurricanes tend to follow upper-air winds and developed the concept of steering current. Norton applied this concept with some intuition, and short-term track predictions improved as a result during the next ten years, reducing the number of fatalities.

Norton also effectively communicated warnings to the public. Whenever a hurricane threatened the state, he transmitted hurricane information to more than twenty Florida radio stations in a strong, clear, calm, and folksy manner, sometimes for two days without rest. As a result, he became the most trusted hurricane forecaster in the region. Norton was also an active public awareness advocate, giving many after-dinner talks on hurricane preparation. He also worked with government officials and engineers to improve building codes.

Due to minimal human resources at the busy Jacksonville office, in 1943 this forecast center was moved to Miami to establish a joint hurricane warning service with the U.S. Air Corps and the U.S. Navy. Norton continued as the meteorologist in charge during this transition. In 1949, he received the Department of Commerce Silver Medal in recognition of his outstanding hurricane work.

Norton suffered from high blood pressure, which resulted in severe migraine headaches in the last decade of his life. However, he kept working against his doctor's advice, partly because the general public wanted to know his personal predictions. After a twelve-hour stint forecasting for Hurricane Hazel on October 9, 1954, Norton suffered a stroke at home and passed away later that day. In 1955, the Miami office was officially designated as the National Hurricane Center (NHC), with Norton's understudy, junior forecaster Gordon Dunn, named as the first director. However, most people, including Dunn, recognized Norton as the honorary first director of the NHC.

Katsuyuki V. Ooyama (1929–)

Katsuyuki V. Ooyama (often referred to as "Vic" by colleagues) was born in Japan on March 5, 1929. He obtained a bachelor's of science in physics in 1951 at the University of Tokyo. From 1951 to 1955, he worked at the Japan Meteorological Agency as a forecaster. During 1955, he enrolled at New York University where he earned a master's of science and PhD under the mentorship of Dr. A. K. Blackadar and Dr. B. Haurwitz to study atmospheric processes near the earth's surface. From 1958 to 1962, he participated in remote sensing of the ozone layer. From 1962 to 1973 he was a faculty member at New York University, and from 1973 to 1979 he was a senior research scientist at the National Center for Atmospheric Research. Since 1980, he has been a senior research scientist at NOAA's Hurricane Research Division, developing a three-dimensional weather model.

In 1969, Ooyama was the first scientist to realistically duplicate the basic two-dimensional features of a hurricane using a computer model, which he described in the landmark paper "Numerical Simulation of the Life Cycle of Tropical Cyclones." In the same paper, Ooyama also performed the first multiscale perturbation analysis to explain how tropical genesis can occur based on the growth of small disturbances. This analysis launched a decade of theoretical research on tropical genesis and hurricane development based on similar mathematical theories. Ooyama is the recipient of the Meisinger Award (1968), the Fujiwhara Award (1971), and several Distinguished Authorship awards from the Environmental Research Laboratory of NOAA.

Henry Piddington (1797–1858)

Henry Piddington was the first person to coin the word "cyclone" for rotary storms (from the Greek words *kyklon* and *kyklos* meaning "moving in a circle, whirling around" and also "coiling of the snake"). He was also a prolific writer about these weather systems. Born in Uckfield, East Sussex, England, Piddington began his career in the mercantile marines, eventually becoming a ship commander in the British East India Company. In the late 1820s, Piddington became the curator of the Calcutta Museum, and during the next two decades he wrote articles for the *Journal of the Asiatic Society* on iron ore, minerals, plants, and soil chemistry. While in Calcutta, Piddington also became the president of the Marine Court and held the positions of officiating secretary, assistant secretary, and curator of the Geological Department of the Asiatic Society of Calcutta.

In the mid-1840s, Piddington became interested in rotary storms after reading William Reid's (see biography in this chapter) book *Law of Storms*. Recalling past experiences with such storms, especially one that shipwrecked Piddington's vessel from which he narrowly escaped death, Piddington began collecting as many ship logs as possible that involved cyclones. Based on this data, Piddington published twenty-five memoirs and a variety of books and journal articles. His most famous book has the mind-boggling title *The Sailor's Horn-book for the Law of Storms: Being a Practical Exposition of the Theory of the Law of Storms, and Its Uses to Mariners of All Classes in All Parts of the World, Shown by Transparent Cards and Useful Lessons*. This book, typically called *The Sailor's Horn Book,* taught mariners how to navigate around the circulation of a hurricane; it describes how the barometer changes as a storm approaches and also describes typical storm tracks throughout the world. This book identifies the storm's right side as the "dangerous" semicircle (because the storm's translation speed is the same direction as the circulation, causing faster winds) and the left side as a "safe" semicircle (where the storm motion is opposite of the circulation). Mariners utilize these rules even today.

Piddington also performed research on ocean waves and the storm surge. He studied storm surges in India and warned against a port being built south of Calcutta, which was constructed anyway. Three years later, the port was destroyed by a hurricane.

Piddington went blind in 1855 and died in 1858.

William Redfield (1789–1857)

William Redfield made the first comprehensive analysis of the cyclonic rotation of hurricanes. Redfield was born on March 26, 1789, and after his seafarer father's death he apprenticed as a saddle and harness maker in 1803. He completed his apprenticeship in 1810, and after some traveling in Ohio he made saddles and ran a store in Connecticut for a decade. Stimulated by an encounter with Fulton's steamboat during the Ohio trip, in 1822 Redfield began a career as a marine engineer and transportation promoter with a steamboat on the Connecticut River. He moved to New York in 1824 after the steamboat operations expanded to the Hudson River. In 1829, Redfield served as the superintendent of the steamboat company, in which he promoted a towboat operation for barges. He was also involved in railroad promotion and planning.

Redfield was a self-taught scientist with an insightful and remarkable organizational ability. While traveling around Connecticut after a strong hurricane in September 1821, he noted that trees and corn in some areas were pointing toward the northwest, while those a few miles away were pointing to the southeast. He hypothesized that storms (and not just hurricanes) have a counterclockwise rotation, which he confirmed ten years later when two more storms hit the New York area. A chance meeting with Professor Denison Olmsted of Yale, who thereafter constantly urged Redfield to print his theories, led Redfield to publish an article in the *American Journal of Science* in 1831 that concluded: "This storm exhibited in the form of a great whirlwind." This paper brought Redfield instant recognition, and sea captains provided Redfield with ship logs from which he published updated theories on storm rotation for the next twenty-five years in British and American journals. These later theories, explaining the source of rotation and storm development, were sometimes incorrect or incomplete and were the source of many lively discussions among scientists. However, Redfield correctly noted the following observations about cyclones and hurricanes:

1. That winds do not rotate in horizontal circles but spiral inward toward the center.
2. That cyclones rotate counterclockwise in the Northern Hemisphere and clockwise in the Southern Hemisphere, with wind speeds increasing toward the center.
3. That cyclones move at a variable rate.
4. That hurricanes generally form deep in the tropics and travel west-northwest with the tropical trade wind flow, eventually turning back to the east as they propagate away from the tropics.
5. That hurricane size is large, frequently with a diameter greater than 1,000 miles.
6. That pressure decreases with increasing rapidity near the center of a hurricane.

Stimulated by his son's paper on fossil fish in 1836, Redfield also developed an interest in paleontology, publishing seven papers on the subject between 1838 and 1856 that established him as America's first specialist on fossil fish. He transformed the Association of American Geologists and Naturalists into the American Association for the Advancement of Science and was its first president in 1848. Yale University awarded Redfield an honorary degree in 1839; his name is commemorated by one genus and several species of fish and by Mount Redfield in the Adirondacks.

Redfield died on February 12, 1857.

William Reid (1791–1858)

William Reid was one of the first researchers on hurricanes and the author of a key book on the subject titled the *Law of Storms*. Born in Kinglassie, Reid was educated at the Royal Military

Academy at Woolwich, England. He entered the British army as an engineer in 1809, participated in many military engagements, and survived several conflict-induced injuries (including one diagnosed as fatal). His life is characterized by a lifetime of public service and scientific achievement.

After the wars, Reid was appointed major of engineers to restore government buildings after a hurricane hit Barbados in 1831. Upon seeing the destruction, Reid sought additional information about previous hurricanes in hopes of understanding them better. He collected detailed information about a number of hurricanes while in Barbados, corresponded with William Redfield (see biography in this chapter), and developed theories on the circulation and movement of hurricanes. He spent the next few years writing papers about hurricanes and three editions of his book *The Law of Storms*. This book contained rules for mariners on how to avoid the most dangerous quadrants of hurricanes.

During 1838, Reid was appointed governor of Bermuda, and in this position he aspired to improve the country's agriculture, education, and general welfare. In 1847 he established the first hurricane warning display system when the barometer indicated the approach of a hurricane. He returned to England in 1848 after the government refused to support the removal of a judge who oppressed the people. In 1849 he was appointed commanding engineer at Woolwich and was appointed governor of Malta in 1851 during the Russian war, in which he was promoted to general. In 1851 he also received the Order of Knighthood from the queen.

Reid died on October 31, 1858.

Lewis Fry Richardson (1881–1953)

Lewis Fry Richardson, a British mathematician and meteorologist, was born on October 11, 1881, in Newcastle. He was educated at Durham College and at Cambridge University, from which he graduated in 1903. In 1913 he became superintendent of the meteorological observatory at Eskdalemuir, Scotland. From 1920 to 1929 he was head of the physics department at Westminster Training College. He then became the principal of Paisley Technical College, Scotland, where he remained until his retirement in 1940.

In 1922 Richardson published the classic book *Weather Prediction by Numerical Procedures,* which described how meteorological equations can be approximated to forecast the weather and also described the first attempt at numerical forecasting. This effort was very laborious since there were no electronic calculators or computers then, requiring many tedious hand calculations and a large support staff. In fact, Richardson estimated that an operational numerical forecast for 2,000 locations would require 64,000 people just to carry out the calculations! Richardson attempted a six-hour pressure forecast for just one location, and unfortunately it produced a grossly inaccurate prediction of 145 mb. It is now known that this failure was due to an improper approximation method and incorrect formulation of the equations. This forecast failure, along with the incredible amounts of labor involved, discouraged further numerical modeling efforts for decades until computers were invented.

Nevertheless, Richardson's attempt at applying mathematics to weather forecasting was truly visionary. As computer speed increased, and as understanding of the atmosphere and its governing equations improved, computer forecasts improved dramatically, eventually becoming the most important tool for hurricane forecasts (and weather forecasts in general). Richardson was also the first to envision a roomful of people performing mathematical calculations to predict the weather. Today, the staff at the National Centers for Environmental Prediction performs such a task on supercomputers.

Richardson worked on other aspects of meteorology, including turbulence studies in which a fractional value called the Richardson number, using wind and temperature values, is named after him. Richardson also attempted to apply a mathematical framework to the causes of war, publishing several papers on the subject.

Richardson died on September 30, 1953, at Kilmun, Scotland.

Herbert Riehl (1915–1997)

Widely regarded as the father of tropical meteorology, Herbert Riehl was born in Munich, Germany, on March 30, 1915. He immigrated to the United States in 1933 and became a citizen in 1939. In 1942, he received master's degree in meteorology from New York University and taught in the Department of Meteorology at the University of Washington. When the U.S. Army Air Corps (Air Force) and the University of Chicago established the

Institute of Tropical Meteorology on the campus of the University of Puerto Rico in 1943 to provide weather research in equatorial regions, Riehl was a member of the team. He served as director of the Institute in 1945–1946. He earned a PhD from the University of Chicago in 1947 and remained on staff there until 1960, when he moved to Fort Collins, Colorado, and established the Department of Atmospheric Science at Colorado State University. He served as department head until 1968 and as a professor until 1972. Even after Riehl left, Colorado State University continued its reputation as a premier research institute on tropical meteorology.

In 1972, Riehl accepted the post of director of the Institute of Meteorology and Geophysics at the Free University of Berlin. In 1976, he moved to Boulder to join the staffs of the National Center for Atmospheric Research (NCAR) and Co-operative Institute for Research in Environmental Sciences (CIRES), a joint laboratory of the federal government and the University of Colorado, as senior scientist. He held those posts until 1989. After retirement, Dr. Riehl remained active in tropical meteorology research as a consultant in both North and South America. One of his passions was mountain climbing. He climbed many mountains in Colorado and the United States and major peaks throughout the world, including Kilamanjaro and Mont Blanc.

Riehl profoundly influenced tropical meteorology with landmark papers and books on hurricane genesis, motion, intensification, wind shear interaction, and energetics. During his career he received many honors, including the AMS Meisinger Award in 1948, the Losey Award of the American Institute for Aeronautics and Astronautics in 1962, and the Carl-Gustav Rossby Research Medal in 1979. In May 1997, Riehl's contributions to tropical meteorology were honored at the 22nd Conference on Hurricanes and Tropical Meteorology of the American Meteorological Society.

Riehl died on June 1, 1997.

Carl-Gustav Arvid Rossby (1898–1957)

Carl-Gustav Rossby was born on December 28, 1898, in Stockholm and educated at the University of Stockholm. In 1919 he joined the Geophysical Institute at Bergen, Norway, one of the world's main centers for meteorological research at that time, as an apprentice forecaster on Vilhelm Bjerknes' team. In 1922, he joined the Swedish Meteorological and Hydrological Institute (SMHI), where he established a pilot balloon network and served as bench forecaster with four other meteorologists, analyzing maps and issuing nationwide forecasts that were read over the newly established radio network. He also served two summers as forecaster on a sailing ship, gave lectures to the crew, and took pilot balloon observations.

In 1926, the Swedish-American Foundation and SMHI sponsored Rossby to study dynamics and meteorology in the United States. Because the Weather Bureau's observation network was superior to Europe's, Rossby decided to stay in the United States. Rossby originally worked in Washington, D.C., performing research, pioneering aviation forecasting techniques, and teaching the Bergeron methods of forecasting. Rossby then was employed by the Scandinavian-American Foundation to develop an aviation forecast system for a flight between San Francisco and Los Angeles; this system was later transitioned to the Weather Bureau. He then joined academia, first as a professor at the Massachusetts Institute of Technology, then to found a meteorology program at the University of Chicago. He returned to Sweden in 1948 and founded the Institute of Meteorology.

Rossby was one of the most eminent meteorologists of the twentieth century. Although he never actually performed any hurricane research, his influence was so encompassing in meteorology that his accomplishments must at least be mentioned. He had an aptitude for theoretical research but often focused on practical applications of his work. He made major contributions to our understanding of air masses and air movements. In the late 1930s Rossby simplified the equations that Richardson attempted to solve by noting that certain terms could be neglected for large-scale weather. From these simplified equations, he also derived an expression for large undulations in the upper atmosphere, now known as Rossby waves. Because Rossby's research laid the foundation for many future scientific advances in weather, he is known as the Father of Modern Meteorology.

Rossby wave motion is linked to the north-south variation of planetary vorticity and the upper atmosphere's westerly winds. *Vorticity* is the rate an imaginary paddle wheel placed in fluid would rotate due to curved flow or horizontally sheared flow. Because the earth's rotation varies with latitude, planetary vorticity varies poleward. Wave motion also depends on wavelength, with shorter waves propagating faster to the east than long waves. Very long waves will actually move backwards against the westerly flow, called retrograding. Because a hierarchy of wavelengths exists in any situation, atmospheric flow evolution can become complicated.

Other types of Rossby waves exist, such as in hurricanes. Due to their curved flow and rapid variation in wind speed away from the storm center, storm vorticity varies radially. The result is a special class of Rossby waves known as *vortex Rossby waves*. Recent research shows that hurricane spiral bands are explained by the properties of vortex Rossby waves.

Rossby died on August 19, 1957, in Stockholm, Sweden.

Joanne Simpson (1923–)

Joanne Simpson was the first woman meteorologist to earn a PhD and has made many contributions to cloud physics and hurricanes. The daughter of an editor of the *Boston Herald* who reported on aviation as a hobby, Simpson was fascinated by flying and earned her student pilot's license at age 16. Since flying is weather-dependent, Simpson subsequently developed an interest in meteorology. She earned a bachelor of science degree in meteorology in 1943 at the University of Chicago. She then taught meteorology to aviation cadets and military forecasters while she pursued her master of science degree, which she completed in 1945. Unable to obtain a fellowship for a PhD in an era when women were strongly discouraged from such aspirations, Simpson became a physics and meteorology instructor at the Illinois Institute of Technology. Because faculty could take classes for free, she took many of her graduate courses there after taking the minimum required at the University of Chicago. In 1947, Herbert Riehl lectured on aircraft observations of wind flow and cloud structure in the tropics. Fascinated by this new field of tropical meteorology, Simpson ended up completing her PhD work in 1949 at the University of Chicago with Riehl as her advisor.

Riehl and Simpson wrote several landmark papers about hurricane structure, hurricane energetics, the thermodynamic structure of the tropics, and key concepts regarding the role of the tropical general circulation. Simpson left Illinois Tech in 1951 to become a research meteorologist at the Woods Hole Oceanographic Institute to learn more about weather over the ocean.

During this period, she constructed some of the first mathematical models of clouds based on buoyancy-driven motion (a novel idea at the time) and flew into clouds to validate her computations. In 1954 she won a Guggenheim fellowship to work in England, and in 1955 she was an honorary lecturer at the Imperial College in London.

In 1960, Simpson served as a professor at UCLA. In 1962, she wrote a chapter in volume 1 of the landmark book *The Sea*. This book presented the most comprehensive treatment to date of the coupling between the ocean and the atmosphere. Simpson's chapter dealt with the complexity of how large- and small-scale atmospheric features interact, and in particular how difficult it would be to explicitly compute the individual contributions of small-scale clouds and turbulence on large-scale weather patterns. This was a factor leading to the concept of parameterization, in which the net effect of nonmeasurable small-scale weather features is computed in terms of large-scale (measurable) weather variables. Simpson also showed that much energy in the *InterTropical Convergence Zone* is transferred through narrow clouds with strong updrafts, the so-called hot tower concept.

After five major hurricanes made landfall on the eastern United States in 1954 and 1955, Congress established the National Hurricane Research Project and named Simpson as an advisor. There she met the first director of the project, Robert Simpson (see next section), whom she married in 1965. From 1965 to 1974, she was director of NOAA's Experimental Meteorology Laboratory in Coral Gables, Florida, and participated in attempts to modify clouds and hurricanes using cloud-seeding techniques (see chapter 2 about Project STORMFURY). In 1974, she again accepted a professorship, this time at the University of Virginia. In 1979 she went to the NASA Goddard Space Flight Center as head of the Severe Storm Branch; she has remained at Goddard ever since. She served as the project scientist for the Tropical Rainfall Measuring Mission from 1986 until launch in 1997 and is now chief scientist for meteorology. Simpson has published 150 reviewed papers, been on many scientific committees, and been an editor on several journals. She has won numerous awards for her achievements, including the *Los Angeles Times* Woman of the Year in 1963; the American Meteorological Society's Meisinger Award; and the American Meteorological Society's highest honor, the Carl-Gustav Rossby Research Medal. She is a former president of the American Meteorological Society.

Robert Simpson (1912–)

Robert Simpson's experience with hurricanes began when he was a 6-year-old boy in Corpus Christi, Texas, when a category 4 hurricane hit the region in 1919 (see chapter 4). Simpson's family, with Robert hanging on his father's back, swam through the rapidly rising water three blocks to the only high building downtown, the courthouse, for shelter, where they then watched the hurricane decimate Corpus Christi. However, initially Simpson's career was not in science, but music and architecture. He served as an apprentice architect until the Great Depression, when he lost his job. He then attended Southwestern University at Georgetown as first chair in its orchestra but ended up earning a degree in physics and mathematics. Afterward, he embarked on graduate studies at Emory University and earned a masters in physics, but he was unable to obtain employment and instead became a music director in the Corpus Christi school system. However, he wanted to apply his science education, so he took Civil Service exams to qualify for government work and was offered a job with the Weather Bureau in Brownsville. To advance in the Weather Bureau, he became a weather observer in Swan Island and eventually earned employment at the New Orleans office. To advance further, Simpson took classes at an aviation cadet program designed to create a pool of 10,000 forecasters in the early years of World War II, then was sent to the Miami hurricane office to be a forecaster under Grady Norton. In 1945, Simpson taught classes to the Air Force School of Tropical Meteorology in Panama.

Simpson's experiences in Panama and Miami sparked his interest in tropical meteorology and hurricanes. He started publishing papers, which led to a research assignment at the Weather Bureau in Washington, D.C., in 1947. During 1945–1947, Simpson also participated in the first reconnaissance flights into hurricanes. Based on a 1946 flight, Simpson published a paper documenting the vertical structure of the hurricane eye. He also published several observational and theoretical papers on the eye and eyewall thermodynamic properties. Simpson took part in hundreds of flights during his career.

To further his career advancement, Simpson accepted a management position for the Weather Bureau offices in Hawaii. He established the Mauna Loa summit laboratory to take ozone measurements (used as a tracer) to study upper-level cyclones. A few years later, through a chance encounter with a government astronomer while on vacation, Simpson was able to expand this facility into the now-famous Mauna Loa Observatory by agreeing to also measure carbon dioxide. These carbon dioxide measurements today provide the most conclusive evidence on the increase of this greenhouse gas. Simpson returned to Washington, D.C., in 1952, when several tornado incidents inspired him to establish the first national radar network.

After the East Coast was hit by a series of hurricanes (Carol, Edna, and Hazel), Simpson convinced Congress to fund the National Hurricane Research Program (now the Hurricane Research Division), for which he became its first director in 1955. In 1959, Simpson went to the University of Chicago to finish his PhD, which had been interrupted by World War II. His advisor, Herbert Riehl (see biography in this chapter), obtained permission from the U.S. Navy to accompany a flight through Hurricane Donna in 1960. Riehl noticed during this flight that the jet experienced icing, and the idea that perhaps a hurricane's structure and development could be modified with artificial ice nuclei was born. This ultimately culminated in the hurricane modification program known as Project STORMFURY (see chapter 2), in which Simpson was the chief proponent and first director.

In 1960, Simpson was assigned as deputy director in the Washington, D.C., office. The accuracy of plane data was questionable and required more research. Therefore, Simpson approved the establishment of the National Severe Storms Laboratory in Norman, Oklahoma, for radar development and aircraft programs. Today, this laboratory is a leading research organization on severe weather.

In 1968, Simpson became director of the National Hurricane Center (NHC), where he remained until 1973. Simpson established a small research and development unit at the NHC that developed many forecast applications. Simpson created the hurricane specialist position, which has forecast responsibilities during the hurricane season and research and public service in the remainder of the year. As NHC director, Simpson collaborated with a consulting engineer (Herbert Saffir) to devise a scale to quantify hurricane damage based on five intensity categories, known as the Saffir-Simpson Hurricane Scale. In 1974, Simpson founded Simpson Weather Associates, where he continued contract work on matters related to hurricanes. In 1981, he coauthored with Herbert Riehl the important book *The Hurricane and Its Impact* (see chapter 6).

Simpson's career ambitions unfortunately led to his first marriage ending in 1948. However, he later married researcher Joanne Malkus (now Joanne Simpson). His marriage with Joanne Simpson "has meant more to me both professionally and personally than any other factor in my life," says Simpson.

Chris Velden (1956–)

Chris Velden has been a leader in the development of satellite applications to meteorology, with the primary focus on hurricane forecasting. Velden received his bachelor's of science in 1979 from the University of Wisconsin-Stevens Point in natural science with a minor in physics. This was followed by a master's of science in meteorology at the University of Wisconsin-Madison in 1982 dealing with satellite images of hurricanes in the microwave spectrum. For much of his career afterward he has remained at the University of Wisconsin to continue pursuing satellite research, except for a visiting scientist position at the Australian Bureau of Meteorology in 1987–1988. Currently Velden is a physical science senior researcher at the Cooperative Institute for Meteorological Satellite Studies in the Space Science and Engineering Center located at the University of Wisconsin-Madison.

Velden and his colleagues have developed and automated cloud-drift wind data derived from satellites, which is invaluable in data-sparse regions, particularly in tropical oceans. Cloud-drift winds are computed by tracking individual cloud elements and deducing their motion. These winds are used to identify steering currents and to visualize atmospheric features that affect tropical cyclone genesis and intensification. This data is being assimilated in computer models, resulting in improved weather forecasts and hurricane track predictions. For this effort, Velden received an award from the AMS in 1997. Velden and his team have also developed an objective satellite-based scheme that estimates a tropical cyclone's intensity. This technique is now operational at the National Hurricane Center and is supplementing the current subjective satellite scheme known as the Dvorak technique. For this work, Velden and his colleagues won the AMS Banner Miller Award in 2000. His team at CIMSS is now focussing on a microwave-based method to augment this objective method, with promising results. Velden has participated in numerous meteorological field experiments and has published more than sixty journal articles, approximately twenty as lead author.

Benito Viñes (1837–1893)

In his day, Father Benito Viñes was the greatest authority on hurricanes. Also known as the "Hurricane priest," Viñes contributed to both theory and forecasting techniques and laid the foundation for today's observational and hurricane warning network. Viñes was born on September 19, 1837, in Poboleda, Spain. A fugitive from revolutionary Spain with training in physics, Viñes came to Cuba from France in 1870 to serve as director of the Magnetic and Meteorological Observatory of the Royal College of Belen in Havana, which had been established by Jesuit priests in 1857. After arriving, Viñes investigated ways to help people prepare for hurricanes by studying twelve years of records at the observatory, old newspaper accounts, and eyewitness accounts. His documentation was detailed and exhaustive, with hourly weather observations from 4 a.m. to 10 p.m. every day of pressure, temperature, moisture, and cloud height. He also measured wind direction and speed at the surface and inferred wind aloft from cloud motion. After hurricane disasters, Viñes rummaged through debris and questioned survivors, meticulously documenting everything.

Based on this work, Viñes was the first to develop a methodology for locating a hurricane's center based on cloud motion aloft and sea swells. Implicit in this work was a remarkably accurate deduction of a hurricane's three-dimensional wind flow. Viñes was also the first to forecast hurricane movement based on the motion of high clouds that flow out from the storm center. He also developed a methodology for predicting when a hurricane will recurve to the north and for predicting average storm motion based on the latitude and time of year. This effort was augmented through the development of an observational network consisting of hundreds of observers along the Cuban coastline so that routine hurricane warnings could be issued. Routine forecasts began in 1875, with a pony express system distributing observations and forecasts between villages.

Viñes issued his first printed forecast on September 11, 1875, stating accurately that an intense hurricane would strike southern Cuba two days later. Many lives were saved, and a legend was born. Later that year he issued another successful forecast. On September 14, 1876, Viñes warned of a dangerous hurricane offshore, and the captain of the ship *Liberty* ignored him, only to sail directly into the path of the storm and sink. Viñes's success garnered funding and complimentary services. In 1877 he expanded his operation with more observation posts, supported by the Havana Chamber of Commerce and several private companies. Telegraph companies didn't charge him, and railroads and steamships provided free passage. The railroads even provided a special express if immediate transportation was needed.

In September 1877 Viñes accurately predicted that a hurricane would pass south of Puerto Rico but hit Santiago de Cuba. The same year Viñes published his most famous work, *Practical Hints in Regard to West Indian Hurricanes*, which quickly became standard learning material for mariners. In 1888, Viñes developed the Antilles cyclonoscope, similar to Henry Piddington's horn cards, designed to detect the center of a hurricane far away for mariners.

The revered Father Viñes died on July 23, 1893, in Havana, Cuba, three days after he mailed to the U.S. Weather Bureau another exceptional technical report about the three-dimensional wind flow of hurricanes, titled "Cyclonic Circulation and the Translatory Movement of West Indian Hurricanes."

Hugh E. Willoughby (1945–)

Hugh Willoughby's impact on hurricane research consists of an interesting combination of high-level theory and observational analysis. Willoughby obtained a bachelor's of science in geophysics geochemistry from the University of Arizona in 1967, followed by a master's of science in 1969 from the Naval Postgraduate School. As a commissioned navy officer, he served as a flight meteorologist in the Airborne Early Warning Squadron One from 1970 to 1971. From 1971 to 1974, Willoughby was a faculty member of the Naval Academy, where he taught meteorology, oceanography, geology, and computer science to midshipmen. Willoughby left active duty as a lieutenant to pursue a PhD in atmospheric science at the University of Miami, which he earned in 1977.

While completing the PhD, Willoughby began working at NOAA's Hurricane Research Division (HRD) in 1975. He used the data collected on the HRD reconnaissance flights to study hurricane structure and to formulate theoretical ideas. He first observed that hurricanes go through a natural (but temporary) weakening process in which a new eyewall forms outside the original eyewall. The outer eyewall chokes off inflow to the inner eyewall, causing it to dissipate. The outer eyewall then propagates inward, replacing the original eyewall. This process, known as the *concentric eyewall cycle,* lasts twelve to thirty-six hours and is associated with temporary weakening of the hurricane. The ramifications of this result was that hurricanes experience internal processes that influence intensification and that any perceived man-made changes to hurricanes in Project STORMFURY may have been the result of natural internal evolution instead.

Willoughby has made more than 400 research and reconnaissance flights into the eyes of typhoons and hurricanes in his career. He has held the position of G. J. Haltiner visiting research chair at the Naval Postgraduate School (January–July 1991); was a visiting research scientist at the Bureau of Meteorology Research Centre in Melbourne, Australia (June–July 1988); and was a visiting lecturer at the Shanghai Typhoon Institute (December 1985). He has published dozens of articles on hurricanes and is a fellow of the American Meteorological Society. He also has served as the director of the Hurricane Research Division. He currently is a faculty member at Florida International University in Miami, Florida.

6

Data, Opinions, and Letters

This chapter provides data, opinions, and letters relevant to hurricanes. Data is presented in a tabular format concerning record events and annual figures since 1900. The opinion articles present topics including global warming and an attempt to control hurricanes in Project STORMFURY. Letters from survivors of hurricane landfalls in the 1700s and 1800s are included at the end of the chapter. Updates of these tables will be available on www.drfitz.net.

Data

TABLE 6.1
The thirty deadliest land-falling tropical storms and hurricanes for the U.S. mainland, 1900–2004

Ranking	Hurricane	Year	Category	Deaths
1	Texas (Galveston)	1900	4	8,000*
2	Florida (Lake Okeechobee)	1928	4	2,500*
3	Florida (Keys)/South Texas	1919	4	600†
4	New England	1938	3‡	600
5	Florida (Keys)	1935	5	408
6	Audrey (Southwest Louisiana/North Texas)	1957	4	390
7	NE United States	1944	3‡	390§
8	Louisiana (Grand Isle)	1909	4	350
9	Louisiana (New Orleans)	1915	4	275
10	Texas (Galveston)	1915	4	275
11	Camille (Mississippi/Louisiana)	1969	5	256
12	Florida (Miami/Pensacola), Mississippi, Alabama	1926	4	243
13	Diane (NE United States)	1955	1	184
14	Southeast Florida	1906	2	164
15	Mississippi/Alabama/Florida (Pensacola)	1906	3	134
16	Agnes (NE United States)	1972	1	122
17	Hazel (South Carolina/North Carolina)	1954	4‡	95
18	Betsy (Southeast Louisiana/Southeast Florida)	1965	3	75
19	Carol (NE United States)	1954	3‡	60
20	Floyd (Mid-Atlantic and Northeastern U.S.)	1999	2	57
21	Southeast Florida/Louisiana/Mississippi	1947	4	51
22	Donna (Florida/Eastern U.S.)	1960	4	50
23	Georgia/South Carolina/North Carolina	1940	2	50
24	Carla (Texas)	1961	4	46
25	Allison Texas/ Louisiana/Mississippi/Alabama	2001	TS**	41
26	Texas (Velasco)	1909	3	41
27	Texas (Freeport)	1932	4	40
28	South Texas	1933	3	40
29	Hilda (Louisiana)	1964	3	38
30	Southwest Louisiana	1918	3	34
Addendum: Pre-1900, California, and Caribbean Islands				
	"Chenier Caminada" (Louisiana)	1893	4	2,000
	"Sea Islands" (South Carolina/Georgia)	1893	3	1,000–2,000
	"Brunswick" (Georgia/South Carolina)	1881	2	700
	San Felipe (Puerto Rico)	1928	4	312
	U.S. Virgin Islands, Puerto Rico	1932	2	225
	Donna (St. Thomas, Virgin Islands)	1960	4	107
	Southern California	1939	TS††	45
	Eloise (Puerto Rico)	1975	TS††	44

(continues)

TABLE 6.1
(Continued)

Ranking	Hurricane	Year	Category	Deaths
Addendum: Most Deadly Hurricanes in the Western Hemisphere				
	The Great Hurricane (Martinique, St. Eustatius, Barbados) plus sinking of ships	1780	Unknown[§§]	22,000
	Mitch (Central America, including Honduras and Nicaragua)	1998	1	11,000 to 18,000[‡‡]
	Texas (Galveston)	1900	4	8,000[*]
	Fifi (Honduras)	1974	2	8,000
	Dominican Republic	1930	4	8,000
	Flora (Haiti, Cuba)	1963	4 in Haiti, 3 in Cuba	7,200
	Martinique (Point Petre Bay)	1776	Unknown	>6,000

[*]May actually been as high as 10,000 to 12,000.
[†]More than 500 of these lost on ships at sea; 600–900 estimated deaths.
[‡]Moving more than 30 miles an hour.
[§]Some 344 of these lost on ships at sea.
[**]TS = tropical storm.
[††]Only of tropical storm intensity.
[‡‡]Exact death toll count unknown at press time.
[§§]Unknown = intensity not sufficiently known to establish category.
Source: Hebert, Jarrell, and Mayfield (1997); updated based on Landsea (2005).

TABLE 6.2
The thirty-six costliest land-falling hurricanes for the U.S. mainland for 1900–2004

Ranking	Hurricane	Year	Category	Damage (U.S.)
1	Andre (SE Florida/SE Louisiana)	1992	5*	$26,500,000,000
2	Charley (SE Florida)	2004	4	$15,000,000,000
3	Ivan (Alabama, NW Florida)	2004	3	$14,200,000,000
4	Frances (SE Florida)	2004	2	$8,900,000,000
5	Hugo (South Carolina)	1989	4	$7,000,000,000
6	Jeanne (SE Florida)	2004	3	$6,900,000,000
7	Allison (Texas, Louisiana, Mississippi, Alabama)	2001	TS†	$5,408,000,000
8	Floyd (Mid-Atlantic and Northeastern U.S.)	1999	2	$4,500,000,000
9	Isabel (North Carolina, Virginia, New York, New Jersey, Maryland)	2003	2	$3,370,000,000
10	Fran (North Carolina)	1996	3	$3,200,000,000
11	Opal (NW Florida, Alabama)	1995	3	$3,000,000,000
12	Georges (Florida Keys, Mississippi, Alabama)	1998	2	$2,310,000,000
13	Frederic (Alabama, Mississippi)	1979	3	$2,300,000,000
14	Agnes (Northeastern U.S.)	1972	1	$2,100,000,000
15	Alicia (N Texas)	1983	3	$2,000,000,000
16	Bob (North Carolina, Northeast U.S.)	1991	2	$1,500,000,000
17	Juan (Louisiana)	1985	1	$1,500,000,000
18	Camille (Mississippi, Alabama)	1969	5	$1,420,700,000
19	Betsy (Florida, Louisiana)	1965	3	$1,420,500,000
20	Elena (Mississippi, Alabama, NW Florida)	1985	3	$1,250,000,000
21	Gloria (Eastern U.S.)	1985	3‡	$900,000,000
22	Diane (Northeastern U.S.)	1955	1	$831,700,000
23	Bonnie (North Carolina, Virginia)	1998	2	$720,000,000
24	Erin (Central and NW Florida, SW Alabama)	1995	2	$700,000,000
25	Allison (N Texas)	1989	TS§	$500,000,000
26	Alberto (NW Florida, Georgia, Alabama)	1994	TS§	$500,000,000
27	Frances (Texas)	1998	TS§	$500,000,000
28	Eloise (NW Florida)	1975	3	$490,000,000
29	Carol (Northeastern U.S.)	1954	3§	$461,000,000
30	Celia (S Texas)	1970	3	$453,000,000
31	Carla (Texas)	1961	4	$408,000,000
32	Claudette (N Texas)	1979	TS§	$400,000,000
33	Gordon (S and Central Florida, North Carolina)	1994	TS§	$400,000,000
34	Donna (Florida, Eastern U.S.)	1960	4	$387,000,000
35	David (Florida, Eastern US)	1979	2	$320,000,000
36	New England	1938	3§	$306,000,000
Addendum: Non-Atlantic or non-Gulf Coast Systems				
4	Georges (U.S. Virgin Islands, Puerto Rico)	1998	3	$3,600,000,000
10	Iniki (Kauai, Hawaii)	1992	Unknown**	$1,800,000,000

(continues)

TABLE 6.2
(Continued)

Ranking	Hurricane	Year	Category	Damage (U.S.)
09	Marilyn (U.S. Virgin Islands/Eastern Puerto Rico)	1995	2	$1,500,000,000
09	Hugo (U.S. Virgin Islands, Puerto Rico)	1989	4	1,000,000,000
09	Hortense (Puerto Rico)	1996	4	$500,000,000
09	Olivia (California)	1982	TD[††]	$325,000,000
09	Iwa (Kauai, Hawaii)	1982	Unknown	$312,000,000

Costs are not adjusted for inflation, personal property increases, and population changes. Note that 2004 hurricanes are only estimates at time of publication and could change.

*Reclassified as category 5 in 2002.
[†]TS = tropical storm.
[‡]Moving more than 30 miles an hour.
[§]Only of tropical storm intensity but included because of high damage.
[**]Unknown = intensity not sufficiently known to establish category.
[††]TD — Only a tropical depression.
Source: Hebert, Jarrell, and Mayfield (1997); updated by author.

TABLE 6.3
Costliest land-falling hurricanes of 1900–2004, adjusted to 2003 dollars by inflation, personal property increases, and coastal county population changes

Rank	Hurricane	Year	Category	Damage (U.S.)
1	SE Florida, Alabama	1926	4	$98,051,000,000
2	Andrew (SE Florida, Louisiana)	1992	5	$44,878,000,000
3	N Texas (Galveston)*	1900	4	$36,096,000,000
4	N Texas (Galveston)*	1915	4	$30,585,000,000
5	SW Florida	1944	3	$22,070,000,000
6	"New England" (New York, Rhode Island)	1938	3	$22,549,000,000
7	SE Florida, Lake Okeechobee	1928	4	$18,708,000,000
8	Betsy (SE Florida/Louisiana)	1965	3	$16,863,000,000
9	Donna (Florida, Eastern U.S.)	1960	4	$16,339,000,000
10	Charley (SE Florida)	2004	4	$15,000,000,000
11	Camille (Mississippi, Louisiana, Virginia)	1969	5	$14,870,000,000
12	Agnes (NW Florida, Northeastern U.S.)	1972	1	$14,515,000,000
13	Ivan (Alabama, NW Florida)	2004	3	$14,200,000,000
14	Diane (Northeastern U.S.)	1955	1	$13,875,000,000
15	Hugo (South Carolina)	1989	4	$12,718,000,000
16	Carol (Northeastern U.S.)	1954	3	$12,291,000,000
17	SE Florida, Louisiana, Alabama	1947	4	$11,266,000,000
18	Carla (N and Central Texas)	1961	4	$9,587,000,000
19	Hazel (South Carolina, North Carolina)	1954	4	$9,545,000,000
20	Frances (SE Florida)	2004	2	$8,900,000,000
21	Northeastern U.S.	1944	3	$8,763,000,000
22	SE Florida	1945	3	$8,561,000,000
23	Frederic (Alabama, Mississippi)	1979	3	$8,534,000,000
24	SE Florida	1949	3	$7,918,000,000
25	S Texas*	1919	4	$7,253,000,000
26	Jeanne (SE Florida)	2004	3	$6,900,000,000
27	Alicia (N Texas)	1983	3	$5,501,000,000
28	Allison (Texas/Louisiana/Mississippi/Alabama)	2001	TS†	$5,408,000,000
29	Floyd (Mid-Atlantic and Northeastern U.S.)	1999	2	$5,264,000,000
30	Celia (S Texas)	1970	3	$4,526,000,000
31	Dora (NE Florida)	1964	2	$4,215,000,000
32	Fran (North Carolina)	1996	3	$4,201,000,000
33	Opal (NW Florida, Alabama)	1995	3	$4,068,000,000
34	Isabel (North Carolina, Virginia, New York, New Jersey, Maryland)	2003	2	$3,370,000,000

Costs for 2004 hurricanes are estimates only at time of publication and could change.
*Hurricanes included from years 1900–1924 using simplifying assumptions to extend the normalization methodology to 1900.
†TS = tropical storm.
Source: Pielke and Landsea (1998); updated from Landsea (2005) and by author.

TABLE 6.4
Distribution of top nine (of thirty) insured property catastrophe losses from 1970 to 2002

Hazard	Total insured loss (Billions of 2002 U.S. dollars)	Fraction of top 30 largest losses
U.S. hurricane	44	0.31
U.S. earthquake	17	0.12
U.S. tornado	7	0.05
European wind	21	0.15
European flood	7	0.05
Japanese typhoon	12	0.08
Japanese earthquake	3	0.02
Man-made	24	0.17
Caribbean hurricane	5	0.04

Note that $19 billion of the $24 billion in insured losses from man-made catastrophes were caused by the September 11, 2001, attack on the World Trade Center. However, total insured losses may exceed $40 billion when liability risks are included. Of U.S. hurricane losses, $20 billion was due to Hurricane Andrew (1992). Note that tropical cyclones (typhoon and hurricanes) account for 43 percent of the total losses worldwide.
Source: Murnane (2004).

TABLE 6.5
Median and mean losses by the three phases of the El Niño cycle: La Niña, El Niño, and neither (neutral)

	Median ($ millions)	Mean ($ millions)
La Niña	3,292	5,887
Neutral	927	6,979
El Niño	152	2,056

This table shows that hurricane damage is strongly modulated by the phase of El Niño, with increased losses during La Niña and reduced losses during El Niño.
Source: Pielke and Landsea (1999).

TABLE 6.6
Major U.S. hurricanes at landfall during 1900–2004

Ranking	Hurricane	Year	Category (at Landfall)	Pressure (Millibars)
1	Florida (Keys)	1935	5	892
2	Camille (Mississippi, SE Louisiana, Virginia)	1969	5	909
3	Andrew (SE Florida, SE Louisiana)	1992	5*	922
4	Florida (Keys), S Texas	1919	4	927
5	Florida (Lake Okeechobee)	1928	4	929
6	Donna (Florida, Eastern U.S.)	1960	4	930
7	Texas (Galveston)	1900	4	931
7	Louisiana (Grand Isle)	1909	4	931
7	Louisiana (New Orleans)	1915	4	931
7	Carla (N and Central Texas)	1961	4	931
11	Hugo (South Carolina)	1989	4	934
12	Florida (Miami, Pensacola), Mississippi, Alabama	1926	4	935
13	Hazel (South Carolina, North Carolina)	1954	4†	938
14	SE Florida, SE Louisiana, Mississippi	1947	4	940
15	N Texas	1932	4	941
15	Charley (SW Florida)	2004	4	941
17	Gloria (Eastern U.S.)	1985	3††	942
17	Opal (NW Florida, Alabama)	1995	3†	942
19	Audrey (SW Louisiana, N Texas)	1957	4§	945
19	Texas (Galveston)	1915	4§	945
19	Celia (S Texas)	1970	3	945
19	Allen (S Texas)	1980	3	945
23	New England	1938	3†	946
23	Frederic (Alabama, Mississippi)	1979	3	946
23	Ivan (Alabama, NW Florida)	2004	3	946
26	"Great Atlantic" (Northeastern U.S.)	1944	3†	947
26	South Carolina, North Carolina	1906	3	947
28	Betsy (SE Florida, SE Louisiana)	1965	3	948
28	SE Florida, NW Florida	1929	3	948
28	SE Florida	1933	3	948
28	S Texas	1916	3	948
28	Mississippi, Alabama	1916	3	948
33	Diane (North Carolina)	1955	3**	949
33	S Texas	1933	3	949
35	Beulah (S Texas)	1967	3	950
35	Hilda (Central Louisiana)	1964	3	950
35	Gracie (South Carolina)	1959	3	950
35	Central Texas	1942	3	950
35	Jeanne (SE Florida)	2004	3	950
40	SE Florida	1945	3	951
40	Bret (S Texas)	1999	3	951
42	Tampa Bay, Florida	1921	3	952
42	Carmen (Central Louisiana)	1974	3	952
44	Edna (New England)	1954	3†	954

(continues)

TABLE 6.6
(Continued)

Ranking	Hurricane	Year	Category (at Landfall)	Pressure (Millibars)
44	SE Florida	1949	3	954
44	Fran (North Carolina)	1996	3	954
47	Eloise (NW Florida)	1975	3	955
47	King (SE Florida)	1950	3	955
47	Central Louisiana	1926	3	955
47	SW Louisiana	1918	3	955
47	SW Florida	1910	3	955
52	North Carolina	1933	3	957
52	Florida Keys	1909	3	957
54	Easy (NW Florida)	1950	3	958
54	N Texas	1941	3	958
54	NW Florida	1917	3	958
54	N Texas	1909	3	958
54	Mississippi, Alabama	1906	3	958
59	Elena (Mississippi, Alabama, NW Florida)	1985	3	959
60	Carol (Northeastern U.S.)	1954	3†	960
60	Ione (North Carolina)	1955	3	960
60	Emily (North Carolina)	1993	3	960
63	Alicia (N Texas)	1983	3	962
63	Connie (North Carolina, Virginia)	1955	3	962
63	SW Florida, NE Florida	1944	3	962
63	Central Louisiana	1934	3	962
67	SW and NE Florida	1948	3	963
68	NW Florida	1936	3	964

Hurricanes may have had greater intensities at other times before landfall.
*Reclassified as category 5 in 2002.
†Moving more than 30 miles an hour.
‡Highest category justified by winds.
§Classified category 4 because of estimated winds.
**Cape Fear, North Carolina, area only; was a category 2 at final landfall.
Source: Hebert, Jarrell, and Mayfield (1997); updated statistics from http://www.nhc.noaa.gov.

TABLE 6.7
Estimated annual deaths and damages (1900–1995)

Year	Deaths	Unadjusted Damage ($ millions)	Adjusted Damage ($ millions)	Year	Deaths	Unadjusted Damage ($ millions)	Adjusted Damage ($ millions)
1900	8,000	$30	$790	1948	3	$18	$113
1901	10	$1	$26	1949	4	$59	$370
1902	0	Minor	Minor	1950	19	$36	$222
1903	15	$1	$26	1951	0	$2	$11
1904	5	$2	$53	1952	3	$3	$16
1905	0	Minor	Minor	1953	2	$6	$33
1906	298	$3	$79	1954	193	$756	$4,180
1907	0	$0	$0	1955	218	$985	$5,348
1908	0	$0	$0	1956	19	$27	$139
1909	406	$8	$211	1957	400	$152	$758
1910	30	$1	$26	1958	2	$11	$55
1911	17	$1	$26	1959	24	$23	$116
1912	1	Minor	Minor	1960	65	$396	$2,006
1913	5	$3	$79	1961	46	$414	$2,101
1914	0	$0	$0	1962	3	$2	$10
1915	550	$63	$1,660	1963	10	$12	$59
1916	107	$33	$723	1964	49	$515	$2,564
1917	5	Minor	Minor	1965	75	$1,445	$6,996
1918	34	$5	$71	1966	54	$15	$70
1919	287	$22	$278	1967	18	$200	$900
1920	2	$3	$30	1968	9	$10	$43
1921	6	$3	$38	1969	256	$1,421	$5,647
1922	0	$0	$0	1970	11	$454	$1,699
1923	0	Minor	Minor	1971	8	$213	$747
1924	2	Minor	Minor	1972	122	$2,100	$6,924
1925	6	Minor	Minor	1973	5	$3	$9
1926	269	$112	$1,415	1974	1	$150	$396
1927	0	$0	$0	1975	21	$490	$1,191
1928	1,836	$25	$315	1976	9	$100	$233
1929	3	$1	$12	1977	0	$10	$22
1930	0	Minor	Minor	1978	36	$20	$38
1931	0	$0	$0	1979	22	$3,045	$5,210
1932	0	$0	$0	1980	2	$300	$463
1933	63	$47	$701	1981	0	$25	$36
1934	17	$5	$68	1982	0	Minor	Minor
1935	414	$12	$163	1983	22	$2,000	$2,751
1936	9	$2	$28	1984	4	$66	$88
1937	0	Minor	Minor	1985	30	$4,000	$5,197
1938	600	$306	$3,864	1986	9	$17	$21
1939	3	Minor	Minor	1987	0	$8	$10
1940	51	$5	$66	1988	6	$9	$11
1941	10	$8	$98	1989	56	$7,670	$8,640
1942	8	$27	$286	1990	16	$57	$63

(continues)

TABLE 6.7
(Continued)

Year	Deaths	Unadjusted Damage ($ millions)	Adjusted Damage ($ millions)	Year	Deaths	Unadjusted Damage ($ millions)	Adjusted Damage ($ millions)
1943	16	$17	$169	1991	16	$1,500	$1,637
1944	64	$165	$1,641	1992	24	$26,500	$28,687
1945	7	$80	$773	1993	4	$57	$59
1946	0	$5	$41	1994	38	$973	$973
1947	53	$136	$935	1995	29	$3,723	$3,582

Both unadjusted and inflation-adjusted values are shown in the mainland United States from land-falling hurricanes and tropical storms during 1900–1995. The adjusted values are based on 1994 dollars using the U.S. Department of Commerce Implicit Price Deflator for Construction. "Minor" applies to years with less than $1 million in unadjusted damage.
Source: Hebert, Jarrell, and Mayfield (1997).

TABLE 6.8
Progress of the average Atlantic hurricane season (1944–1996)

Number	Tropical storms	Hurricanes	Category 3 or Greater
1	July 11	August 14	September 4
2	August 8	August 30	September 28
3	August 21	September 10	—
4	August 30	September 24	—
5	September 7	October 15	—
6	September 14	—	—
7	September 23	—	—
8	October 5	—	—
9	October 21	—	—

Average chronological date upon which each tropical storm, hurricane, and major hurricane forms. This will exhibit considerable variability from year to year. However, in general most tropical storms and hurricanes occur from late August to early October. Only occasionally will a major hurricane (category 3 or better) occur in June or July.
Source: National Hurricane Center, http://www.nhc.noaa.gov.

TABLE 6.9
Number of hurricanes by category to strike the United States during 1900–2004

Category*	Number of Hurricanes
5	3
4	15
3	50
2	41
1	64
Total	173

This shows that on average two major hurricanes make landfall in the United States every three years and that when all categories are combined, an average of five hurricanes strike the United States every three years.
*Major hurricanes (categories 3, 4, and 5) total 68.
Source: Hebert, Jarrell, and Mayfield (1997); updated from http://www.nhc.noaa.gov.

TABLE 6.10
List of category 5 Atlantic hurricanes since 1886

Number	Storm Name	Maximum Winds	Date Attained (UTC)	Landfall as Category 5
1	Not named	160 mph	September 13, 1928	Puerto Rico
2	Not named	160 mph	September 5, 1932	Bahamas
3	Not named	160 mph	September 3, 1935	Florida Keys
4	Not named	160 mph	September 19, 1938	—
5	Not named	160 mph	September 16, 1947	Bahamas
6	Dog	185 mph	September 6, 1950	—
7	Easy	160 mph	September 7, 1951	—
8	Janet	175 mph	September 28, 1955	Mexico
9	Cleo	160 mph	August 16, 1958	—
10	Donna	160 mph	September 4, 1960	—
11	Ethel	160 mph	September 15, 1960	—
12	Carla	175 mph	September 11, 1961	—
13	Hattie	160 mph	October 30, 1961	—
14	Beulah	160 mph	September 20, 1967	—
15	Camille	190 mph	August 17, 1969	Mississippi
16	Edith	160 mph	September 9, 1971	Nicaragua
17	Anita	175 mph	September 2, 1977	—
18	David	175 mph	August 30, 1979	Dominica
19	Allen	190 mph	August 4–5, 7, 8–9, 1980	—
20	Gilbert	185 mph	September 14, 1988	Mexico
21	Hugo	160 mph	September 15, 1989	—
22	Andrew	165 mph	August 24, 1992	South Florida
23	Mitch	155–180 mph	October 26–28, 1998	—
24	Isabel	155–165 mph	September 11–14, 2003	—
25	Ivan	165 mph	September 8, 11, and 12, 2004	—

To qualify as a category 5 hurricane on the Saffir-Simpson scale, maximum sustained winds must exceed 155 mph. Through 2004, only twenty-five Atlantic storms have reached this intensity, and only ten were of category 5 strength at landfall. Of these twenty-five, only three made U.S. landfall: the 1935 Florida Keys hurricane; Hurricane Camille, which hit the Mississippi coast in 1969; and Hurricane Andrew, which devastated south Florida. This table lists all known category 5 Atlantic hurricane since records began in 1886.

Interesting Category 5 Facts

- Hurricane Allen reached category 5 intensity three times along its path through the southern Caribbean and Gulf of Mexico: twice these periods were of twenty-four-hour duration, and the third lasted 18 hours.
- With the exception of Camille, no category 5 hurricanes have ever existed north of 30°N or south of 14°N.
- Areas that have never experienced a land-falling hurricane of category 5 intensity include the U.S. East Coast, Cuba, Jamaica, and most of the Windward or Leeward Islands.

Source: National Climatic Data Center, http://www.ncdc.noaa.gov. Updated by author.

World Records

Most Intense Ever

870 mb (over 180 mph) in Supertyphoon Tip, western North Pacific Ocean, October 12, 1979; reports from JTWC indicate that the twelve most intense cyclones on record occurred in the western North Pacific Ocean.

Most Intense, Atlantic Ocean

888 mb in Hurricane Gilbert, in September 1988. Although the pressure is almost 20 mb higher than western Pacific storms, the maximum sustained winds are comparable at over 180 mph.

Fastest Intensification

100 mb (from 976 to 876 mb) in just under twenty-four hours in Supertyphoon Forrest, western North Pacific Ocean. Estimated surface winds increased from 75 mph to 173 mph during this period.

Extreme Storm Surge

42 feet in the Bathurst Bay Hurricane, Northern Australia, 1899; this value was derived from reanalysis of debris sightings and

eyewitness reports and, as a result, is controversial. But clearly a phenomenal storm surge occurred.

Highest Wave

112 feet from USS *Ramapo* in the western North Pacific, 6–7 February, 1933; also in the western North Pacific, 82 feet on 26 September, 1935. Hurricane Ivan produced an Atlantic record 91-foot wave on September 15, 2004, in the Gulf of Mexico.

Highest Rainfall

All have occurred at La Reunion Island.

12 hours	45 inches at Foc-Foc (7,511 feet altitude) in Tropical Cyclone Denise, 7–8 January 1966
24 hours	71 inches at Foc-Foc in Tropical Cyclone Denise, 7–8 January 1966
48 hours	96 inches at Aurere (3,083 feet altitude), 8–10 April 1958
72 hours	126 inches at Grand-Ilet (3,773 feet altitude) in Tropical Cyclone Hyacinthe, 18–27 January 1980
10 days	221 inches at Commerson (7,610 feet altitude) in Tropical Cyclone Hyacinthe, 18–27 January 1980

Largest

683-mile radius of gale-force winds (35 mph) in Supertyphoon Tip, western North Pacific Ocean, October 12, 1979.

Smallest

25-mile radius of gale force winds (35 mph) for Tropical Cyclone Tracy, Darwin Australia, December 24, 1974. A storm this small is called a midget tropical cyclone.

Smallest Eye

4-mile radius from radar in Tropical Cyclone Tracy, Darwin, Australia, December 24, 1974; 5-mile radius of Supertyphoon Tip (1979).

Largest Eye

115-mile radius from radar of Typhoon Carmen as it passed over Okinawa, western North Pacific, August 20, 1960; 115-mile radius from radar of Typhoon Winnie, western North Pacific, August 17, 1997; 56-mile radius from aircraft reconnaissance, Tropical Cyclone Kerry, Coral Sea, February 21, 1979.

Longest Lived

Hurricane/Typhoon John lasted 31 days as it traveled across both the northeastern and northwestern parts of the Pacific Ocean during August and September 1994. In the Atlantic, the longest lived was Hurricane Ginger, which lasted 28 days in 1971.

Most Deaths

As quoted from Holland (1993a), "The death toll in the infamous Bangladesh Cyclone of 1970 has had several estimates, some wildly speculative, but it seems certain that at least 300,000 people died from the associated storm surge in the low-lying deltas." Also, several disastrous floods from land-falling typhoons in the Yangtze River valley occurred in the mid-1850s and resulted in many millions of deaths.

Costliest

Hurricane Andrew (1992) in Florida and Louisiana is estimated to have caused at least $25 billion in damage.

Source: Holland (1993a); Landsea (2005); Lander (1999); Anon. (2005).

TABLE 6.11
Record number of storms by ocean basin

Basin	Total Storms or Greater with Sustained Winds of 39 mph or More			Hurricanes, Typhoons, and Severe Tropical Cyclone with Sustained Winds of 73 mph or More		
	Most	Least	Average	Most	Least	Average
Atlantic	19	4	10.6	12	2	5.9
NE Pacific	27	8	16.3	16	4	9.0
NW Pacific	35	17	26.7	24	9	16.9
N Indian	11	2	5.4	5	0	2.2
SW Indian	18	7	13.3	11	2	6.7
SE Indian	13	1	7.3	8	0	3.6
SW Pacific	18	4	10.6	12	0	4.8
N Hemisphere	76	39	58.7	47	24	33.7
S Hemisphere	38	19	29.0	22	7	14.5
Globally	106	68	87.7	64	36	48.3

Based on data from 1968–2003 for all Northern Hemisphere ocean basins except the Atlantic Ocean, and from 1968/1969 to 1989/2003 for the Southern Hemisphere. Starting in 1944, systematic aircraft reconnaissance was commenced for monitoring tropical storms and hurricanes in the Atlantic, hence these records are valid since 1944 in this basin. Atlantic records also contain a storm similar to hurricanes called subtropical cyclones.
Source: Landsea (2005).

TABLE 6.12
More specific record information for the Atlantic Ocean from 1944 to 2004
(coincident with the reconnaissance flight era)

Category	Maximum	Minimum
Tropical storms/hurricanes	19 (1995)*	4 (1983)
Hurricanes	12 (1969)	2 (1982)
Major hurricanes (category 3 or more)	7 (1950)	0 (many times, 1994 last)
U.S. land-falling tropical storms/hurricanes	8 (1916)	1 (many times, 1991 last)
U.S. land-falling hurricanes	6 (1916, 1985)	0 (many times, 2001 last)
U.S. land-falling major hurricanes	3 (1909, 1933, 1954, 2004)	0 (many times, 2003 last)

Because of highly populated coastlines along the U.S. East Coast and Gulf Coast, data with good reliability extends back to around 1899 for land-falling storms. Thus, the landfall records include the time period 1899–2004.
*The year 1933 is recorded as being the most active of any Atlantic basin season on record (reliable or otherwise) with twenty-one tropical storms and hurricanes. However, since reconnaissance flights were not conducted at that time, it is excluded from this table.
Source: Landsea (2005).

TABLE 6.13
Number of U.S. land-falling hurricanes for each state during 1900–2004, stratified by the Saffir-Simpson scale

Area	Category Number					All Categories (1, 2, 3, 4, 5)	Major Hurricanes (Categories 3, 4, 5)
	1	2	3	4	5		
U.S. (Texas to Maine)	65	41	50	15	3	174	68
Texas	13	9	10	6	0	38	16
N Texas	7	3	3	4	0	17	7
Central Texas	3	2	1	1	0	7	2
S Texas	3	4	6	1	0	14	7
Louisiana	10	5	8	3	1	27	12
Mississippi	1	2	5	0	1	9	6
Alabama	5	1	6	0	0	12	6
Florida	19	18	19	6	2	64	27
NW Florida	10	8	8	0	0	26	8
NE Florida	2	7	0	0	0	9	0
SW Florida	7	4	6	3	1	21	10
SE Florida	5	11	8	3	1	28	12
Georgia	1	4	0	0	0	5	0
South Carolina	8	4	2	2	0	16	4
North Carolina	11	7	10	1*	0	29	11
Virginia	3	1	1*	0	0	5	1*
Maryland	0	1*	0	0	0	1*	0
Delaware	0	0	0	0	0	0	0
New Jersey	1*	0	0	0	0	1*	0
New York	3	1*	5*	0	0	9	5*
Connecticut	2	3*	3*	0	0	8	3*
Rhode Island	0	2*	3*	0	0	5*	3*
Massachusetts	2	2*	2*	0	0	6	2*
New Hampshire	1*	1*	0	0	0	2*	0
Maine	5*	0	0	0	0	5	0

Some of these hurricanes may have hit land more than once, and each hit is counted individually. As a result, state totals will not equal U.S. totals, and Texas and Florida totals will not necessarily equal sum of sectional totals. The data is shown starting from Texas eastward, then up the East Coast.
*Hurricanes moving faster than 30 mph.
Source: Hebert, Jarrell, and Mayfield (1997); updated from http://www.nhc.noaa.gov.

Deadliest U.S. Tropical Storms and Hurricanes during 1970–1998

Historically, most deaths associated with tropical storms and hurricanes were caused by storm surge. The Hurricane Camille tragedy made emergency officials and the general public more conscientious about the hurricane storm surge. As evacuation procedures have improved in the United States, fatalities due to the storm surge have dramatically diminished. In fact, most deaths in tropical storms and hurricanes are now associated with inland flooding, as shown in the tables below.

TABLE 6.14
Deadliest top ten U.S. tropical storms and hurricanes during 1970–1998

Rank	Name	Year	Dead
1	Agnes	1972	125
2	Alberto	1994	34
3	Amelia	1978	33
4	Celia	1970	25
5	Andrew	1992	23
6	Fran	1996	22
7	Hugo	1989	17
8	Alicia	1983	13
8	Chantal	1989	13
8	Charley	1998	13

This is the post-Camille period where deaths from the storm surge are fewer due to better hurricane awareness and preparation. Fatalities from heart attacks, house fires, and vehicle accidents during the storms are considered indirect and are excluded from the database.

TABLE 6.15
Causes of 1970–1998 tropical storm and hurricane deaths

Cause of Death	Number of Deaths	Percentage of 517 Total Deaths (Rounded to Whole Number)
Drowned	415	80%
Wind*	66	13%
Tornado	20	4%
Other†	16	3%

This is the post-Camille period where deaths from the storm surge are fewer due to better hurricane awareness and preparation.
*From building failures, airborne debris, etc.
†Nine of these deaths are from lightning, hypothermia, and downed aircraft. Seven are from unknown causes.

TABLE 6.16
Causes of tropical storm and hurricane deaths by drowning between 1970 and 1998

Cause	Drowned (415 Total Deaths)		Percentage of 517 Total Deaths
	Number	Percentage	
Freshwater	292	70%	57%
Shoreline*	62	15%	12%
Offshore†	52	13%	10%
Storm Surge	5	1%	1%
Other	4	1%	.8%

Note that most deaths are from inland flooding, not the storm surge, in the post-Camille period.
*Rough surf, rip currents, large swells.
†Beyond breakers to 50 nautical miles from shore.

TABLE 6.17
Location of U.S. tropical storm and hurricane deaths between 1970 and 1998

Location	Number of Deaths (517 Total)	Percentage of 517 Total Deaths
Inland county	295	57%
Coastal county*	128	25%
Offshore†	55	11%
Unknown	39	7%

*Includes storm surge, large swells, and rip current areas.
†Beyond breakers to 50 nautical miles from shore.
Source: Rappaport, Fuchs, and Lorentson (1998).

Opinions

Excerpts from "Project STORMFURY: A Scientific Chronicle, 1962–1983," by Willoughby et al., 1985.

This article chronicles Project STORMFURY, in which reconnaissance flights attempted to weaken hurricanes by seeding them with silver iodide. Please refer to chapter 1 for a discussion on Project STORMFURY.

Willoughby et al. explain the motivation for Project STORMFURY:

> The years 1954 and 1955 each brought three major hurricanes to the East Coast of the United States. Hurricanes Carol, Edna, Hazel, Connie, Diane, and Ione together destroyed more than six billion dollars in property (adjusted to 1983) and killed nearly 400 U.S. citizens. Six hurricane landfalls in two years seemed to call for some sort of governmental action. In the spring of 1955, the U.S. Congress appropriated substantially more money for hurricane research. After the 1955 hurricane season, it mandated that the U.S. Weather Bureau establish the National Hurricane Research Project (NHRP), which was to become the National Hurricane Laboratory in 1964 and the Hurricane Research Division of the Atlantic Oceanographic and Meteorological Laboratory in 1983. The mission of NHRP was fourfold: to study the formation of hurricanes; to study their structure and dynamics; *to seek means for hurricane modification;* and to seek means for improvement of forecasts. The third objective was, in time, to provide justification for Project STORMFURY.

For a variety of political and scientific reasons, funding for STORMFURY was threatened in the late 1970s. Below Willoughby et al. list five criterion needed to justify the continuation of STORMFURY, and they then explain why it was eventually abandoned.

> STORMFURY had to be viable in five different respects [to justify continuation]:
>
> *Political.* Governments had to be willing to accept the risk of a public outcry if a seeded hurricane (or typhoon) devastated a coastal region. This outcry and its legal consequences might arise even if human intervention had no effect at all on the hurricane.
>
> *Operational.* The aircraft, instrumentation, and personnel to do the seeding and to document the result had to be available.

Microphysical. Convection in hurricanes had to contain enough supercooled water for seeding to be effective.

Dynamic. The hurricane vortex had to be sufficiently labile for human intervention to change its structure.

Statistical. The experiment had to be repeatable, and the results had to be distinguishable from natural behavior.

STORMFURY itself, however, had two fatal flaws: it was neither microphysically nor statistically feasible. Observational evidence indicates that seeding in hurricanes would be ineffective because they contain too little supercooled water and too much natural ice. Moreover, the expected results of seeding are often indistinguishable from naturally occurring intensity changes. By mid-1983, none (or perhaps only one: dynamic feasibility) of the five conditions for development of an operational hurricane amelioration strategy could be met, and Project STORMFURY ended. Its lasting legacies are the instrumented aircraft and two decades of productive research. Of the four original objectives for hurricane research set down in 1955 at the establishment of NHRP, three—understanding formation, understanding structure and dynamics, and improvement of forecasts—remain areas of active investigation.

Source: Willoughby et al. (1985).

Excerpts from the 1996 IPCC Report on Whether Global Warming Is Influencing Global Hurricane Activity

The Intergovernmental Panel on Climate Change (IPCC) is a group of more than 200 scientists that originally met in 1990 to discuss issues related to possible man-induced climate change. Based on these and subsequent meetings in 1992, 1994, and 1996, the IPCC issued conclusions regarding global warming based on current data and the state of the science. The 1996 IPCC report states the following:

> The state-of-the-art for [hurricane simulations in numerical models] remains poor because: (i) tropical cyclones cannot be adequately resolved in general circulation models; (ii) some aspects of ENSO [El Niño-Southern Oscillation] are not well-simulated in general circulation models; (iii) other large-scale changes in the atmospheric general circulation which could affect tropical cyclones cannot yet be discounted; and (iv) natural variability of tropical cyclones is very large, so small trends are likely to be lost in the noise.

This report also concludes that

> It is not possible to say whether the frequency, area of occurrence, mean intensity, or maximum intensity of tropical cyclones will change.
>
> Source: Houghton et al. (1996).

Excerpt from "Tropical Cyclones and Global Climate Change: A Post-IPCC Assessment," by Henderson-Sellers et al. 1998

As a follow-up to the 1996 IPCC report, a group of tropical cyclone experts issued some more definitive statements regarding any relationships to possible global warming and tropical cyclone activity in 1998:

> Our knowledge has advanced to permit the following summary.
>
> - There are no discernible global trends in tropical cyclone number, intensity, or location from historical data analyses;
> - Regional variability, which is very large, is being quantified slowly by a variety of methods;
> - Empirical methods do not have skill when applied to tropical cyclones in greenhouse conditions;

- Global and mesoscale model-based predictions for tropical cyclones in greenhouse conditions have not yet demonstrated prediction skill;
- There is no evidence to suggest any major changes in the area or global location of tropical cyclone genesis in greenhouse conditions;
- Thermodynamic "upscaling" models seem to have some skill in predicting maximum potential intensity (MPI); and
- These thermodynamic schemes predict an increase in MPI of 10%–20% for a doubled CO_2 climate but the known omissions [from MPI theory] all act to reduce these increases.

Source: Henderson-Sellers et al. (1998).

Letters

Alexander Hamilton and the 1772 St. Croix Hurricanes

It was hurricanes that indirectly brought Alexander Hamilton to America. Hamilton experienced two hurricanes on August 3 and September 3, 1772, on St. Croix in the West Indies. His description of this experience in a letter to his father so impressed local planters that they took up a collection to send him to King's College (now Columbia University) in New York. He was in America when the first rumblings of revolution occurred, and the rest is history.

The following is an excerpt of Hamilton's letter to his father:

> Good God! What horror and destruction! It is impossible for me to describe it or for you to form any idea of it. It seemed as if a total dissolution of nature was taking place. The roaring of the sea and wind, fiery meteors flying about in the air, the prodigious glare of almost perpetual lightning, the crash of falling houses, and the ear-piercing shrieks of the distressed were sufficient to strike astonishment into Angels. A great part

of the buildings throughout the island are leveled to the ground; almost all the rest very much shattered, several persons killed and numbers utterly ruined—whole families roaming about the streets, unknowing where to find a place of shelter—the sick exposed to the keenness of water and air, without a bed to lie upon, or a dry covering to their bodies, and out harbors entirely bare. In a word, misery, in its most hideous shapes, spread over the whole face of the country.

Source: Hughes (1976, 1987).

The Great Hurricane of 1780

One of the most destructive hurricanes of the 1700s struck the West Indies in October 1780, killing around 22,000 people. This storm, called "The Great Hurricane" by many in that century, also inflicted great losses on the fleets of both Britain and France, who were planning assaults on each other to claim the Antilles Islands during the American Revolution.

The first news of the hurricane disaster was published in the England journal *The Political Magazine* in a letter from Major General Vaughan, commander-in-chief of British forces in the Leeward Island, dated October 30, 1780:

> I am much concerned to inform your Lordship, that this island was almost entirely destroyed by a most violent hurricane, which began on Tuesday the 10th instant. And continued almost without intermission for nearly forty-eight hours. It is impossible for me to attempt a description of the storm; suffice it to say, that few families have escaped the general ruin, and I do not believe that 10 houses are saved in the whole island: scarce a house is standing in Bridgetown; whole families were buried in the ruins of their habitations; and many, in attempting to escape, were maimed and disabled: a general convulsion of nature seemed to take place, and an universal destruction ensued. The strongest colours could not paint to you Lordship the miseries of the inhabitants: on the one hand, the ground covered with mangled bodies of their friends and relations, and on

the other, reputable families, wandering through the ruins, seeking for food and shelter: in short, imagination can form but a faint idea of the horrors of this dreadful scene.

British Admiral George Rodney was in New York during the hurricane, and when he visited Barbados he incorrectly concluded that the storm must have been accompanied by an earthquake to achieve such total destruction: In describing the ruin, Rodney writes:

> The whole face of the country appears an entire ruin, and the most beautiful island in the world has the appearance of a country laid waste by fire, and sword, and appears to the imagination more dreadful than it is possible for me to find words to express.

Source: Ludlum (1963).

Eyewitnesses Account of Hurricane Landfall on the Florida Panhandle in 1843

The following is an excerpt from the September 15, 1843, *Commercial Gazette* describing the damage on Port Leon (about 30 miles south of today's city of Tallahassee near St. Marks):

> Our city is in ruins! We have been visited by one of the most horrible storms that it ever before devolved upon us to chronicle. On Wednesday [13 September] about 11 o'clock A.M. the wind lulled and the tide fell, the weather still continued lowering. At 11 at night, the wind freshened, and the tide commenced flowing, and by 12 o'clock it blew a perfect hurricane, and the whole town was inundated. The gale continued with unabated violence until 2 o'clock, the water making a perfect breach ten feet deep over our town. The wind suddenly lulled for a few minutes, and then came from southwest with redoubled violence and blew until daylight.
>
> Every warehouse in the town was laid flat with the ground except one, Messers Hamlin & Shell's, and a

part of that also fell. Nearly every dwelling was thrown from its foundations, and many of them crushed to atoms. The loss of property is immense. Every inhabitant participated in the loss more or less. None have escaped, many with only the clothes they stand in. St. Marks suffered in like proportion with ourselves. But our losses are nothing compared with those at the lighthouse. Every building but the lighthouse gone—and dreadful to relate fourteen lives lost! And among them some of our most valued citizens. We cannot attempt to estimate the loss of each individual at this time, but shall reserve it until our feelings will better enable us to investigate it.

Source: Ludlum (1963).

The Last Island Disaster

On August 10, 1856, a violent hurricane struck Isle Derniere (Cajun for "Last Island"), a pleasure resort southwest of New Orleans. The highest points were under 5 feet of water. The resort hotel and surrounding gambling establishments were destroyed, and the island was cut in half by the storm. Crew members on the ferry boat *Star* rode out the storm and rescued survivors seen floating among the debris. A total of about forty people survived on the *Star;* the rest—hundreds of people—were killed. The following letter was sent to the *Daily Picayune* (the New Orleans newspaper) from a survivor of this event and published on August 14, 1856:

> Dear Pic.—You may have heard ere this reaches you of the dreadful catastrophe which happened on Last Island on Sunday the 10th inst. As one of the sufferers it becomes my duty to chronicle one of the most melancholy events, which have ever occurred. On Saturday night, the 9th inst., a heavy northeast wind prevailed, which excited the fears of a storm in the minds of many; the wind increased gradually until about ten o'clock Sunday morning, when there existed no longer any doubt that we were threatened with imminent

danger. From that time the wind blew a perfect hurricane; every house upon the island giving way, one after another, until nothing remained. At this moment everyone sought the most elevated point on the island, exerting themselves at the same time to avoid the fragments of buildings, which were scattered in every direction by the wind. Many persons were wounded; some mortally. The water at this time (about 2 o'clock P.M.) commenced rising so rapidly from the bay side, that there could no longer be any doubt that the island would be submerged. The scene at this moment forbids description. Men, women, and children were seen running in every direction, in search of some means of salvation. The violence of the wind, together with the rain, which fell like hail, and the sand blinded their eyes, prevented many from reaching the objects they had aimed at.

At about 4 o'clock, the Bay and Gulf currents met and the sea washed over the whole island. Those who were so fortunate as to find some object to cling to, were seen floating in all directions. Many of them, however, were separated from the straw to which they clung for life, and launched into eternity; others were washed away by the rapid current and drowned before they could reach their point of destination. Many were drowned from being stunned by scattered fragments of the buildings, which had been blown asunder by the storm; many others were crushed by floating timbers and logs, which were removed from the beach, and met them on their journey. To attempt a description of this sad event would be useless. No words could depict the awful scene which occurred on the night between the 10th and 11th inst. It was not until the next morning the 11th, that we could ascertain the extent of the disaster. Upon my return, after having drifted for about twenty hours, I found the steamer *Star*, which had arrived the day before, and was lying at anchor, a perfect wreck, nothing but her hull and boilers, and a portion of her machinery remaining. Upon this wreck the lives of a large number were saved. Toward her each one directed his path as he was recovered from the deep, and was welcomed with tears by his fellow-sufferers, who had been so fortunate as to escape. The scene was heart-rending; the good fortune of many an individual in being saved, was blighted by the news of the loss of a father, brother, sister, wife or some near relative.

Source: Jennings (1970); Ludlum (1963).

7
Directory of Organizations

This chapter describes some organizations relevant to hurricane research and forecasting. Other organizations are included because of their importance to meteorology as a whole or because of their commitment to disaster mitigation or assistance. A description of each group's function and history is included, as well as the address, point of contact, and website. Relevant publications are also listed. Updates of websites and addresses will also be provided on http://www.drfitz.net.

American Meteorological Society

The American Meteorological Society (AMS) is a nonprofit, professional society whose objectives are the development and dissemination of knowledge of the atmospheric and related oceanic and hydrologic sciences and the advancement of its professional applications. Charles Franklin Brooks of the Blue Hill Observatory in Milton, Massachusetts, founded the AMS in 1919.

Its initial membership came primarily from the U.S. Signal Corps and U.S. Weather Bureau. Its initial publication, the *Bulletin of the American Meteorological Society*, was meant to serve as a supplement to the *Monthly Weather Review*, which, at the time, was published by the U.S. Weather Bureau. However, during the 1930s and 1940s, because of the key role meteorologists played in support of military activities, as well as associated technology developments, the AMS substantially grew in numbers and in purpose.

Carl-Gustav Rossby served as president of the AMS for 1944 and 1945 and developed the framework for its first scientific journal, the *Journal of Meteorology,* which later split into the two current AMS journals: the *Journal of Applied Meteorology* and the *Journal of the Atmospheric Sciences.*

This role as a scientific and professional organization serving the atmospheric and related sciences continues today. The AMS now publishes eight scientific journals and an abstract journal, in addition to the *Bulletin,* and sponsors and organizes more than a dozen scientific conferences each year, including a biannual Hurricane and Tropical Meteorology Conference where scientists and forecasters present the latest research on these storms. The AMS has published almost fifty monographs in its continuing series, as well as many other books and educational materials. The journals are directed toward an audience with at least bachelor of science degrees in meteorology, and many are written for advanced degrees. However, the *Bulletin* (provided with annual dues) strives toward general public readability, and many of the educational materials are also quite readable. Internet access and CD-ROMS of all journals and some other publications are available.

The AMS administers two professional certification programs—the Radio and Television Seal of Approval and the Certified Consulting Meteorologist programs—and also offers an array of undergraduate scholarships and graduate fellowships to support students pursuing careers in the atmospheric and related oceanic and hydrologic sciences. Local AMS chapters for many cities and universities also exist, providing a means for community interaction among those interested in meteorology and professional meteorologists.

Professionals in the field of atmospheric or related oceanic or hydrologic sciences holding a baccalaureate or higher degree can join the AMS under the classification of "member." Those who do not meet this criterion but are interested in joining the AMS can apply under the classification of "associate member." Classifications for students also exist. Annual dues are required, with many optional purchases detailed on the AMS website.

The address is:

American Meteorological Society
45 Beacon Street

Boston, MA 02108-3693
Phone: (617) 227-2425
URL: http://www.ametsoc.org

Educational Material and Publications

For professional meteorologists, the following journals are published by the AMS: *The Bulletin of the American Meteorological Society; Weather and Forecasting; Journal of Climate; Monthly Weather Review; Journal of Physical Oceanography; Journal of the Atmospheric Sciences; Journal of Applied Meteorology; Journal of Atmospheric and Oceanic Technology; Journal of Hydrometeorology; Earth Interactions;* and *Meteorological and Geostrophysical Abstracts*. Conference papers are also available for download in the "Conferences" section of the website, and presentation recordings with their corresponding PowerPoint files are also available.

In addition to its journals, the AMS publishes many monographs on weather topics. The following partially or exclusively discuss hurricanes: *Historical Essays on Meteorology; Early American Hurricanes: 1492–1870; Tropical Cyclones: Their Evolution, Structure, and Effects; Glossary of Weather and Climate; The 1938 Hurricane;* and *Cloud Systems, Hurricanes, and the Tropical Rainfall Measuring Mission (TRMM): A Tribute to Dr. Joanne Simpson*. Another authoritative book, containing more than 12,000 meteorological terms and definitions based on input from an editorial board of forty-one distinguished scientists, is the *Glossary of Meteorology*.

The AMS's educational program supports students at all education levels (K–12, college, and graduate school). For college students, the AMS offers Online Weather Studies, a course with electronic and printed components. The course focuses on the study of weather as it happens by utilizing Internet-delivered learning materials. The course is available for licensing by colleges and universities. The AMS offers approximately fifty minority and other undergraduate scholarships and also offers graduate fellowships to the nation's future meteorologists, oceanographers, hydrologists, and climatologists each year. Students interested in becoming meteorologists are encouraged to read the online AMS documents *Careers in Atmospheric Research and Applied Meteorology* and *A Career Guide for Atmospheric Sciences*, in the "Student Resources" section.

To promote weather education at the elementary and secondary school levels, Project ATMOSPHERE has trained more than

70,000 teachers and provided instructional material for classes. DataStreme Atmosphere Project, a distance-learning weather course based on Internet-delivered information, is also available. Equivalent oceanography classes—the Maury Project and the DataStreme Ocean Project—are also offered. Another distance-learning class, the DataStreme Water in the Earth System (WES) Project, teaches about the global water cycle and related issues. Other countries also use these programs, including Canada, Australia, Great Britain, Argentina, Croatia, Japan, Mexico, South Africa, and Switzerland. The AMS book *Hands-on Meteorology* is an excellent source of ideas for teachers to use in the classroom.

American Red Cross and the International Federation of Red Cross and Red Crescent Societies

The American Red Cross is a humanitarian organization, led by volunteers, that provides relief to victims of disasters and helps prevent, prepare for, and respond to emergencies. It is the largest volunteer emergency service organization in the United States, with numerous chapters nationwide, as well as regional Blood Service and Tissue Service centers. The Red Cross is not a U.S. government agency; it is a private, nonprofit organization relying on charitable donations for funding.

The Red Cross was founded by Henri Dunant in June 1859 when he visited the battlefields of Solferino in Italy and encountered thousands of wounded and dying with no medical care. He organized a group to assist and wrote a book about the experience afterward. This promoted a movement that authorizes medical personnel, wearing a red cross on a white armlet to distinguish them as noncombatants, to assist wounded soldiers. A second component of the movement required governments to sign treaties that protected these workers and fostered the formation of national volunteer societies to carry out his mission. During this movement, a volunteer named Clara Barton cared for soldiers during the American Civil War. After the Civil War, Barton visited Europe and learned of Dunant's movement. She returned to the United States deeply committed to persuading U.S. participation in this movement. In 1864, the United States and many other countries signed such a treaty, known as the Geneva Convention.

It established that "wounded or sick combatants, to whatever nation they belong, shall be collected and cared for" by personnel wearing the Red Cross emblem, and that persons and facilities bearing the symbol are protected from attack. On May 21, 1881, Barton founded the American Red Cross, and in 1900 the U.S. Congress chartered the organization to provide services to the U.S. military forces and to victims at home and abroad. In a typical year, the Red Cross responds to more than 66,000 natural and man-made disasters including hurricanes, floods, earthquakes, tornadoes, fires, hazardous materials spills, civil disturbances, explosions, and transportation accidents.

The American Red Cross is part of the larger International Federation of Red Cross and Red Crescent Societies (IFRC), consisting of different Red Cross organizations in various countries. The IFRC is the largest humanitarian network in the world with a presence in almost every country. The IFRC's mission is to improve the lives of vulnerable people. Vulnerable people are those at greatest risk from situations that endanger their survival or threaten their capacity to live with an acceptable level of social and economic security and human dignity. Often, these are victims of natural disasters, poverty brought about by socioeconomic crises, refugees, and victims of health emergencies. The IFRC carries out relief operations to assist victims of disaster, and combines this with development work to strengthen the capacities of its member national societies. The IFRC's work focuses on four core areas: promoting humanitarian values, disaster response, disaster preparedness, and health and community care.

For more information in the United States, contact the local Red Cross or the U.S. headquarters at:

American Red Cross National Headquarters
2025 E Street N.W.
Washington, DC 20006
Phone: (202) 303-4498
URL: http://www.redcross.org

Educational Material and Publications

The American Red Cross publishes several pamphlets related to hurricane preparedness and hurricane relief. Available online are: *Hurricanes . . . Unleashing Nature's Fury!*; *Are You Ready for A Hurricane?*; *Against the Wind: Protecting Your Home from Hurricane Wind*

Damage; Your Family Disaster Plan; Your Family Disaster Supplies Kit; Chain Saw Safety Fact Sheet; Using a Generator When Disaster Strikes; and *Safety Information for Short-Term Power Outages or "Rolling Blackouts."* Some of these pamphlets are also available from the National Weather Service. Two coloring books for children can be ordered for a nominal fee: *Jason and Robin's Awesome Hurricane Adventure* and *After the Storm Coloring Book.* Hurricane tracking charts can also be purchased. The American Red Cross sells videos on hurricane preparedness, including *Against The Wind; Home Preparedness for Hurricanes; Before the Wind Blows; Hurricane Information Guide for Coastal Residents;* and (for children) *Jason and Robin's Awesome Hurricane Adventure.* Lesson plans known as the Master of Disaster curriculum kit are available for teachers of K–8 for disaster safety.

For international information, contact the IFRC at:

International Federation of Red Cross and Red Crescent Societies
P.O. Box 372 CH-1211
Geneva 19 Switzerland
Phone: +41 22 730 42 22
Fax: +41 22 733 03 95
URL: http://www.ifrc.org

Educational Material and Publications

The IFRC publishes a variety of documents on the subjects of emergency response and disaster relief, disaster preparedness, and health and social welfare. An *Annual Report* and *World Disaster Report* are published every year. The *Red Cross Red Crescent* magazine is published quarterly. This magazine tackles humanitarian issues and features the activities of Red Cross and Red Crescent societies as they seek to help vulnerable people in their different environments.

Cooperative Institute for Meteorological Satellite Studies at the University of Wisconsin-Madison

The Cooperative Institute for Meteorological Satellite Studies (CIMSS) was established in 1980 to formalize and support cooperative research between the National Oceanic and Atmospheric Administration's (NOAA) National Environmental Satellite, Data, and Information Service (NESDIS) and the University of Wisconsin-Madison's Space Science and Engineering Center. Sponsorship and membership of the CIMSS was expanded to include the National Aeronautics and Space Administration (NASA) in 1989. During the 1980s, a need emerged for joint federal-university research centers to support the NOAA weather research program. CIMSS was established to focus on the development and testing of the operational utility of new weather satellite observing systems to improve weather forecasts.

The CIMSS develops techniques for using weather satellite observations to improve forecasts of severe storms, including tornadoes and hurricanes. One notable achievement has been the development and automation of cloud-drift winds, which are invaluable in data-sparse regions, particularly in tropical oceans. Cloud-drift winds are computed by tracking individual cloud elements and deducing their motion. These winds are used to identify steering currents and to visualize atmospheric features that affect tropical cyclone genesis and intensification. This data is being assimilated into computer models to improve weather forecasts, including hurricane track predictions. The CIMSS also conducts research on satellite techniques for estimating tropical storm and hurricane intensity. One research area uses the Advanced Microwave Sounding Unit (AMSU) on board the NOAA polar-orbiting satellites to measure a storm's upper-level warm core; this can indirectly quantify the spatial distributions of winds as well as a storm's central surface pressure. The CIMSS has also developed an objective Dvorak technique using satellite data that supplements the subjective Dvorak technique currently being used at operational hurricane forecast centers worldwide.

In addition, the CIMSS participates in satellite instrument design, international field programs, and computer visualization of satellite measurements. For example, software called the Man computer Interactive Data Access System (McIDAS) displays, enhances, and animates satellite data and can also overlay the imagery with meteorological measurements. McIDAS is still actively used today by many governmental and university institutions, including the NHC.

The CIMSS's address is:

The Cooperative Institute for Meteorological Satellite Studies
1225 West Daytona Street
Madison, WI 53706
Phone: (608) 263–7435
Fax: (608) 262–5974
URL: http://cimss.ssec.wisc.edu

Educational Material and Publications

Their website is the best source for information about hurricane research at CIMSS. Under the "Tropical Cyclones" link, worldwide satellite images of hurricanes are displayed in a variety of formats, as are cloud-drift winds. Infrared, visible, water vapor, and AMSU satellite images are available. Tropical cyclone forecasters from around the world use this state-of-the-art website.

Federal Emergency Management Agency

The Federal Emergency Management Agency (FEMA) is a former independent agency that became part of the new Department of Homeland Security in March 2003. Since its founding in 1979, FEMA's mission is to reduce loss of life and property and protect our nation's critical infrastructure from all types of hazards (including hurricanes) through a centralized, comprehensive, risk-based, emergency management program of mitigation, preparedness, response, and recovery. FEMA has about 2,500 full-time employees and nearly 5,000 standby disaster assistance employees who help out after disasters. Often FEMA works in partnership with other organizations that are part of the nation's emergency management system. These partners include state and local emergency management agencies, twenty-seven federal agencies, and the American Red Cross. During hurricane threats, FEMA provides coordination efforts to federal, state, and local governments.

FEMA is called in to help when the president declares a disaster. Disaster areas are declared after hurricanes, tornadoes, floods, earthquakes, or other similar events strike a community. The governor of the state must ask for help from the president before FEMA can respond. FEMA helps provide disaster victims with temporary housing if their homes are damaged or destroyed and furnishes loans for uninsured disaster victims during major disaster declarations.

FEMA is actively involved in disaster mitigation. Services include:

1. Evaluation of emergency evacuation plans and shelters.
2. Training and outreach programs to increase public awareness of disaster hazards (including hurricanes).
3. The National Flood Insurance Program, an important program since homeowners insurance *does not* cover flood damage (see chapter 2).
4. Grants in the form of matching funds to states, local governments, and property owners to implement long-term mitigation measures after a major disaster declaration. Examples include elevating properties that repeatedly flood from rain or storm surges; buy-outs, administered through state or community programs, of properties that repeatedly flood and are deemed no longer suitable for residence or development; improvement of drainage ditches in flood-prone areas; installation or improvements of flood pumps; retrofitting homes through second-story conversion, which entails cutting the roof off the house, building a second story on top of the existing first story, placing the roof back on the house, and gutting the bottom floor; and dry flood-proofing in which the exterior of a building is completely sealed to prevent the entry of floodwaters.
5. Funding for HAZards US (HAZUS) software, which provides potential loss estimates for a variety of earthquake, flood, and hurricane scenarios. Current scientific and engineering knowledge is coupled with the latest Geographic Information Systems (GIS) technology to produce estimates of hazard-related damage before or after a disaster occurs. With regard to hurricanes, HAZUS estimates potential damage and loss to residential, commercial, and industrial buildings. It also allows users to estimate direct economic loss, poststorm shelter needs, and building debris on the Atlantic and Gulf Coast. Future versions will include the capability to estimate wind effects in island territories, storm surge, indirect

economic losses, casualties, and impacts to utility and transportation lifelines and agriculture.

FEMA's headquarters is located at:

Federal Emergency Management Agency
Federal Center Plaza
500 C. Street S.W.
Washington, D.C. 20472
URL: http://www.fema.gov

FEMA has ten regional offices and two area offices. Each region serves several states, and regional staff works directly with the states to help plan for disasters, develop mitigation programs, and meet needs when major disasters occur. These regions are shown on the home page of FEMA's website. For more immediate responses to emergency management issues, contact the nearest regional office.

Educational Material and Publications

FEMA's website contains extensive information and numerous documents on hurricane preparedness, hurricane recovery issues, flood insurance, and flood protection, with many in electronic form for downloading to a computer. Too many exist to list them all here, and new ones become available routinely. A few of the many publications are *Hurricanes: Floods and Flood Safety Tips for Coastal and Inland Flooding; Safety Tips for Hurricanes; Hurricane Awareness: Action Guidelines for Senior Citizens; Hurricane Awareness: Action Guidelines for School Children; Against the Wind: Protecting Your Home from Hurricane Wind Damage; After a Flood: The First Steps; How the NFIP Works; Things You Should Know about Flood Insurance; Avoiding Hurricane Damage: A Checklist for Homeowners;* and *Hurricane Preparedness in Marinas*. Publications, videos, CDs, and coloring books on hurricanes and disaster preparedness are also available for children. The FEMA website contains a HAZUS section that provides detailed information on this software. The publication *Battling Hazards with a Brand New Tool* is also available on HAZUS. The software is free and can be ordered through a form on the website, and its manuals can be downloaded off the website.

Geophysical Fluid Dynamics Laboratory

The Geophysical Fluid Dynamics Laboratory (GFDL) is a NOAA research laboratory. The GFDL conducts research of the physical processes that govern the behavior of the atmosphere and the oceans as complex fluid systems. These systems can then be modeled mathematically, and their phenomenology can be studied by computer simulation methods. The GFDL focuses on predictability and climate issues at all scales in the atmosphere and ocean, as well as the air-sea interface. It conducts leading-edge research on many topics of great practical value, including weather and hurricane forecasts, El Niño prediction, and global warming. The GFDL also collaborates with Princeton University on these topics.

The GFDL has developed a hurricane model that provides forecast track and intensity guidance to the National Hurricane Center. A similar version is used by the U.S. Navy for the western Pacific Ocean and is called the GFDN model. GFDL's model, on average, produces the smallest track errors with respect to most other operational models.

The GFDL is continuously upgrading the hurricane model. Recently, ocean model coupling has been added to the hurricane model so that important wind and ocean temperature feedbacks are incorporated. Hurricanes mix the upper ocean, resulting in colder water beneath the hurricane that affects storm intensity. Computer upgrades at the National Centers for Environmental Prediction (discussed later in this chapter) have also allowed the hurricane model to simulate higher-resolution features, which will improve the model's forecasts. Better physics and land algorithms (after a hurricane makes landfall) are in development.

The GFDL is also investigating hurricane sensitivity to climate change. In particular, should global warming occur, the GFDL is exploring whether, on average, hurricanes will become stronger or more numerous.

The GFDL's address is:

Geophysical Fluid Dynamics Laboratory/NOAA
Princeton Forrestal Campus
201 Forrestal Road
Princeton, NJ 08542-0308
Phone: (609) 452-6500

Fax: (609) 987-5063
URL: http://www.gfdl.gov

Mail correspondence should be sent to P.O. Box 308.

Educational Material and Publications

The GFDL's website is the best source for information about hurricane research at GFDL. Topics include hurricanes and climate change, information and statistics on the hurricane model, and animations of well-known hurricanes. One section of the website contains bibliographies of researchers' abstracts going back as far as 1965.

Hurricane Hunters

The 53rd Weather Reconnaissance Squadron, known as the Hurricane Hunters of the U.S. Air Force Reserve, is the only Department of Defense organization flying into tropical storms and hurricanes on a routine basis. Moreover, they are the only organization conducting operational hurricane reconnaissance in the world. The 53rd Weather Reconnaissance Squadron (WRS) is a component of the 403rd Wing located at Keesler Air Force Base in Biloxi, Mississippi. A detailed history on hurricane reconnaissance is provided in chapter 4.

The mission of the Hurricane Hunters is to recruit, organize, and train assigned personnel to perform aerial weather reconnaissance. During the hurricane season, from June 1 to November 30, they provide surveillance of tropical disturbances and hurricanes in the Atlantic (west of 55°W), Caribbean, and Gulf of Mexico for the National Hurricane Center in Miami, Florida. They also may provide aerial weather reconnaissance for the Central Pacific Hurricane Center in Honolulu, Hawaii.

From November 1 through April 15, the unit also flies winter storms off both coasts of the United States in support of the National Centers for Environmental Prediction. These missions are flown at high altitude (30,000 feet) and can be just as challenging as the hurricane missions, with turbulence, lightning, and icing. These hurricane reconnaissance flights provide the most accurate measurements of the storm's location, central pressure, wind speeds, and other critical data that satellites cannot measure (or can only estimate). Even though the flights are expensive, the information provided to forecasters, emergency preparedness officials, and the general public more than compensates for the cost. Studies have shown that the plane observations improve forecast accuracy by about 25 percent. The planes also have detected sudden, dangerous changes in hurricane intensity and movement that are currently very difficult to detect by satellite alone. A typical hurricane warning costs an estimated $192 million due to preparation, evacuation, and lost commerce. Narrowing the warning area can save $640,000 per mile or more. Furthermore, as coastal populations continue to grow, evacuation decisions need to be made earlier.

The reservists bring a wide variety of experience with them. Their primary occupations include airline pilot, meteorologist, teacher, electrician, computer systems operator, law enforcement officer, firefighter, medical doctor, veterinarian, and businessperson. While most live in the Gulf Coast area, the reservists are drawn from fifteen states and frequently travel at their own expense to perform their duties. Reservists schedule their annual two-week training tour to coincide with the peak of the storm season in the summer. Reservists may also be given short notice to participate in additional storm missions, with the generous cooperation of their civilian employers.

Colocated with the National Hurricane Center is a small group of U.S. Air Force Reserve civilian personnel assigned to the 53rd WRS. The supervisory meteorologist of the unit serves as Chief, Aerial Reconnaissance Coordination, All Hurricanes, better known as CARCAH. These personnel are responsible for coordinating Department of Commerce requirements for hurricane data, tasking weather reconnaissance missions, and monitoring all data transmitted from the weather reconnaissance aircraft.

To arrange for an interview, tour, or speaker or to obtain more general information on the Hurricane Hunters, call the 403rd Wing Public Affairs or write to:

403rd Wing Public Affairs
701 Fisher Street, Suite 103
Keesler AFB, MS 39534-2572
Phone: (228) 377-2056

Fax: (228) 377-0755
URL: http://www.hurricanehunters.org

Members of the media can obtain permission to fly on one of the reconnaissance flights into a hurricane. Please note that only media members are allowed to accompany the Hurricane Hunters, and all other requests will be turned down.

Educational Material and Publications

The Hurricane Hunters' website is the best source of information. It contains a "Frequently Asked Questions" section, a cyberflight into a hurricane with detailed descriptions of each phase of the flight, how to decipher reconnaissance weather data code such as the "vortex" messages, historical information on the Hurricane Hunters, and photos. They also publish a pamphlet titled *53rd Weather Reconnaissance Squadron: Hurricane Hunters*.

Hurricane Research Division

The Hurricane Research Division (HRD) is a part of the Atlantic Oceanographic and Meteorological Laboratory (AOML) within the National Oceanic and Atmospheric Administration (NOAA) of the U.S. Department of Commerce. The HRD is NOAA's primary focus for studies of tropical cyclones and tropical weather systems. A key component of HRD activity is its annual field program of flights aboard NOAA's research aircraft (two WP-3D turboprops for analysis up to 30,000 feet and a Gulfstream IV-SP jet for analysis between 30,000 and 45,000 feet) flown by NOAA's Aircraft Operations Center (AOC) (see discussion later in this chapter).

The HRD is engaged in several hurricane research topics. These investigations involve theoretical studies, computer modeling, and the collection and examination of measurements taken in actual hurricanes. The ultimate goal is to improve scientific understanding of how and why hurricanes form, strengthen, and dissipate, which will lead to improved forecasting of tropical weather and mitigation of damage from the destructive power of hurricanes.

The HRD's main research topics are:

1. The improvement of hurricane track forecasts, particularly with the use of dropsondes, which measure the vertical profile of wind, temperature, and moisture around a hurricane. The HRD is currently investigating the impact of dropsondes on NCEP weather models and is also exploring new techniques for better track prediction, such as "targeted observations" and "ensemble forecasting" (see chapter 2).
2. The improvement of hurricane intensity forecasts. By analyzing data acquired during the aircraft field missions, HRD scientists are studying air-sea interaction under hurricanes and structural changes in the eyewall region. The HRD considers transmitting the aircraft data real-time to the NHC a top priority and is developing techniques to transmit real-time Doppler radar winds and hurricane surface winds. The HRD is also improving statistical schemes and weather models used to predict hurricane intensity. Finally, the HRD is investigating how mineral dust, dry air, and wind surges from Africa affect Atlantic tropical cyclone genesis and intensity change.
3. To conduct hurricane climate studies. The frequency with which hurricanes occur varies across all timescales, from weekly to centennial. HRD personnel are conducting studies to increase understanding of these variations. The HRD also is developing schemes to improve the predictability of hurricane occurrence and intensity over monthly, seasonal, interannual, decadal, and longer timescales.
4. To address the impact of hurricanes at landfall. The HRD is developing methodologies to improve FEMA's emergency preparedness software, how buildings and vegetation impact hurricane winds, the extratropical transition of hurricanes, and the physics on why hurricanes weaken inland. A major HRD research priority is to improve precipitation forecasting for tropical storms and hurricanes. Drowning from inland flooding in land-falling tropical storms and hurricanes is the leading cause of death by these storms in the past thirty years in the United States.

These are lofty goals, and therefore the HRD collaborates with many organizations such as NASA and several universities to attack these issues.

HRD's address is:

NOAA/AOML/HRD
4301 Rickenbacker Causeway
Miami, FL 33149
Phone: (305) 361-4400
Fax: (305) 361-4402
URL: http://www.aoml.noaa.gov/hrd

Educational Material and Publications

The HRD website is loaded with information about the organization's research activities and general hurricane information. The website contains probably the most popular hurricane link on the Internet, "Frequently Asked Questions" about hurricanes. Another link gives detailed advice on storm shutters. The AOML produces a bimonthly newsletter, *AOML Keynotes,* that can also be viewed on the website in pdf format (go to http://www.aoml.noaa.gov and look in the "Keynotes" link). HRD scientists have published numerous articles in AMS journals and other publications; a list is available on the website. See the section "National Weather Service" for a detailed listing of other related NOAA pamphlets.

The HRD website also provides many unique data sets, all of which are free to download. These include:

1. The historical HURricane DATabase (HURDAT). This contains the best determination of positions and intensities for past Atlantic and East Pacific tropical storms and hurricanes and is therefore also sometimes called the "best track" data set. The Atlantic best track data set is quite extensive, going back to the year 1851, and is used for analyzing past hurricane tracks as well as climate studies.
2. The H*Wind data set. The HRD has developed a computer program that integrates wind data in and around a hurricane from a variety of platforms into a single wind analysis reduced to the surface. This product is transmitted real-time to the NHC and is also used in poststorm analysis and research.
3. Dropsonde data from the hurricane research flights.
4. Radar images and upper-level wind graphics during the hurricane research flights.

5. Flight videos.
6. More technical data for scientists, such as cloud microphysics data, bathythermograph data, and stepped frequency microwave radiometer data.

The instruments used to derive these data sets are described on the HRD website as well as on the AOC website. Please refer to the AOC section below for more information. Chapter 4 chronicles HRD's history and field programs in more detail.

Joint Typhoon Warning Center

The Joint Typhoon Warning Center (JTWC) provides tropical cyclone forecasting support to the U.S. military and to allies within the Pacific and Indian Ocean basins under the auspices of the Naval Pacific Meteorology and Oceanography Center in Hawaii. The JTWC was founded May 1, 1959, when the U.S. commander-in-chief of Pacific forces directed that a single tropical cyclone warning center be established for the western North Pacific region. The JTWC is a combined U.S. Air Force/Navy organization in which the air force provides forecasters and the navy provides the facility, equipment, and forecasters.

The mission of the JTWC includes:

1. Continuous monitoring of all tropical weather activity in the Northern and Southern Hemispheres from the west coast of the Americas to the east coast of Africa, the issuance of Tropical Cyclone Formation Alerts when tropical cyclone development is anticipated, and the issuance of warnings for all significant tropical cyclones in these regions. This covers all tropical cyclone regions except the Atlantic.
2. Poststorm analysis of tropical cyclones occurring within the western North Pacific and North Indian Oceans.
3. Cooperation with the Naval Research Laboratory in Monterey, California, on operational evaluation of tropical cyclone models and forecast aids and the development of new techniques to support operational forecast requirements.

For the first forty years, the JTWC was located at Nimitz Hill, Guam. Due to the Base Realignment and Closure (BRAC) legislation of 1995, the JTWC moved from Guam to Hawaii in 1999. The address is:

Naval Pacific Meteorology and Oceanography Center/
Joint Typhoon Warning Center
Box 113
Pearl Harbor, HI 96860-5050
Phone: (808) 474-5301
URL: http://www.npmoc.navy.mil

Forecasters at the Naval Pacific Meteorology and Oceanography Center use this information to help the navy prepare for hurricane events, with particular attention on ship routing and whether to deploy the fleet (sortie) from naval bases. Deploying ships is expensive and time-consuming, costing more than $1 million in fuel and several hundreds of thousands of dollars in staff costs, so such decisions are made meticulously. (Incidentally, the Naval Atlantic Meteorology and Oceanography command serves a similar function for Atlantic hurricanes but uses guidance from the NHC.)

Educational Material and Publications

Every year, the JTWC staff prepares an *Annual Tropical Cyclone Report* summarizing tropical cyclone activity in the Pacific and Indian Ocean basins. These annual reports are available on their website. In addition, a historical data set containing positions and intensities for West Pacific and Australian tropical cyclones (commonly called besttrack data sets) can be downloaded for free. The JTWC's website also contains current typhoon information and forecasts. It also provides current satellite, radar, and weather analysis data as well as model forecasts.

National Aeronautics and Space Administration

The National Aeronautics and Space Administration (NASA) was established by Congress in the National Aeronautics and Space Act of 1958 "to provide for research into problems of flight within and outside the earth's atmosphere, and for other purposes." While NASA certainly contains a rich space flight history, the organization also conducts research on remote sensing, climate topics, and weather topics (including hurricanes). Different

NASA organizations, spread throughout the country, participate in weather, hurricane, and satellite research. The most active organizations include Goddard Space Flight Center, Langley Research Center, and the Jet Propulsion Laboratory, but other NASA affiliations also have weather-related projects. NASA also participates in weather research field programs. With regard to hurricane research, the Convection And Moisture EXperiments (CAMEX) are devoted to the study of hurricane tracking, structure, and cloud physics using NASA-funded aircraft remote-sensing instrumentation.

NASA plays a major role in operational and research satellite technology and has been the leading agency for conducting research on cutting-edge satellite technology since TIROS in 1960 (see chapter 4). Generally, between fifteen and twenty NASA research satellites are orbiting the earth. This decade, NASA deployed the next generation of satellite research through the concept of the Earth Observing System (EOS), in which a constellation of satellites (Aqua, Terra, and Aura) provide a near-simultaneous view of the major earth components (land, oceans, atmosphere, ice, and life) in unprecedented detail. Other notable satellites used in hurricane research (see chapter 4 for more details) include the Tropical Rainfall Measuring Mission (TRMM), which is a "flying rain gauge" and contains several sensors for detecting moisture and rainfall, and QuikSCAT, which measures surface winds over the data-sparse ocean. This data is used experimentally by forecasters and for research by scientists. NASA also continues to design, develop, and launch the nation's civilian operational environmental satellites, in both polar and geostationary orbits, by agreement with NOAA. NOAA then assumes control of these satellites after activation and provides the resulting data on a continuous basis for weather prediction and other services.

The address of NASA headquarters is:

NASA Headquarters
Washington, DC 20546-0001
Phone: (202) 358-0000
URL: http://www.nasa.gov

Educational Material and Publications

There is no shortage of weather publications and satellite pictures at NASA websites. The main NASA website contains links discussing all earth science activities. The reader is also referred to the Jet Propulsion Laboratory, Goddard, and Langley websites for further browsing.

National Centers for Environmental Prediction

The National Centers for Environmental Prediction (NCEP) was established in 1958 as the National Meteorological Center. Since the center's beginning, operational weather forecasting has transformed from an infant discipline into a mature science. Under the auspices of the National Oceanic and Atmospheric Administration's National Weather Service, the NCEP provides operational forecast products worldwide and is the starting point for nearly all weather forecasts in the United States. It is comprised of nine national forecast centers. Each center has a specific responsibility for a portion of the NCEP products and services suite. Seven of the centers provide direct products to users, while two of the centers provide essential support through developing and running complex computer models of the atmosphere on supercomputers, the world's fastest computers. One of the centers providing direct forecast products is the Tropical Prediction Center, which contains the National Hurricane Center.

The NCEP's address is:

National Centers for Environmental Prediction
5200 Auth Road
Camp Springs, MD 20746

To contact the NCEP, the website http://www.ncep.noaa.gov provides a contact section. For phone calls, it is probably best to call the NOAA Office of Public Affairs (see NOAA section for more details).

Educational Material and Publications

A brochure titled *National Centers for Environmental Prediction* describes the individual function of the nine centers. See the section "National Weather Service" for a detailed listing of other related NOAA pamphlets. The NCEP website provides links to all nine forecast centers. In addition to the "Tropical Prediction Center" link, valuable centers to review online during hurricane threats are the Hydrometeorological Prediction Center (HPC) and the NCEP Central Operations (NCO). The HPC specializes in rain forecasts and predicts heavy precipitation potential during hurricane threats. The NCO link shows weather model products (see chapter 2 for details). The National Hurricane Center analyzes the suite of computer models (which typically show slightly different hurricane track predictions) to make their best forecast judgments. By looking at the NCO weather model graphics, you could make your own hurricane forecast!

National Climatic Data Center

The National Climatic Data Center (NCDC) is one of several agencies in the National Environmental Satellite and Data Information Service (NESDIS). NESDIS administers the development and use of all operational civilian satellite-based environmental remote sensing systems and the national and international acquisition, processing, dissemination, and exchange of environmental data. The NCDC is one of the more popular NESDIS entities because it contains the world's largest archive of weather data and documentation (going back 150 years). NCDC receives a staggering amount of weather data, over 200 gigabytes (the equivalent of 72 million pages) per day. Data sources include satellites, radar, aircraft, ships, radiosonde, and other unique meteorology platforms as well as National Weather Service products. The NCDC also maintains one of the four World Data Centers in which data are exchanged among countries by cooperative agreement. The NCDC produces numerous publications involving weather data of hourly, three-hourly, daily, and monthly time frames.

The NCDC provides a wide range of services including consultations, subscriptions, publications, copies of original records, certifications, specialized climate studies, and other climate-related activities. Services are delivered on a variety of media, including online (Internet) access, CD-ROMs, magnetic tape, floppy disks, computer tabulations, maps, and publications. There is a charge for most services, with a discount for online access. However, some of the more popular data sets are available at the website for free.

The address of NCDC is:

National Climatic Data Center
Federal Building
151 Patton Avenue

Asheville, NC 28801-5001
Phone: (828) 271-4800
Fax: (828) 271-4876
Email: ncdc.info@noaa.gov
URL: http://www.ncdc.noaa.gov

Educational Material and Publications

Many data sets and products related to hurricanes are available from the NCDC website. A valuable and popular book, *Tropical Cyclones of the North Atlantic Ocean, 1871–1992*, provides historical track information and statistical summaries. For information about a specific hurricane, the publication *Storm Data* is highly recommended. *Storm Data* documents significant U.S. storms (including hurricanes) and contains statistics on property damage and human casualties. In addition, special reports on some notable hurricanes are available for free download. Graphics showing historical land-falling hurricanes is on the website. A publication titled *Products and Services Guide* contains detailed information about all NCDC products, data sets, and publications.

National Oceanographic and Atmospheric Administration

The National Oceanographic and Atmospheric Administration (NOAA) heads the U.S. weather and ocean programs, coastal and marine resource programs, and related research programs. NOAA is subdivided into the following five branches: the National Weather Service (NWS); the National Ocean Service; the National Marine Fisheries Service; the National Environmental Satellite, Data and Information Service (NESDIS); and NOAA Research. The NWS is actively involved in hurricane forecasting; NESDIS provides satellite information and archives weather data; and NOAA Research (through the Office of Oceanic and Atmospheric Research) contains several laboratories where weather research is conducted, including the Hurricane Research Division and the Geophysical Fluid Dynamics Laboratory. Many of the organizations are described in this chapter.

NOAA also participates in the Global Learning and Observations to Benefit the Environment (GLOBE) program, which is a worldwide network of students, teachers, and scientists working together to study and understand the environment. GLOBE students make environmental observations at or near their schools and report their data through the Internet to the scientists, who provide feedback to the students about their data.

NOAA can be contacted at:

NOAA Public and Constituent Affairs
U.S. Department of Commerce
14th Street & Constitution Ave., N.W.
Room 6217
Washington, DC 20230
Phone: (202) 482-6090
Fax: (202) 482-3154
URL: http://www.noaa.gov

Educational Material and Publications

There are numerous links related to weather education on NOAA's home page as well as links to its branches, which contain more specialized information. As you browse this website, you will find satellite photos, graphics, tables, and news releases. NOAA Public Affairs publishes a series of brochures known as the *NOAA Backgrounder* that contain detailed information about specific NOAA organizations, weather topics, and ocean topics. See the section "National Weather Service" for a detailed listing of other related NOAA publications (available for download in pdf format).

National Weather Association

The mission of the National Weather Association (NWA) is to help, support, and promote excellence in operational meteorology and related activities. The NWA started as a nonprofit organization, incorporated in Washington, D.C., in 1975 mainly to serve individuals interested in operational meteorology and related activities. It has grown to more than 2,800 members, 50 corporate members, and more than 250 subscribers including many colleges, universities, and weather service agencies. There are no restrictions on membership except age. Applicants should be 18 years old or older. Members receive publications (see below), discounts on buying NWA publications, and discounted registration fees for the NWA Annual Meeting. The NWA has an annual awards program and offers the NWA Seal of Approval to be earned by radio and television weathercasters to promote standards and quality broadcasting.

To join the NWA, print out the application form available from its website and send it to the NWA office, or request an application from the NWA office at:

NWA Office
1697 Capri Way
Charlottesville, VA 22911-3534
Phone or Fax: (434) 296-9966
Email: NatWeaAsoc@aol.com
URL: http://www.nwas.org

One can also join by simply sending into the NWA office a short letter of request with your full name, mailing address, and other information that is helpful to the NWA: name of employment organization or name of school attending, phone numbers for home, office, and fax, and email address. A check or money order in U.S. funds is also required to cover the annual dues. As of this writing, first-year dues are $25 ($12.50 for full-time students). Follow-on year dues are $28 ($14 for full-time students).

Educational Material and Publications

Members receive the quarterly journal *National Weather Digest* and a monthly newsletter. Nonmembers may also purchase these publications. While the NWA publications are not as technical as the AMS publications, in general they are written for operational meteorologists who have had at least a few meteorology classes and some forecasting experience, and only those with a meteorology degree will really appreciate the articles.

However, many of the books available from the NWA are nontechnical and suitable for all people interested in a variety of weather topics. A monograph on satellites contains sections on hurricane imagery: *Satellite Imagery Interpretation for Forecasters*. A 25-mm slide show (with a script) titled *Polar Orbiter Satellite Imagery Interpretation* provides examples of polar orbiter satellite imagery of all weather phenomenon, including hurricanes.

The NWA also provides grants to K–12 teachers to help improve the education of their students in meteorology.

National Weather Service

The mission of the National Weather Service (NWS) is to provide weather, hydrologic, and climate forecasts and warnings for the United States, its territories, adjacent waters, and ocean areas for the protection of life and property and the enhancement of the national economy. TV weathercasters and private meteorology companies prepare their forecasts using the basic forecast and weather observation data the NWS issues continuously all day. In addition, NWS data and products are used by private meteorologists for the provision of all specialized services, including consultant work and environmental impact studies.

Another interesting component of the NWS is the Meteorological Development Laboratory (MDL), which develops and implements scientific techniques into NWS operations. One aspect of this branch is the development of storm surge forecasts. Another aspect is the development of short-term forecasts of severe weather events, including hurricanes, using Doppler radar, satellites, and other state-of-the-art platforms. Their website is http://www.nws.noaa.gov/mdl/.

The NWS Headquarters' address is:

National Weather Service Headquarters
1325 East-West Highway
Silver Spring, MD 20910
URL: http://www.nws.noaa.gov (also called http://weather.gov)

The best method for contacting the NWS is through the NWS website, which contains an interactive form (in the "Contact Us" section) for general inquiries. You can also call the nearest NWS office for local information. The website shows your nearest NWS office and provides a contact phone number.

Educational Material and Publications

Numerous pamphlets and brochures are available on the NWS website. You can also order them through your local NWS office, the Red Cross, or the NOAA Outreach Unit at:

NOAA Office of Public and Constituent Affairs
Outreach Unit
1305 East-West Highway

Silver Spring, MD 20910
Phone: (202) 482-6090

Publications relevant to hurricanes include *Floods . . . The Awesome Power!*; Flash Flood (wallet card); *Hurricane Flooding: A Deadly Inland Danger; Red Cross: Are You Ready for a Flood or Flash Flood?; River and Flood Program (Hydrologic Services Program); The Hidden Danger: Low Water Crossing; Survival in a Hurricane; Hawaiian Hurricane Safety Measures with Central Pacific Tracking Chart; Hurricanes . . . Unleashing Nature's Fury; Hurricane! A Familiarization Booklet; Red Cross: Are You Ready for a Hurricane? A Mariner's Guide to Marine Weather Services: Coastal, Offshore, and High Seas; NOAA Weather Radio Pamphlet; NOAA Weather Radio . . . The Voice of the National Weather Service; Red Cross: Atlantic Hurricane Tracking Map; Pacific Hurricane Tracking Map; Atlantic Hurricane Names: 2003–2008;* and *Pacific Hurricane Names: 2003–2008.*

The website provides a comprehensive education section. Available commercial videos, slides, books, posters, charts, software are listed along with the vendors who sell them. Classroom ideas and experiments are presented in an excellent manner.

Complete and detailed weather information is available at the NWS-sponsored website http://weather.gov. Current weather warnings, forecasts, weather maps, radar, river information, and satellite imagery are shown. Animations of radar and satellite images are also available. Links to specialized forecast services are provided for aviation, hydrology, severe weather, weather impacts on fires, climate, and hurricanes. For instance, the link on hurricanes goes to the National Hurricane Center website. Another link is provided to the National Centers for Environmental Prediction to access weather model graphics.

The Meteorological Development Laboratory (MDL) produces an online monthly progress. The NOAA technical report *Sea, Lake, and Overland Surges from Hurricanes* contains documentation about the MDL hurricane storm surge model called SLOSH.

Naval Research Laboratory

The Naval Research Laboratory (NRL) conducts a broadly-based multidisciplinary program of scientific research and advanced technological development directed toward maritime applications of new and improved materials, techniques, equipment, systems, and ocean, atmospheric, and space sciences and related technologies. Congress commissioned the NRL in 1923 for the Department of the Navy. Today it is a field command under the chief of naval research and has thousands of personnel (nearly half of these PhDs) who address basic research issues concerning the navy's environment of sea, sky, and space. The NRL's parent organization is the Office of Naval Research (ONR) in Arlington, Virginia. The ONR coordinates, executes, and promotes the science and technology programs of the U.S. Navy and Marine Corps through universities, government laboratories, and nonprofit and for-profit organizations.

The NRL has a number of major sites and facilities. The largest facility is located at the Stennis Space Center in Bay St. Louis, Mississippi. Others include a facility at the Naval Postgraduate School in Monterey, California, and the Chesapeake Bay Detachment in Maryland. Additional sites are located in Maryland, Virginia, Alabama, and Florida. The unit responsible for atmospheric research is the Marine Meteorology Division in Monterey.

The address of the Monterey facility is:

Naval Research Laboratory
Marine Meteorology Division
7 Grace Hopper Ave.
Monterey, CA 93943-5502
Phone: (408) 656-4721
Fax: (408) 656-4314
URL: http://www.nrlmry.navy.mil/

NRL also has a public affairs officer who can be contacted at (202) 767-2541.

Educational Material and Publications

A web page under the "Project Demonstration" link of http://www.nrlmry.navy.mil is a popular site for hurricane enthusiasts. It contains global real-time satellite images of hurricanes and other weather phenomenon displayed in a variety of imagery formats, some of which are experimental. Products include cloud-motion winds, surface winds, rainfall estimates, microwave images (which show storm structure better), and high-resolution visible, infrared, and water vapor imagery. Satellite animations of these products are also online and are perhaps the best in the world. A unique product is the merging of both geostationary and polar orbiter products for composite pictures centered over a hurricane. Warning and track forecasts relevant to the times of the satellite data sets are shown for all storms. Hurricane forecasters from around the world use this state-of-the-art website.

The NRL publishes an annual review that contains short articles covering a variety of research areas. It also publishes *The NRL Fact Book* every two years; this is primarily a reference source for information on NRL's organizational structure, capabilities, and points of contact. An online publication titled *Tropical Cyclone Forecaster's Reference Guide* presents a technical but readable document on all facets of tropical cyclones. A list of all books, technical reports, and journal articles is available on the website.

A major concern of the U.S. Navy during hurricane threats is whether vessels should remain at port or evade the storm at sea. This decision is based on the circumstances of the hurricane, the facilities of the port, and the capabilities of the vessel and crew. Guidelines for all major ports are provided to the navy (and probably useful to all mariners) in the following publications, most of which are online and also available by CD-ROM: *Hurricane Havens Handbook for the North Atlantic Ocean; Typhoon Havens Handbook for the Western Pacific and Indian Oceans;* and *Typhoon Haven Evaluations for Brisbane, Cairns, Mackay, Townsville, Darwin, Fremantle, Stirling, Geraldton, and Bunbury.*

Copies of NRL reports and technical publications can be obtained from:

Ruth H. Hooker Research Library
4555 Overlook Ave., S.W.
Washington, DC 20375
(202) 767-READ (7323)
URL: http://infoweb2.nrl.navy.mil/

NOAA Aircraft Operations Center

The NOAA Aircraft Operations Center (AOC) provides specially equipped aircraft in support of NOAA's research, including hurricane research missions. The AOC was created in 1983 to consolidate the aviation assets operated by NOAA, including jet, propeller, helicopter, and amphibious aircraft. The aircraft: (1) collect scientific data essential to NOAA's weather research on hurricanes (by the Hurricane Research Division), tornadoes, winter storms, air pollution, and other atmospheric and oceanographic issues; (2) provide coastal and aeronautical charting and photogrammetry; and (3) conduct mammal, fish, and marine sanctuary surveys. The AOC supports all NOAA line offices and also participates in joint projects with other government agencies such as NASA, the Office of Naval Research, the Naval Research Laboratories, the U.S. Geological Survey, the U.S. Corps of Engineers, and the National Science Foundation.

AOC pilots are members of the NOAA Corps and are among the few pilots in the world who are trained and qualified to fly into hurricanes at dangerously low altitudes. These officers also occupy a few of the key managerial and staff positions within the AOC. The NOAA Corps makes up approximately 30 percent of the personnel within AOC, with the remaining staff comprised of civilian government employees whose background include flight engineer, aircraft maintenance technician, meteorologist, electronic engineer, electronic technician, computer specialist, and administrative and support personnel.

The aircraft used for hurricane research, reconnaissance, and surveillance include the Lockheed WP-3D 43 Orions and a G-IV Gulfstream Jet. The P-3s are the flagship aircraft of the AOC and are among the most advanced airborne environmental research platforms flying today. Built in 1975 and 1976, their cruising speed is 374 mph with a maximum range varying from 2,645 to 3,450 miles depending on their altitude. Since these are turboprop aircraft, the maximum altitude they can attain is 27,000 feet. They can carry up to twenty personnel, including up to ten scientists or observers.

The planes are essentially flying weather stations, each with a variety of proven and experimental aircraft sensors, radar instruments, and expendable probes. The following weather fields are measured or inferred from instrumentation on one or both aircraft: (1) rainfall and cloud distribution from the airborne radars; (2) winds above 3,000 feet in three dimensions from the airborne radars (requires two planes since one plane gives only two-dimensional winds); (3) flight-level temperature, dewpoint, wind, and solar and terrestrial radiation; (4) atmospheric electrification; (5) cloud particles and aerosols; (6) remotely sensed rain rate and surface winds using a C-band SCATerometer (C-SCAT) and a Stepped Frequency Microwave Radiometer (SFMR); (7) wind profiles below 3,000 feet using the Imaging Wind and Rain Airborne Profiler (IWRAP), a C- and Ku-band conically scanning Doppler radar; (8) camera and video equipment to capture the atmosphere and ocean environment on film; (9) ocean waves using a Scanning Radar Altimeter (SRA); (10) wind and heat fluxes using the Best Atmospheric Turbulence (BAT) probe; (11) moisture fluxes from a humidiometer probe; (12) sea spray droplet spectra from a Cloud Imaging Probe (CIP); and (13) sea-surface temperature in clear regions from a downward-looking, dual-frequency infrared radiometer.

The following have also been measured from deployed instruments in recent field programs: (1) vertical profiles of temperature, relative humidity, and wind using GPS dropsondes; (2) surface pressure from dropsondes; (3) vertical profiles of ocean temperature using Airborne Expendable BathyThermographs (AXBTs); (4) vertical profiles of ocean currents using Airborne Expendable Current Probes (AXCPs); (5) vertical profiles of ocean salinity and temperature using Airborne Expendable Conductivity Temperature Depth (AXCTD) probes; and (6) air temperature, ocean waves, sea surface temperature, and water temperature profiles from drifting buoys or submergible floats.

In 1997, the Gulfstream IV was added to the AOC's fleet. Since the Gulfstream is a jet, it can reach altitudes of 45,000 feet; it is also faster than the P-3s (with a cruising speed of 506 mph) and can fly farther (with a maximum range of 4,700 miles). The jet's primary mission is hurricane surveillance and will complement the work of the P-3s by flying into the steering currents of hurricanes up to 45,000 feet. The objective is to release dropsondes at multiple points around hurricanes from 40,000 feet and above. The dropsondes measure atmospheric conditions that surround and steer a hurricane. This data, which are ingested into hurricane forecast computer models, improves hurricane track forecasts and provide a better scientific understanding of hurricane intensification.

Amusingly, the planes are named after the Muppet characters (with the permission of Jim Henson Productions). Over a decade ago, one P-3 had become such a maintenance problem that it earned the nickname "The Pig" in the maintenance and operations departments. In 1987, a new crew chief and flight engineer, with some hard work, corrected many of the problems, cleaned the plane to a sparkling finish, and then redubbed the plane "Miss Piggy" because of that Muppet character's fussy nature about her appearance. Later, the second P-3 was renamed after Miss Piggy's best friend, Kermit. From 1991 to 1992, Jim Henson Productions worked with the AOC to develop the character art that now adorns the fuselage of each P-3. When NOAA purchased the Gulfstream, it was designated "Gonzo" because of its large, nonstandard nose radome. (Muppet characters are copyrighted by Jim Henson Productions.)

The planes are flown by members of the NOAA Corps, the smallest of the seven uniformed services of the United States. Originally founded for ship surveying missions in 1917 by Congress, the NOAA Corps consists of 200–300 officers, most of whom have a bachelor of science in engineering, mathematics, physics, computer science, meteorology, oceanography, biology, or a related discipline. Some have degrees in business or administration. Officers operate NOAA ships, fly NOAA aircraft, lead mobile field parties, conduct diving operations, manage research projects, and serve in staff positions. Recruits experience no military drill or stringent physical training but undergo demanding ship-handling exercises with classroom instruction in leadership, seamanship, navigation, and military protocol.

The AOC is located in Tampa, Florida. The address is:

NOAA/Aircraft Operations Center (AOC)
P.O. Box 6829
MacDill Air Force Base, FL 33608-0829
Phone: (813) 828-3310
Fax: (813) 828-3266
URL: http://www.aoc.noaa.gov

Individuals interested in a NOAA Corps career should contact:

NOAA Corps Commissioned Personnel Center
1315 East-West Highway, Room 12100
Silver Spring, MD 20910-3282
Phone: 1-800-299-6622
Email: NOAACorps.Recruiting@noaa.gov
URL: http://www.noaacorps.noaa.gov

Educational Material and Publications

A variety of brochures are available from the AOC, including an annual report, additional aircraft information, historical information, and general AOC facts. Additional information about the aircraft and its instrumentation can also be obtained from the

Hurricane Research Division's home page, http://www.aoml.noaa.gov/hrd. More information about the GPS dropwindsonde can be found at the University Corporation for Atmospheric Research Atmospheric Technology Division's website, http://www.atd.ucar.edu. The brochure *Commissioned Officer Corps* is available from the NOAA Corps and provides more information on the organization.

Tropical Prediction Center

The Tropical Prediction Center (TCP), where the National Hurricane Center is located, is a branch of the National Centers for Environmental Prediction (NCEP). The mission of the TPC is to save lives, mitigate property loss, and improve economic efficiency by issuing the best watches, warnings, forecasts and analyses of hazardous tropical weather and by increasing understanding of these hazards. Through international agreement, the TPC has responsibility within the World Meteorological Organization to generate and coordinate tropical cyclone analysis and forecast products for twenty-four countries in the Americas and the Caribbean and for the waters of the North Atlantic Ocean, Caribbean Sea, Gulf of Mexico, and the eastern North Pacific Ocean. TPC products are distributed through a close working relationship with the media and emergency management communities.

To fulfill this mission, the TPC is comprised of three branches: the National Hurricane Center (NHC), the Tropical Analysis and Forecast Branch (TAFB), and the Technical Support Branch (TSB).

National Hurricane Center

The NHC maintains a continuous watch on hurricanes over the Atlantic, Caribbean, Gulf of Mexico, and the eastern Pacific from May 15 through November 30. The NHC prepares and issues forecasts, watches, and warnings within text advisories and graphical products. Although many countries issue their own warnings, they generally base them on direct discussions with, and guidance from, the NHC. During other parts of the year, the NHC conducts an extensive outreach and education program, training U.S. emergency managers and representatives from many other countries affected by hurricanes. The NHC is also actively involved in public awareness programs.

Tropical Analysis and Forecast Branch

The TAFB provides year-round marine weather analysis and forecast products over the tropical and subtropical waters of the eastern North and South Pacific and the North Atlantic basin. The branch also produces satellite-based weather interpretation and rainfall estimates for the international community. In addition, the TAFB provides support to the NHC when additional forecasters are necessary and also provides hurricane position and intensity estimates from the Dvorak technique.

Technical Support Branch

The TSB provides support for TPC computer and communications systems, including the McIDAS satellite data processing systems. The TSB also maintains a small applied research unit that develops tools for hurricane and tropical weather analysis and prediction. In addition, the TSB has a storm surge group that provides information for developing evacuation procedures for coastal areas.

The TPC is located on the campus of the Florida International University in southwestern Dade County in Miami, Florida. The address is:

Tropical Prediction Center
11691 S.W. 17th Street
Miami, FL 33165-2149
Phone: (305) 229-4470
URL: http://www.nhc.noaa.gov

The TPC has a public affairs officer who can be contacted at (305) 229-4404.

Educational Material and Publications

Numerous hurricane topics are available either as direct links or in pdf format in the brochure section of the website. Topics include general hurricane information and hurricane preparedness issues. A variety of technical reports have been published (see website) and are available from NTIS (see chapter 8). A popular technical report written by three NHC forecasters is *The Deadliest, Costliest, and Most Intense United States Hurricanes of this Century (and Other Frequently Requested Hurricane Facts)*, and an abbreviated but updated version of this report is available on the website. Each year NHC forecasters publish annual hurricane summaries in *Weatherwise*, the *Mariner's Weather Log*, *WMO's Annual Global Tropical Cyclone Summary*, and the *Monthly Weather Review* (please see chapter 8 for details on ordering these publications). In addition, all *Monthly Weather Review* hurricane summaries are available online in pdf format in the history archive section. See the section "National Weather Service" for a detailed listing of other related NOAA pamphlets. The publication *Mariner's Guide for Hurricane Awareness* is a very informative publication on hurricanes in general, with a focus on boating issues, and is available online in pdf format.

The website itself contains TAFB weather products, NHC advisories (see the glossary), abundant hurricane education material, archives of previous years' advisories, verification of previous NHC forecasts, and hurricane statistics (some are reproduced in chapter 6). Previous tropical storm and hurricane track graphics are online. The popular HURricane DATabase (HURDAT), which contains six-hourly center locations and intensities of historical tropical storms and hurricanes, can be downloaded from the NHC website. Hurricane tracking charts can also be downloaded.

The Weather Channel

Started in 1982, the Weather Channel has been the only cable TV media outlet providing twenty-four-hour weather coverage and special weather programs seven days a week. The Weather Channel also provides weather forecasts to some newspapers and radio stations. During the hurricane season, a Tropical Update is provided about ten minutes before each hour. The Tropical Update shows satellite photos of all tropical disturbances, tropical storms, and hurricanes in the Atlantic and East Pacific; the western North Pacific also is sometimes shown. During hurricane landfall situations, coverage expands, including frequent discussions with the Weather Channel's hurricane experts and live reports from meteorologists near the expected area of landfall. Current weather information and forecasts are available on its website, http://weather.com. In addition, weather-related news and seasonal features are online. During the U.S. hurricane season, the "Tropical Update" website link provides satellite pictures, past storm track, forecast track, storm surge forecast, and wind distribution, as well as a summary from a Weather Channel meteorologist.

The Weather Channel's headquarters are in Atlanta, Georgia. It does not want phone and address information released publicly. However, its website http://weather.com contains an interactive section for general inquiries, resume submission, etc., in the "Contact Us" link.

Educational Material and Publications

The Weather Channel's online bookstore sells general weather books and videos. *The Weather Channel: The Improbable Rise of a Media Phenomenon* is about the history of this cable network. The Weather Channel also airs "The Weather Classroom" on Mondays and Thursdays at 4:00 through 4:30 A.M. EST, appropriate for grades 5–10. Each episode includes three separate weather-related topics lasting eight minutes each. Educators can tape the program and use it in their classrooms. The Weather Channel has also set up an education website at http://www.weatherclassroom.com. Students can issue their own weather forecasts and record weather observations or obtain information on careers in meteorology. An online weather encyclopedia and glossary are also available. The Weather Channel presents two thirty-minute shows every day between 8 P.M. and 9 P.M. EST called "Storm Stories," and often the episode is on a particular hurricane.

World Meteorological Organization

The World Meteorological Organization (WMO) is the successor to the International Meteorological Organization (IMO), founded in 1873 to coordinate global research on weather and climate. In 1947, the IMO met in Washington, D.C., and decided to restructure as a specialized agency of the United Nations. This agency was named the World Meteorological Organization and was located with the United Nations in Geneva, Switzerland. It began operations in 1951.

The purposes of the WMO are to facilitate international cooperation in the establishment of networks of stations for making meteorological, hydrological, and other observations and to promote the rapid exchange of meteorological information, the standardization of meteorological observations, and the uniform publication of observations and statistics. It also furthers the application of meteorology to aviation, shipping, water problems, agriculture, and other human activities; promotes operational hydrology; and encourages research and training in meteorology. The 187-member organization, consisting of countries and territories, provides the authoritative scientific voice on the state and behavior of the earth's atmosphere and climate.

The WMO is structured in eight technical commissions covering atmospheric sciences, basic systems, climatology, hydrology, instruments and methods of observations, aeronautical meteorology, marine meteorology, and agricultural meteorology. Members are grouped in six regional associations: Africa, Asia, South America, North and Central America, the Southwest Pacific, and Europe. The WMO's governance consists of a congress of members that meets every four years and an executive council, consisting of a subset of thirty-seven members, that conducts business yearly between congresses. Leadership consists of an elected president, three vice presidents, and an appointed secretary-general. The secretary-general directs a professional staff called the secretariat. The leadership positions have a tradition of maintaining balance between the developing and developed countries.

The WMO administers several programs, including the important World Weather Watch (WWW). The mission of the WWW is to coordinate and monitor observational and telecommunication facilities all over the world so that every country has available the information it needs to provide weather services on a day-to-day basis. The WWW implements standardization of measuring methods and techniques, common telecommunication procedures, and the presentation of observed data and processed information in a manner that is understood by all, regardless of language.

The WWW is also involved in long-term planning and research. One component of the WWW is the Tropical Cyclone Programme (TCP). The main purpose of the Tropical Cyclone Programme is to assist WMO members, through an internationally coordinated program, in their efforts to mitigate tropical cyclone disasters. Under the coordination of the WWW, this program is designed to assist more than fifty countries in areas vulnerable to tropical cyclones, to minimize destruction and loss of life by improving forecasting and warning systems, and to encourage members to establish national disaster prevention and preparedness measures. Another contribution from this program has been several important technical publications.

For general information, the WMO address is:

World Meteorological Organization
7 bis Avenue de la Paix
SP 2300
1211 Geneva 2
Switzerland
Phone: +41 22 730 8111
Fax: + 41 22 730 8181
Email: wmo@wmo.int
URL: http://www.wmo.int

Publication orders in the United States and Canada must be made through the American Meteorological Society (see the AMS website publication section for details). Other countries need to order directly from the WMO.

Educational Material and Publications

The WMO produces many publications. With respect to hurricanes, the technical reports "Global Perspectives on Tropical Cyclones" and "Global Guide to Tropical Cyclone Forecasting" are strongly recommended. The WMO issues an annual report as well as an annual summary of global tropical cyclone activity for all five WMO regions (see chapter 3 for details). A variety of reports are available that cover workshops, conferences, forecast manuals, regional WMO operational plans, training courses, WMO observation code, telecommunications of weather data, weather data management, and disaster prevention/preparedness. Decadal progress reports have been prepared, such as *Twenty Years of Progress and Achievement of the WMO Tropical Cyclone Programme* (1980–1999) and *A Decade of Progress: The World Meteorological Organization in the 1990s and the New Century*. Many of these reports are available online in Microsoft Word or html format. In particular, the "Global Guide to Tropical Cyclone Forecasting" is a popular link. The WMO also provides a link to websites of the Regional Specialized Meteorological Centres (RSMCs) and the Tropical Cyclone Warning Centres (TCWCs).

8

Print, Nonprint, and Internet Resources

This chapter describes a number of publications on hurricanes, separated by topical sections. These books range from elementary to general to technical in nature and focus on topics such as individual storms, hurricane preparedness, and societal impact of hurricanes.

Almost all recent (after 1990) U.S. government publications and technical reports can be ordered through the National Technical Information Service (see information at end of chapter). Should your library not contain a particular book, try having your librarian order it through an interlibrary loan from another library. A good starting point is the NOAA library, which contains a user-friendly interface for listing books by author, subject, title, or keyword on its website (see http://www.lib.noaa.gov for details; click on its "Library Catalog" link to begin the reference search).

Updates of Internet links are provided on the author's website at http://www.drfitz.net.

Print Resources

Young Adult (Elementary) Books on Hurricanes

These books contain basic information about hurricanes supplemented with good pictures. They are ideal for students and other readers looking for a short, easy to read book on hurricanes. The author S. Lee also provides some science projects.

Sharon M. Carpenter and T. G. Carpenter. *The Hurricane Handbook: A Practical Guide for Residents of the Hurricane Belt.* Tailored Tours Publications, 1993. Inc. 128 pp. ISBN 0-96312-414-5.

S. Hood and J. Barkan. *Hurricanes!* Simon and Schuster Children's, 1998, 64 pp. ISBN 0-68982-017-8.

P. Lauber. *Hurricanes: Earth's Mightiest Storms.* New York: Scholastic, 1996, 64 pp. ISBN 0-59047-4065.

S. Lee. Hurricanes. New York: Franklin Watts, 1993, 63 pp. ISBN 0-53115-665-6.

Mariner Books

S. Dashew and L. Dashew. *Mariner's Weather Handbook.* Tucson, AZ: Beowulf, 1998, 594 pp. ISBN 0-96580-282-5.

This book provides forecasting and boating tactics used by professional routers and ocean-racing navigators. It includes techniques for reading the ocean for weather changes as well as analyzing weather maps. All facets of weather are covered. Quick reference checklists and executive summaries are also provided in the page margins. With regard to hurricanes (and the reason this source is included in this chapter), this handbook gives one of the best descriptions of using the Dvorak technique of estimating hurricane intensity from satellite images. In addition, it provides basic information on hurricane climatology, radio broadcast schedules for hurricane forecasts worldwide, and how to navigate around a hurricane.

E. J. Holweg. *Mariner's Guide for Hurricane Awareness in the North Atlantic Basin.* Miami, FL: National Hurricane Center, 2000, 72 pp. Available at http://www.nhc.noaa.gov/marinersguide.pdf.

This electronic book, available from the National Hurricane Center website, is a thorough source for mariners on navigating during the hurricane season as well as an excellent reference for general readers interested in hurricanes. Chapter 1 discusses hurricane basics, including definitions, formation, structure, and observations at sea. Chapter 2 is on hurricane motion, genesis regions, and general tracks in different ocean basins. Chapter 3 discusses NHC products and methods for obtaining information at sea by radio, radio fax, high frequency (HF) transmission, medium frequency (MF) transmission, very high frequency (VHF) transmission, and the National Weather Service weather radio. Chapter 4 gives guidance on how to avoid hurricanes at sea.

General Books on Hurricanes

M. Allaby. *A Chronology of Weather.* New York: Facts on File, 1998, 154 pp. ISBN 0-81605-321-0.

The largest part of this book consists of two chronological accounts. One lists some of the major weather disasters (i.e., hurricanes, tornadoes, droughts, etc.) by the years in which they occurred, from 3200 B.C. through 1997. The other chronological account lists important developments in the understanding of weather and improvements in forecasting. This book also contains thirty simple weather experiments that can be done at home to elucidate basic weather principles.

Peter R. Chaston. *Hurricanes!* Kearney, MO: Chaston Scientific, 1996, 182 pp. ISBN 0-96451-722-1.

This book, written by a meteorologist, covers many aspects of hurricanes, including life cycles, structure, the Saffir-Simpson scale, wind destruction and storm surge damage, and spectacular hurricane stories from survivors. It also includes 100 photos from reconnaissance flights, weather satellites, and radars as well as easy to understand graphics. The history of hurricanes is chronicled from Columbus's time through today. Maps detail the tracks of hurricanes, and the naming of these storms and the hurricane category scale of intensity are explained.

D. E. Fisher. *The Scariest Place on Earth.* New York: Random House, 1994, 250 pp. ISBN 0-679-42775-9.

Dr. David Fisher is a professor of cosmochemistry and director of the Environmental Science Program at the University of Miami, and he personally experienced Hurricane Andrew's landfall. This book is written in a three-topic fashion, jumping back and forth from one topic to another. These three topics include historical accounts and facts on hurricanes, a description of hurricanes, and a chronological account of his Andrew experience. Fisher also presents a reasonably balanced discussion on whether or not global warming will increase the number and intensity of hurricanes. However, some readers may not appreciate his conclusion that global warming is already occurring.

J. Fishman and R. Kalish. *The Weather Revolution: Innovations and Imminent Breakthroughs in Accurate Forecasting.* New York: Plenum, 1994, 276 pp. ISBN 0-30644-764-9.

This excellent book discusses the history of weather-related technological advances, including computer models, satellites, hurricane forecasting, severe weather forecasting, and long-term weather prediction. The authors write about complex topics in a thorough but understandable fashion.

R. Pielke Jr. and R. Pielke Sr., eds. *Storms,* vol. 1. London: Routledge, 2000, 563 pp. ISBN 0-41517-239-X.

This book summarizes the current state of knowledge of storms and forecasting procedures, including their hazards from both a scientific and social standpoint. Part 2 of this book is devoted to hurricanes, with several articles written by experts on seasonal predictions, the WMO Tropical Cyclone Programme, warning and mitigation procedures, and the impact of these storms on Australia, China, and India.

J. R. Fleming, ed. *Historical Essays on Meteorology.* Boston, MA: American Meteorological Society, 1996, 617 pp. ISBN 1-87822-017-9.

This is a book on modern meteorology history. It chronicles achievements in theoretical meteorology, computer models, observation tools, cloud physics, hurricanes, convection, lightning, climatology, hydrology, the private sector, and education. Dr. Mark DeMaria, a former researcher at the National Hurricane Center and the Hurricane Research Division, presents in chapter 9 a history of hurricane forecasting from 1920 to the present.

D. M. Ludlum. *Early American Hurricanes, 1492–1870.* Boston, MA: American Meteorological Society, 1989, 198 pp. ISBN 0-93387-616-5.

This book describes, in chronological order, the hurricanes prior to 1870 that either closely approached or actually crossed the Atlantic and Gulf coastlines of the United States. The book is divided into two periods. The first section starts with the voyages of Christopher Columbus and extends through the years of exploration and colonization to the end of 1814. The year 1814 makes a convenient break, for the first federal attempts to institute a national weather observing service started during the War of 1812. With the cessation of hostilities in 1815, an era of worldwide peace emerged with vigorous worldwide trading accompanied by weather logs by merchant marines. The second section spans this period, carries through the Civil War, and extends to 1870, the year the U.S. Signal Corps established its storm warning system. This book's documentation is meticulous, with much detail, and is a must for any weather historian.

B. Sheets and J. Williams. *Hurricane Watch*. New York: Vintage, 2001, 331 pp. ISBN 0-37570-390-X.

Written by a former director of the National Hurricane Center and *USA Today's* weather page editor, this book gives an excellent overview on the growth of hurricane knowledge. It provides in chronological order each major discovery and weather technology breakthrough as well as many tragedies due to the early ignorance of these storms. The book is written in a dramatic, journalistic style woven with facts. The first three chapters cover the first few hundred years through the early 1900s. Chapter 4 is about hurricane reconnaissance and is followed by the evolving of hurricane forecasting and science in the 1950s (chapter 5) and the 1960s (chapter 6). Chapter 7 talks about weather modification experiments in Project STORMFURY, and chapter 8 covers the 1970s and 1980s. Chapter 9 is devoted to how computer modeling has dramatically improved hurricane track forecasts. The book finishes with a chapter on Hurricane Andrew, which Bob Sheets experienced firsthand. This book was recognized by the American Meteorological Society with the Louis J. Battan Award, which honors each year's best general public weather book.

R. Simpson, R. Anthes, M. Garstang, and J. M. Simpson, eds. *Hurricane! Coping with Disaster*. Special Publication Vol. 55. Washington, D.C.: American Geophysical Union, 2002, 360 pp. ISBN 0-87590-298-7 (hardcover), ISBN 0-87590-297-9 (paper).

This book chronicles the century-long struggle to understand the enormous power and devastating impact of hurricanes. Some authors provide a look at recent advances and promising new technologies for tracking and predicting the path of hurricanes. Others examine what has been done and what still could be done to reduce the vulnerability of people and property to this extreme force of nature.

Stories from the men and women who pioneered the effort to understand, track, and cope with hurricanes bring this volume to life. The book has something to offer a wide range of readers who may be interested in the history of scientific discovery, the science and technology of hurricane research, or the societal and economic challenges posed by major disasters.

P. Davies. *Inside the Hurricane.* New York: Henry Holt, 2000, 264 pp. ISBN 0-80506-574-1.

The author spent the 1999 hurricane season with the scientists of the National Hurricane Center and the Hurricane Research Division in Miami as they studied a series of fascinating and intense hurricanes and struggled with budget limitations. Davies supplies a vivid description of the devastation of Hurricane Mitch and Hurricane Floyd and the experience of people in the midst of the storm. Davies flew missions with the NOAA's P-3 hurricane hunter aircraft and gives a good feel for the combination of raw excitement, pure terror, and occasional boredom of these flights.

Popular Weather Magazines

The *Mariner's Weather Log,* a publication of the National Weather Service (NWS), contains articles, news, and information about marine weather events and phenomenon, storms at sea, weather forecasting, the NWS Voluntary Observing Ship (VOS) program, Port Meteorological Officers (PMOs), cooperating ships officers, and their vessels. The magazine provides meteorological information to the maritime community and contains a comprehensive chronicle on marine weather. It also recognizes ships officers for their efforts as voluntary weather observers and allows NWS to maintain contact with and communicate with more than 10,000 shipboard observers (ships officers) in the merchant marine, NOAA Corps, U.S. Coast Guard, U.S. Navy, etc. In addition, the magazine provides annual summaries of hurricane activity for all ocean basins worldwide. The *Mariner's Weather Log* is published three times yearly (April, August, and December) and is distributed to mariners, marine institutions, the shipping industry, scientists, educational and research facilities, educators, libraries, government agencies, and the general public.

Weatherwise, available bimonthly, is a popular magazine intended for weather enthusiasts in the general public. Articles are written about a wide variety of weather-related issues, including hurricanes, with compelling columns and color photographs. In the February/March issue, the previous Atlantic and Pacific hurricane seasons are reviewed.

Hurricane Preparation

Staff of the Miami Herald. *Hurricanes: How to Prepare and Recover.* Kansas City, MO: Andrews and McMeel, 1993, 125 pp. ISBN 0-83621-718-7.

Based on the Hurricane Andrew experience, this book provides in-depth instructions on how to prepare one's home and family for a hurricane onslaught. Instructions for recovery and repairs are also included.

Societal Impact of Hurricanes

R. A. Cook and M. Soltani, eds. *Hurricanes of 1992.* New York: American Society of Civil Engineers, 1994, 808 pp. ISBN 0-78440-046-6.

The year 1992 was unusually costly due to three major hurricanes that hit U.S. coasts or territories: Hurricane Andrew, Hurricane Iniki, and Typhoon Omar, causing more than $35 billion in damages. This proceeding contains papers presented at a conference in Miami, Florida, to disseminate information on wind hazards and managing wind-related disasters based on experiences with these three storms. Topics include (1) wind speeds and wind loads; (2) risk assessment; (3) insurance; (4) damage assessment; (5) building codes; (6) building code implementation and enforcement; (7) coastal structures; (8) manufactured, residential, and commercial structures; (9) essential facilities; and (10) lifelines.

Henry F. Diaz, ed. *Hurricanes: Climate and Socioeconomic Impacts.* Verlag, Germany: Springer, 1997, 292 pp. ISBN 3-54062-078-8.

This book contains reports from a workshop titled "Atlantic Hurricane Variability on Decadal Time Scales: Nature, Causes, and Socioeconomic Impacts" held at the National Hurricane Center during February 9–10, 1995. Climate change issues are discussed along with socioeconomic issues such as private insurance losses and the availability of insurance for coastal residents in the future. This workshop assembled experts with different views on the range and predictability of potential climatic changes and their impact on tropical cyclone activity. The potential of massive economic losses when Atlantic hurricane activity increases to the level observed in the 1940s and 1950s is discussed.

James B. Elsner and A. Birol Kara. *Hurricanes of the North Atlantic.* New York: Oxford University Press, 1999, 496 pp. ISBN 0-19512-508-8.

This book is intended for both hurricane climate researchers and users of hurricane information. Topics include hurricane climatology, a history of the major Atlantic hurricanes, forecasting methods, and methods for analyzing the social and economic impact of hurricanes.

Roger A. Pielke Jr. and Roger A. Pielke Sr. *Hurricanes: Their Nature and Impacts on Society.* Chichester, England: John Wiley, 1997, 279 pp. ISBN 0-47197-354-8.

This book discusses the increased vulnerability of the U.S. coastline to hurricanes due to population increases and property development. It addresses both the scientific and societal impacts of hurricanes and illustrates the economic benefit of hurricane research in an era where scientific research is under pressure to demonstrate a better connection to societal needs. Suggestions for policy implementation are included. A well-done description of hurricanes is contained in the first few chapters.

A good supplementary text, which includes a more basic description of hurricanes, archives of annual Atlantic hurricane tracks, and pictures of the ocean inside hurricanes of different intensities, is:

Roger A. Pielke. *The Hurricane.* London: Routledge, 1990, 228 pp. ISBN 0-41503-705-0.

Technical

R. A. Anthes. *Tropical Cyclones: Their Evolution, Structure, and Effects.* Boston, MA: American Meteorological Society, 1982, 208 pp. ISBN 0-93387-654-8.

Although this book contains somewhat dated material, it presents informative sections discussing the mathematical framework of hurricanes as well as their structure and life cycle. One chapter nicely summarizes the history of hurricane computer simulations. The reader should be warned that this book is quite technical, but it gives a flavor of what true hurricane research is about.

R. L. Elsberry, W. M. Frank, G. J. Holland, J. D. Jarrell, and R. L. Southern. *A Global View of Tropical Cyclones.* Montgomery, CA: Naval Postgraduate School, 1987, 195 pp.

Eighty-three experts in tropical cyclone forecasting, research, and warning strategies from twenty-eight different countries attended the International Workshop on Tropical Cyclones held in Bangkok, Thailand, from November 25 to December 5, 1986. This was the first truly worldwide gathering of tropical cyclone specialists. The result is this book, which, while dated, contains pertinent material that tropical experts still reference. Some parts are technical, but most of the book is quite readable. Chapters are included on tropical cyclone observations, structure, genesis, motion, and their impacts. The final chapter discusses warning and mitigation strategies.

G. R. Foley, H. E. Willoughby, J. L. McBride, R. L. Elsberry, I. Ginis, and L. Chen. *Global Perspectives on Tropical Cyclones.* Geneva, Switzerland: World Meteorological Organization, Report No. TCP-38, 1995, 289 pp.

In many respects, this book is a revision of the book *A Global View of Tropical Cyclones* to reflect developments in tropical cyclone research since 1987. *Global Perspectives on Tropical Cyclones* provides the state of hurricane science through 1994. A chapter on ocean interactions with tropical cyclones has also been added.

This book compliments the World Meteorological Organization's operational forecasting publication *A Global Guide to Tropical Cyclone Forecasting,* edited by Greg Holland (described in this section).

P. J. Hebert, J. D. Jarrell, and M. Mayfield. *The Deadliest, Costliest, and Most Intense United States Hurricanes of this Century (and Other Frequently Requested Hurricane Facts).* NOAA Technical Memorandum NWS TPC-1, National Oceanic and Atmospheric Administration, National Weather Service, National Hurricane Center, Miami, Florida, 1997.

This technical report provides detailed data about the most destructive U.S. hurricanes of the twentieth century. Statistics are included about the most expensive hurricanes, the most intense hurricanes, U.S. landfalls by state and by decade, etc., plus other hurricane facts. Updated versions of this report are contained on the National Hurricane Center website http://www.nhc.noaa.gov.

G. J. Holland. *Global Guide to Tropical Cyclone Forecasting*. Geneva, Switzerland: World Meteorological Organization, Report No. TCP-31, 1993, 347 pp.

The purpose of this book is to complement the theoretical and descriptive content of *A Global View of Tropical Cyclones* and *Global Perspectives on Tropical Cyclones* (described elsewhere in this section). This book is a practical guide to tropical cyclone forecasting. Included are chapters on track forecasting, satellite interpretation, storm surge forecasting, seasonal forecasting, forecast center strategies, and warning procedures. An informative appendix of tables, definitions, and tropical cyclone records is included at the back of the book. This book is also available at the World Meteorological Organization website http://www.wmo.int., as well as http://www. bom.gov.au/brmc (look for additional links under the mesoscale section to find this book).

E. N. Rappaport and J. Fernandez-Partagas. *The Deadliest Atlantic Tropical Cyclones, 1492–1994*. NOAA Technical Memorandum NWS NHC 47, National Oceanic and Atmospheric Administration, National Weather Service, National Hurricane Center, Miami, Florida, 1995.

This technical report contains information on tropical cyclones than have caused twenty-five or more deaths in any part of the Atlantic basin. It also contains information on cyclones that may have caused twenty-five or more deaths but where the total was not quantified by sources. Updated versions of this report are contained on the National Hurricane Center website http://www.nhc.noaa.gov.

General Meteorology Books and Textbooks

Hurricanes embody almost all meteorological processes. For an understanding of hurricanes, a solid background in all atmospheric phenomena is required. Therefore, the interested reader is urged to consult a good general meteorology textbook. A few of the many possible choices are listed below:

C. D. Ahrens. *Meteorology Today.* St. Paul, MN: West Publishing, 1994, 591 pp. ISBN 0-31402-779-3.

E. W. Danielson, J. Levin, and E. Abrams. *Meteorology.* Boston, MA: WCB/McGraw-Hill, 1998, 462 pp. ISBN 0-69721-711-6.

J. M. Nese and L. M. Grenci. *A World of Weather: Fundamentals of Meteorology.* Dubuque, IA: Kendall/Hunt, 1998, 539 pp. ISBN 0-78723-578-4.

J. Williams. *The Weather Book.* Arlington, VA: USA Today, 1992, 212 pp. ISBN 0-67973-669-7.

Encyclopedias and Glossaries on Meteorology

D. Longshore. *Encyclopedia of Hurricanes, Typhoons, and Cyclones.* New York: Checkmark Books, 2000, 372 pp. ISBN 0-81604-291-8.

Organized by topic in alphabetical order, this book discusses famous U.S. hurricanes as well as some well-known international typhoons and tropical cyclones. A discussion of all countries and U.S. states impacted by these storms is provided. Biographies of key figures in hurricane science are included. Meteorological terms are also defined.

S. H. Schneider. *Encyclopedia of Climate and Weather.* New York: Oxford University Press, 1998, 929 pp. ISBN 0-19509-485-9.

In addition to a good general meteorology textbook, a meteorology encyclopedia is useful because hundreds of topics are arranged in alphabetical order with liberal cross-references. This two-volume set is suitable for high school and above, with clear explanations and abundant figures on a variety of climate and weather topics. Each article is written by an authority on the subject. A glossary is provided in the back.

I. W. Geer, ed. *Glossary of Weather and Climate.* Boston, MA: American Meteorological Society, 1996, 272 pp. ISBN 1-87822-021-7.

Written for a general audience, this book contains a glossary of more than 3,000 terms frequently used in discussions and descriptions of meteorological and climatological phenomena. In addition, the glossary includes definitions of related oceanic and hydrologic terms.

Todd S. Glickman, ed. *Glossary of Meteorology*. Boston, MA: American Meteorological Society, 2000, 855 pp. ISBN 1-87822-034-9.

This book attempts to define every important meteorological term found in the literature today and is intended for meteorologists. It presents definitions that are understandable to any meteorologist yet palpable to the specialist. Most are terms from mainstream meteorology, but the book also draws from related disciplines in hydrology, oceanography, geomagnetism, and astrophysics. Some terms come from the basic sciences of physics and chemistry; some are from applicable portions of mathematics, statistics, and electronics; and others are from the folk language of weather lore through many ages. Each definition represents the effort of three or more individuals, at least two of whom are specialists in the subject area. These terms can also be viewed at *http://amsglossary.allenpress.com*.

State Books

John M. Williams and Iver W. Duedall. *Florida Hurricanes and Tropical Storms, 1871–2001*. Gainesville: University Press of Florida, 2002, 176 pp. ISBN 0-81302-494-3.

This book is a comprehensive chronological guide to hurricanes, tropical storms, and near-misses that have impacted Florida since 1871. Additional features include statistics for each hurricane and tropical storm, eyewitness accounts, photos, ten-year tracking charts, and a hurricane preparedness checklist.

Jay Barnes. *Florida's Hurricane History*. Chapel Hill: University of North Carolina Press, 1998, 384 pp. ISBN 0-80784-748-8.

This book provides a detailed record of more than 100 hurricanes that have struck Florida in the last 450 years based on newspaper reports, National Weather Service records, books, and eyewitness accounts. Information on the basics of hurricane structure, formation, naming, and forecasting is included, and there are more than 200 photographs, maps, and illustrations.

Jay Barnes. *North Carolina's Hurricane History*. Chapel Hill: University of North Carolina Press, 1998, 206 pp. ISBN 0-80784-507-8.

This book provides a detailed record of more than fifty hurricanes that have struck North Carolina from 1526 to 1996 based on newspaper reports, National Weather Service records, books, and eyewitness accounts. A chapter is also included on Northeasters, which are nontropical winter cyclones that are near hurricane intensity and impact the U.S. East Coast each winter. Northeasters inflict storm surge and wind damage similar to tropical storms and category 1 hurricanes.

Individual Hurricanes

The Galveston Hurricane of 1900

J. E. Weems. *A Weekend in September.* College Station: Texas A&M University Press, 1997, 192 pp. ISBN 0-89096-390-8.

Weems interviews survivors of the worst natural disaster in U.S. history—the Galveston Hurricane of 1900 that inundated the island with an 8–15-foot storm surge that killed at least 6,000 people on the island and 2,000 farther inland. Efforts to protect Galveston from future hurricanes are also discussed. A 6-mile-long seawall was completed in 1905, and the elevation of the entire city was raised 8 feet to a height of 17 feet. A storm in 1915 tested the newly reshaped island, and although there was some damage, the community survived. Based on this result, the length of the seawall was expanded to 10.5 miles.

E. Larson. *Isaac's Storm.* New York: Crown Publishers, Random House, 1999, 323 pp. ISBN 0-60960-2330.

This book is a combined dramatization and factual account of the 1900 Galveston Hurricane, with the island's chief meteorologist Isaac Cline as the central character. The book delves into Cline's career and character and also exposes the ineptitude, corruption, and arrogance of Weather Bureau administrators in Washington, D.C., who ignored warnings from Cuban meteorologists that a hurricane was in the Gulf of Mexico and instead stated that the storm was headed into the Atlantic; no warning was ever issued from Washington, D.C. This is probably the most factual book on Isaac Cline and the Galveston Hurricane to date and exposes many myths about this tragedy. Also included is brief historical information of other Atlantic hurricanes before 1900.

P. B. Bixel and E. H. Turner. *Galveston and the 1900 Storm*. Austin: University of Texas Press, 2001, 190 pp. ISBN 0-29270-884-X.

The authors draw on survivors' accounts to vividly recreate the storms and its aftermath. They describe how the work of local relief agencies grew into lasting reforms. The book is full of illustrations and pictures to complement the text.

The 1926 Florida Hurricane

L. F. Reardon. *The Florida Hurricane & Disaster, 1926*. Coral Gables, FL: Arva Parks, 1986, 112 pp. ISBN 0-91438-104-0.

This is a reproduction of a survivor's diary, supplemented with pictures, of the 1926 hurricane that hit Miami and killed 243 people. An updated version also includes a description of Hurricane Andrew (1992), which is uniquely appended upside down starting on the back cover.

The 1928 Okeechobee Hurricane

L. W. Will. *Okeechobee Hurricane: Killer Storms in the Everglades*. St. Petersburg, FL: Great Outdoors, 1971, 204 pp. No ISBN.

This book is an eyewitness account of the 1928 Okeechobee Hurricane that killed and injured thousands of people. Most of the deaths were by drowning, but some were due to snakebites as people climbed into trees to escape the floodwaters, only to find hordes of venomous water mocassins also seeking shelter. Although the lake is actually inland, the hurricane forced a storm surge that broke the eastern earthen dike on the southern end of Lake Okeechobee. This calamity occurred within a few miles of a large city and a world-famous resort, yet so isolated was the location that no one knew what had happened for three days afterward. The author also discusses a similar situation from another hurricane that occurred two years earlier and killed about 200 at the Okeechobee town of Moore Haven. Pictures of the aftermath from both of these hurricanes are included. As a result of these disasters, a levee (called the Herbert Hoover Dike in honor of the president who supported its construction) that compares in size to the Great Wall of China was built to prevent further disasters.

R. Mykle. *Killer 'Cane: The Deadly Hurricane of 1928*. Lanham, MD: Cooper Square Press, 2002, 235 pp. ISBN 0-81541-207-X.

This book is a historical novel, written from an omnipresent point of view. Most of the characters actually lived, and the story is built on real-life experiences. Mykle's description of the effects on the residents of Lake Okeechobee from the failure of the dike and subsequent flooding are the climax and highlight of this book, with vivid, horrifying descriptions of the storm and its aftermath. Mykle also estimates the deaths at 3,000, higher than official estimates, pointing out that the search for bodies ended early due to a lack of funding and that many of those killed were migrant African Americans with no available records who were buried in mass graves. Such a death toll would be the second worst from a hurricane after the 1900 Galveston hurricane.

The 1935 Labor Day Hurricane
W. Drye. *Storm of the Century: The Labor Day Hurricane of 1935.* Washington, D.C.: National Geographic, 2002, 326 pp. ISBN 0-79228-010-5.

The most powerful hurricane ever to hit the United States, this category 5 storm with a storm surge possibly 30 feet high killed 400 people, many of them veterans working on a New Deal construction project in the Florida Keys. This book describes how administrators ignored early warnings and were tardy in sending a train to evacuate the workers, resulting in much controversy and political acrimony. Congressional committees tried to assign blame to inadequate forecasts (the prediction was south of the Keys), and the president successfully diverted any blame from the executive branch. This story is told from the viewpoint of storm survivors, relief workers, and government officials. It also describes the difficulties in communications in 1935, the ability of sea captains to read the ocean, and how local people prepared for the storm.

The New England Hurricane (1938)
J. McCarthy. *Hurricane!* New York: American Heritage, 1969, 168 pp. ISBN 0-82810-020-9.

This book provides a detailed account of the 1938 New England Hurricane that made landfall on the south shore of Long Island, New York. This storm surprised New England residents, who were unaccustomed to hurricanes (the last hurricane had hit New England in 1815). Because the hurricane moved at a fast speed of 60 mph, and because the Weather Bureau had thought it would recurve offshore and not hit the United States, there was little preparation time available. Even worse, the storm hit at the equinox when tides are usually their highest. Ten-foot waves atop a 20-foot storm surge killed 600 people and caused immense property damage. Barrier islands were swept so bare that rescue workers used phone company charts to determine where houses once stood. This book includes detailed eyewitness accounts and photos of the storm's aftermath.

E. S. Allen. *A Wind to Shake the World: The Story of the 1938 Hurricane.* Boston, MA: Little, Brown, 1976, 370 pp.

Allen had just started his first job as a newspaper reporter when the storm hit September 21, 1938. In addition to his own experience, the book tells the story of others caught by the storm.

W. E. Minsinger. *The 1938 Hurricane.* Boston, MA: American Meteorological Society, 1988, 128 pp. ISBN 9-99338-216-7.

To mark the fiftieth anniversary of this destructive hurricane, the American Meteorological Society, in cooperation with the Blue Hill Meteorological Observatory, published this detailed historical volume complete with more than 120 archival photographs and weather charts.

R. A. Scotti. *Sudden Sea: The Great Hurricane of 1938.* New York: Little, Brown, 2003, 288 pp. ISBN 0-31673-911-1.

This book provides an account of the storm drawn from newspaper accounts, personal testimony of survivors, and archival sources and is set within the stage of the times. However, it contains a novelist's touch. It also includes the recollections of Katherine Hepburn, whose family had a summer home on the Connecticut shore.

The Great American Hurricane (1944)
The North Atlantic Hurricane. Charleston, SC: Historical Publications, 64 pp.

This storm received its name because it traveled up the Atlantic coast from North Carolina to the Northeast, making landfall in Rhode Island. This publication chronicles the path and destruction of this storm that impacted the whole East Coast.

Hurricane Hazel (1954)
Hurricane Hazel Lashes North Carolina: The Great Storm in Pictures. Charleston, SC: Historical Publications.

This publication shows the damage caused by Hurricane Hazel on the North Carolina shoreline. Hazel is the most destructive storm in North Carolina's history, producing wind gusts of 150 mph and a high storm surge superimposed on the highest ocean tide of the year.

Hurricane Camille (1969)
Miss Camille 1969. Charleston, SC: Historical Publications.

This publication contains photographs of Hurricane Camille's incredible destruction on the Louisiana, Mississippi, Alabama, and Florida Gulf Coast. This is only one of three U.S. hurricanes that made landfall as category 5 storms.

P. D. Hearn. *Miss Camille: Monster Storm of the Gulf Coast.* Hattiesburg: University Press of Mississippi, 2004, 176 pp. ISBN 1-57806-655-7.

This book provides a history of Hurricane Camille, weaving in many historical accounts of those who experienced the storm. The book was motivated by forty-one interviews originally conducted by University of Southern Mississippi's R. W. Pyle, who conducted the survivor interviews a decade after Camille. The book dramatically gives vivid accounts during and after the storm. Some examples include: how a 73-year-old lady left her wheelchair, rode the waves, and survived; how two priests clung to statues of the Virgin Mary and St. Joseph at St. Michael's Catholic Church and survived; and how a man, who lost most of his family, coped with his grief by helping rescue workers recover their bodies and lay them side by side. Hundreds of domesticated animals were killed because no facilities or food existed for their care. The book discusses the emotional trauma in the aftermath of the storm. One woman stepped outside, surveyed the damage, and shot herself. Divorces quadrupled. The book also contains a history of hurricanes and the settlement of the Gulf Coast.

Hurricane Frederic (1979)
Hurricane Frederic. Charleston, SC: Historical Publications. 88 pp.

This publication contains photographs of Hurricane Frederic's impact on southern Alabama. Interviews with those who experienced Frederic's landfall are included.

Hurricane Alicia (1983)
A. Kareem, ed. *Hurricane Alicia: One Year Later.* New York: American Society of Civil Engineers, 1985, 335 pp. ISBN 0-87262-466-8.

Due to its inland passage over downtown Houston, Hurricane Alicia (1983) caused more than $1 billion in damage, providing a future glimpse of what hurricanes can do in heavily developed coastal areas. This proceeding contains papers presented at a conference in Galveston, Texas, devoted to areas of meteorology, structural behavior, hurricane-resistant design, and building codes.

Hurricane Hugo (1989)

Storm of the Century: Hurricane Hugo. Charleston, SC: Historical Publications. 76 pp.

This publication contains photographs of Hurricane Hugo's impact on Charleston, South Carolina. Interviews with those who experienced Hugo's landfall are included. One interview vividly describes a family's struggle with Hugo's storm surge as it broke through their home, causing them to scamper to the attic.

William P. Fox. *Lunatic Wind: Surviving the Storm of the Century.* Chapel Hill, SC: Algonquin, 1992, 197 pp. ISBN 0-94557-542-4.

Fox, who has experience writing fiction and film scripts, documents in a dramatic style the landfall of Hurricane Hugo near Charleston, South Carolina. Based on detailed research and interviews, Fox recreates the night of the storm and the people caught in it. Mixing factual reporting with the storytelling of a novelist, Fox's docudrama recreates the experience of people who survived: two teenage surfers who had to swim for their lives as their beachhouse is destroyed by the storm surge; a shrimp boat captain determined to ride out the storm in the town's harbor; and more than 1,000 people in an evacuation shelter (a high school gym) that is flooded by the storm surge because the shelter is at a lower elevation than local charts had indicated. He also recounts the aftermath of people surviving with no electricity, no phones, and destroyed homes.

Hurricane Bob (1991)

New England's Nightmare: Hurricane Bob. Charleston, SC: Historical Publications, 97 pp.

This publication contains photographs of Hurricane Bob's impact on the northeastern United States. Interviews with those who experienced Bob's landfall are included.

Hurricane Andrew (1992)

W. E. Bailey. *Andrew's Legacy: Winds of Change.* West Palm Beach, FL: Telshare, 1999, 197 pp. ISBN 0-91028-714-7.

Dr. William E. Bailey is an attorney and consultant specializing in insurance industry issues and litigation. In October 1992, Bailey was appointed codirector of the Hurricane Insurance Information Center in Miami, Florida. The center, established right after Hurricane Andrew wreaked its havoc on South Florida, was formed by seven national insurance trade associations and twenty-two major insurance companies, representing the largest segment of property insurers in the state. Its one-year mission was to inform and educate consumers and the media regarding insurance issues in the aftermath of the costliest natural disaster in U.S. history. One of the results of this experience is this book, which discusses implications on possible future hurricane landfalls in metropolitan regions.

Staff of the Miami Herald and El Nuevo Herald. *Hurricane Andrew: The Big One.* Kansas City, MO: Andrews and McMeel, 1992, 160 pp. ISBN 0-83628-012-1.

A humbling, touching collection of photographs displaying the incredible destruction of Hurricane Andrew and its impact on the people of Dade County south of Miami. It is impossible to describe these pictures in words. As one survivor says in the book, "It's something you can't talk about. You just have to see it." Also included are eyewitness accounts and reports in a chronological fashion.

Noorina Mirza and M. Quraishy. *Before and After Hurricane Andrew 1992.* Miami, FL: Kenya Photo Mural, 1992, 96 pp. ISBN 0-96349-620-4.

This is another picture collection of Hurricane Andrew's damage. Much of the book contains interesting comparative pictures of what areas looked like before Andrew and what they looked like immediately afterward.

Florida's Path of Destruction: Hurricane Andrew. Charleston, SC: Historical Publications. 97 pp.

This publication contains more than 300 photographs of Hurricane Andrew's impact on the Bahamas, Miami, and Louisiana.

Interviews with those who experienced its landfall and with weather specialists are included, as are satellite photos.

U.S. Army Corps of Engineers. *Hurricane Andrew Assessment.* Tallahassee, FL: Post, Buckley, Schuh and Jernigan, 1993. No ISBN.

Hurricane Andrew provided the opportunity for the Federal Management Agency's and the Army Corps of Engineers to investigate whether state and local officials used their products, whether their past studies were useful in preparing for Andrew, and which of these past studies were most or least useful. This report addresses these issues.

E. F. Provenzo and A. B. Provenzo. *In the Eye of Hurricane Andrew 1992.* Gainesville: University of Florida Press, 2002, 184 pp. ISBN 0-81302-566-4.

This publication is a collaboration of the authors and undergraduate and graduate students at the University of Miami based on ninety-two interviews. It was originally conceived as a way to cope with the disaster and evolved into a book. This book is divided chronologically, with chapters on the preparation period, during the storm, the immediate aftermath, the weeks after the storm (vividly describing the difficult conditions), and the long-term effects.

Hurricane Iniki (1992)

Hurricane Iniki. Charleston, SC: Historical Publications, 100 pp.

This publication contains photographs of Hurricane Iniki's impact on Kauai in the Hawaii Islands, including unique documentation of the landfall itself since Iniki hit during the day (unlike many other hurricanes, such as Hugo and Andrew, that hit at night).

Hurricane Alberto (1994)

Deadly Waters: Tropical Storm Alberto. Charleston, SC: Historical Publications, 92 pp.

This publication contains photographs of Hurricane Alberto's impact, which produced record-breaking rainfall and floods that took twenty-eight lives in Georgia and two in Alabama. Interviews with those who experienced Alberto are included.

Hurricane Georges (1998)
Hurricane Georges. Charleston, SC: Historical Publications, 52 pp.

This publication contains photographs of Hurricane Georges's impact on the Florida Keys as documented through the *Key West Citizen and Free Press* reporters who live on the Keys. Interviews with those who experienced Georges's landfall are included, such as those who live on "houseboat row" and in the popular Duval Street area.

Historical Hurricane Tracks

C. J. Neumann, B. R. Jarvinen, C. J. McAdie, and J. D. Elms. *Tropical Cyclones of the North Atlantic Ocean, 1871–1992.* Asheville, NC: National Environmental Satellite, Data, and Information Service, National Climatic Data Center, 1993, 193 pp.

This valuable book presents annual tracks for tropical depressions, tropical storms, and hurricanes for the Atlantic Ocean from 1871 to 1992. Also included in these tracks are an additional category of tropical cyclones called "subtropical storms," which are hybrids of tropical and nontropical cyclones. Basic statistical summaries are presented as well.

R. S. Lourensz. *Tropical Cyclones in the Australian Region, July 1909 to June 1980.* Maryborough, Victoria: Bureau of Meteorology, Department of Science and Technology, Australian Government Publishing Service, 1981, 94 pp. ISBN 0-64201-718-2.

This book presents annual tracks for cyclones in the oceanic regions near Australia from 1909 to 1980. Basic statistical summaries and graphics are also presented.

J. Maunder. 1995. *An Historic Overview Regarding the Intensity, Tracks, and Frequency of Tropical Cyclones in the South Pacific during the Last 100 Years, and an Analysis of any Changes in these Factors.* Geneva, Switzerland: World Meteorological Organization, Report No. TCP-37, 1995, 48 pp.

As suggested by the title, this book presents annual tracks for cyclones in the South Pacific for the 1890s–1990s, as well as tabular and statistical information.

National Hurricane Center Annual Summaries

Approximately once a year, the publication *Monthly Weather Review* has a summary of the previous year's Atlantic and East Pacific hurricane activity authored by National Hurricane Center forecasters. Summaries of Atlantic tropical wave activity are also published every few years. The summaries include descriptions of named storms along with pertinent meteorological data and satellite imagery. The *Monthly Weather Review* reports hurricane data back to 1881. Scanned images of all the *Monthly Weather Review* reports are available in pdf format at http://www.nhc.noaa.gov in the "NHC/TPC Archive of Past Hurricane Seasons" link. Similar reviews are also published in the February/March issue of *Weatherwise;* these are less technical and describe the Atlantic and East Pacific hurricane seasons from the preceding year.

Articles in the *Monthly Weather Review* discussing each Atlantic hurricane season since 1960 are as follows:

G. E. Dunn. 1961. The hurricane season of 1960. *Mon. Wea. Rev.*, **89**, 99–108.

G. E. Dunn and Staff. 1962. The hurricane season of 1961. *Mon. Wea. Rev.*, **90**, 107–119.

G. E. Dunn and Staff. 1963. The hurricane season of 1962. *Mon. Wea. Rev.*, **91**, 199–207.

G. E. Dunn and Staff. 1964. The hurricane season of 1963. *Mon. Wea. Rev.*, **92**, 128–138.

G. E. Dunn and Staff. 1965. The hurricane season of 1964. *Mon. Wea. Rev.*, **93**, 175–187.

A. L. Sugg. 1966. The hurricane season of 1965. *Mon. Wea. Rev.*, **94**, 183–191.

A. L. Sugg. 1967. The hurricane season of 1966. *Mon. Wea. Rev.*, **95**, 131–142.

A. L. Sugg and J. M. Pelissier. 1968. The hurricane season of 1967. *Mon. Wea. Rev.*, **96**, 242–250.

A. L. Sugg and P. J. Hebert. 1969. The Atlantic hurricane season of 1968. *Mon. Wea. Rev.*, **97**, 225–239.

R. H. Simpson, A. L. Sugg, and Staff. 1970. The Atlantic hurricane season of 1969. *Mon. Wea. Rev.*, **98**, 293–306.

R. H. Simpson and J. M. Pelissier. 1971. Atlantic hurricane season of 1970. *Mon. Wea. Rev.*, **99**, 269–277.

R. H. Simpson and J. R. Hope. 1972. Atlantic hurricane season of 1971. *Mon. Wea. Rev.*, **100**, 256–267.

R. H. Simpson and P. J. Hebert. 1973. Atlantic hurricane season of 1972. *Mon. Wea. Rev.*, **101**, 323–333.

P. J. Hebert and N. L. Frank. 1974. Atlantic hurricane season of 1973. *Mon. Wea. Rev.*, **102**, 280–289.

J. R. Hope. 1975. Atlantic hurricane season of 1974. *Mon. Wea. Rev.*, **103**, 285–293.

P. J. Hebert. 1976. Atlantic hurricane season of 1975. *Mon. Wea. Rev.*, **104**, 453–465.

M. B. Lawrence. 1977. Atlantic hurricane season of 1976. *Mon. Wea. Rev.*, **105**, 497–507.

M. B. Lawrence. 1978. Atlantic hurricane season of 1977. *Mon. Wea. Rev.*, **106**, 534–545.

M. B. Lawrence. 1979. Atlantic hurricane season of 1978. *Mon. Wea. Rev.*, **107**, 477–491.

P. J. Hebert. 1980. Atlantic hurricane season of 1979. *Mon. Wea. Rev.*, **108**, 973–990.

M. B. Lawrence and J. M. Pelissier. 1981. Atlantic hurricane season of 1980. *Mon. Wea. Rev.*, **109**, 1567–1582.

M. B. Lawrence. 1982. Atlantic season of 1981. *Mon. Wea. Rev.*, **110**, 852–866.

G. B. Clark. 1983. Atlantic season of 1982. *Mon. Wea. Rev.*, **111**, 1071–1079.

R. A. Case and H. P. Gerrish. 1984. Atlantic hurricane season of 1983. *Mon. Wea. Rev.*, **112**, 1083–1092.

M. B. Lawrence and G. B. Clark. 1985. Atlantic hurricane season of 1984. *Mon. Wea. Rev.*, **113**, 1228–1237.

R. A. Case. 1986. Atlantic hurricane season of 1985. *Mon. Wea. Rev.*, **114**, 1390–1405.

M. B. Lawrence. 1988. Atlantic season of 1986. *Mon. Wea. Rev.,* **116**, 2155–2160.

R. A. Case and H. P. Gerrish. 1989. Atlantic hurricane season of 1987. *Mon. Wea. Rev.,* **117**, 939–949.

M. B. Lawrence and J. M. Gross. 1989. Atlantic hurricane season of 1988. *Mon Wea. Rev.,* **117**, 2248–2259.

B. Case and M. Mayfield. 1990. Atlantic hurricane season of 1989. *Mon. Wea. Rev.,* **118**, 1165–1177.

M. Mayfield and M. B. Lawrence. 1991. Atlantic hurricane season of 1990. *Mon. Wea. Rev.,* **119**, 2014–2026.

R. J. Pasch and L. A. Avila. 1992. Atlantic hurricane season of 1991. *Mon. Wea. Rev.,* **120**, 2671–2687.

M. Mayfield, L. A. Avila, and E. N. Rappaport. 1994. Atlantic hurricane season of 1992. *Mon. Wea. Rev.,* **122**, 517–538.

R. J. Pasch and E. N. Rappaport. 1995. Atlantic hurricane season of 1993. *Mon. Wea. Rev.,* **123**, 871–886.

L. A. Avila and E. N. Rappaport. 1996. Atlantic hurricane season of 1994. *Mon. Wea. Rev.,* **124**, 1558–1578.

M. B. Lawrence, B. M. Mayfield, L. A. Avila, R. J. Pasch, and E. N. Rappaport. 1998. Atlantic hurricane season of 1995. *Mon. Wea. Rev.,* **126**, 1124–1151.

R. J. Pasch and L. A. Avila. 1999. Atlantic hurricane season of 1996. *Mon. Wea. Rev.,* **127**, 581–610.

E. N. Rappaport. 1999. Atlantic hurricane season of 1997. *Mon. Wea. Rev.,* **127**, 2012–2026.

R. J. Pasch, L. A. Avila, and J. L. Guiney. 2001. Atlantic hurricane season of 1998. *Mon. Wea. Rev.,* **129**, 3085–3123.

M. B. Lawrence, L. A. Avila, J. L. Beven, J. L. Franklin, J. L. Guiney, and R. J. Pasch. 2001. Atlantic hurricane season of 1999. *Mon. Wea. Rev.,* **129**, 3057–3084.

J. L. Franklin, L. A. Avila, J. L. Beven, M. B. Lawrence, R. J. Pasch, and S. R. Stewart. 2001. Atlantic hurricane season of 2000. *Mon. Wea. Rev.,* **129**, 3037–3056.

J. L. Beven, S. R. Stewart, M. B. Lawrence, L. A. Avila, J. L. Franklin, and R. J. Pasch. 2003. Atlantic hurricane season of 2001. *Mon. Wea. Rev.,* **131**, 1454–1484.

R. J. Pasch, M. B. Lawrence, L. A. Avila, J. L. Beven, J. L. Franklin, and S. R. Stewart. 2004. Atlantic hurricane season of 2002. *Mon. Wea. Rev.* **132**, 1829–1859.

M. B. Lawrence, L. A. Avila, J. L. Beven, J. L. Franklin, R. J. Pasch, and S. R. Stewart. 2005. Atlantic hurricane season of 2003. *Mon. Wea. Rev.* **133**, 1744–1773.

Joint Typhoon Warning Center Annual Tropical Cyclone Summary

The Joint Typhoon Warning Center (JTWC) provides tropical cyclone forecasting support to the U.S. military and to allies within the Pacific and Indian Ocean basins under the auspices of the Naval Pacific Meteorology and Oceanography Center in Hawaii. The JTWC was founded on May 1, 1959, when the U.S. commander-in-chief of the Pacific forces directed that a single tropical cyclone warning center be established for the western North Pacific region. This region stretches from 180°E longitude westward across the western Pacific and Indian Oceans in the Northern Hemisphere and from the oceans near Australia westward to the eastern coast of Africa in the Southern Hemisphere.

Every year, the JTWC staff prepares an annual report summarizing tropical cyclone activity in the region. Included are storm tracks, a history of each storm, and satellite pictures. Tables and graphs present the JTWC's intensity and track forecasts as well as forecast errors. Ongoing research activities are summarized.

Summaries are printed for each year from 1959 to 1995. An example reference for 1994 is:

Staff of the Joint Typhoon Warning Center, *1994 Annual Tropical Cyclone Report,* NAVPACMETOCCENWEST/JTWC, Guam, Mariana Islands, 337 pp. No ISBN.

Starting in 1996, the JTWC stopped printing the annual issue and made the summaries available only in electronic format. In addition, the printed versions have been scanned into an electronic format. All electronic versions are available at http://www.npmoc.navy.mil.

Annual Summary of Australian Cyclones

The *Australian Meteorological Magazine* has a thorough annual summary of tropical cyclones that have occurred in the Australian Southeast Indian and Southwest Pacific basins. For more information about the *Australian Meteorological Magazine*, see http://www.bom.gov.au/amm.

Annual Summary of Cyclones in the North Indian Ocean

The Indian journal *Mausam* carries an annual summary of tropical cyclone activity over the North Indian Ocean.

Annual Summary of Tropical Cyclones for All Ocean Basins

The *Mariner's Weather Log* has articles from all of the global basins in annual summaries. These are descriptive and nontechnical.

Natural Disaster Survey Reports

After every major natural disaster, the National Oceanographic and Atmospheric Administration (NOAA) sends a survey team to determine how effectively its warning and detection system performed and to identify systematic strengths and improvements so that any necessary improvements can be developed and implemented. These Natural Disaster Survey Reports contain useful data, figures, and recommendations not easily found in other publications. A sample of these reports is listed below. Natural Disaster Survey Reports have been prepared for any recent major hurricane affecting the United States, so please contact NOAA for more details.

In addition, occasionally a Committee on Natural Disasters will conduct on-site studies and prepare reports reflecting their findings, as well as make recommendations on the mitigation of natural disasters. Example references of these Natural Disaster Studies for Hurricanes Hugo and Elena are listed below:

Natural Disaster Survey Report. Hurricane Iniki, September 6–13, 1992. U.S. Department of Commerce, National Oceanic and Atmospheric Administration, National Weather Service, Silver Spring, MD, 1993, 54 pp.

Natural Disaster Survey Report. Hurricane Andrew, August 23–26, 1992. U.S. Department of Commerce, National Oceanic and Atmospheric Administration, National Weather Service, Silver Spring, MD, 1993, 131 pp.

Natural Disaster Survey Report. Tropical Storm Alberto, July, 1994. U.S. Department of Commerce, National Oceanic and Atmospheric Administration, National Weather Service, Silver Spring, MD, 1995, 47 pp.

Natural Disaster Survey Report. Hurricane Gilbert, September 3–16, 1988. U.S. Department of Commerce, National Oceanic and Atmospheric Administration, National Weather Service, Silver Spring, MD, 1998. 35 pp.

Natural Disaster Survey Report. Hurricane Frederic, August 29–September 13, 1979. U.S. Department of Commerce, National Oceanic and Atmospheric Administration, National Weather Service, Silver Spring, MD, 1979. 27 pp.

National Research Council. *Natural Disaster Studies: Hurricane Elena, August 29–September 2, 1985.* Washington, D.C.: National Academy Press, 1991, 121 pp. ISBN 0-30904-434-0.

National Research Council. *Natural Disaster Studies: Hurricane Hugo; Puerto Rico, the Virgin Islands, and Charleston, South Carolina, September 17–22, 1989.* Washington, D.C.: National Academy Press, 1994, 276 pp. ISBN 0-30904-475-8.

NCDC *Storm Data* Publication

Storm Data is a list of severe weather observations across the United States and is published on a monthly basis by the National Climatic Data Center. The reports are generated through (1) National Weather Service (NWS) phone calls to damaged regions; (2) reports volunteered to the NWS from emergency management groups (such as the Army Corps of Engineer, the Federal Emergency Management Agency, and the U.S. Geological Survey), law enforcement agencies, the general public, and other credible organization; and (3) information provided to NWS offices by newspaper clippings. Caution is advised when using *Storm Data*, since it is difficult to verify the accuracy of some reports (i.e., general public reports and newspaper clippings).

With regard to hurricanes, detailed documentation of meteorological data, storm surge heights, general statistical information, and storm history can be found when such a storm makes landfall. To find out information about a particular hurricane, you need to know the year and month when the hurricane impacted the United States, then look up that particular monthly version of Storm Data. NWS offices contain these books, as do some universities and libraries. The books can also be ordered from the National Climatic Data Center (see http://www.ncdc.noaa.gov).

How to Order a U.S. Government Technical Report

Almost all recent (after 1990) U.S. government publications and technical reports can be ordered through the National Technical Information Service (NTIS). For example, you could order many of the technical reports from the National Hurricane Center from NTIS. NTIS is located at 5285 Port Royal Road, Springfield, VA 22161. To order a document, call the NTIS sales desk at 1–800–553–6847. For additional information, visit the NTIS website at http://www.ntis.gov.

Nonprint Resources

Videos display the awesome power of hurricanes, augmented by home video footage. This is a rapidly changing market, with certain videos quickly going out of print while others emerge with new storms. Therefore, vendors that currently provide videos are listed below with a brief discussion. Software is also listed, including FEMA's emergency preparedness package, HAZUS.

Hurricane Videos

Individual Hurricane Videos
Richard Horodner's *Hurricane Videos* is possibly the most complete selection of hurricane videos. Richard Horodner is a hurricane chaser, filming many land-falling hurricanes up close. This is possibly the best selection of videos, with individual storms starting from the 1980s to today. As a bonus, he also offers hurricane chase services! His address is private, with contact only available by phone or email.

Richard Horodner's Hurricane Videos
Phone: (772) 978-7673
Email: horodner@aol.com
URL: http://www.hurricanevideo.citymax.com/

General Hurricane Videos

There is a surprising dearth of general videos on hurricanes. In general, only a few stand out as the best on the subject, and they are not available from the original company. A search on the Internet will yield many new and used videos.

One is the PBS NOVA video *Nature's Fury: Hurricanes*. You can also purchase this in a three-video package titled *Nature's Fury*, which not only includes the hurricane footage but other natural phenomenon such as lightning, tornadoes, and earthquakes. This NOVA video contains excellent Hurricane Camille footage as well as interviews of people who were in the storm. A general description of hurricanes is included in this program. NOVA follows hurricane forecasters and researchers as they monitor Hurricane Gilbert, which is the strongest Atlantic storm on record. NOVA shows how NHC forecasters make their forecasts and tags along with the Hurricane Research Division as they fly into Hurricane Gilbert. The aftermath of Hurricane Andrew is also shown.

Another is the Discovery Channel program *Raging Planet: Hurricane*. It provides a general description of hurricanes and footage of historical storms such as Hurricane Andrew, and the 1970 Bangladesh storm that killed 300,000 people is included. A documentary of North Carolina residents and emergency preparedness officials is presented as they prepare for Hurricane Bertha and Hurricane Fran in 1996. Footage of the Hurricane Hunters as they fly into these storms is shown.

A third documentary, *Hurricane Andrew: Ground Zero*, was originally aired by the Learning Channel. It contains mostly human interest stories on this hurricane but does include some scientific background.

A final recommended documentary is *Savage Skies: Monster of the Deep*, which chronicles the events before, during, and after the landfall of Hurricane Andrew on Miami in 1992. Compelling interviews are conducted with National Hurricane Center forecasters and several people who experienced the wrath of the storm as it destroyed their homes.

Good Internet sites for buying these programs include The Weather Shop at http://www.weathershop.com/books-videos.

htm; My Weather Guide at http://shop.myweatherguide.com, an affiliate of amazon.com, that offers weather products in the form of books, instruments, and electronics in addition to videos; and The Teacher's Video Company at http://www.teachersvideo.com.

Educational Hurricane Videos

The Red Cross and FEMA (see chapter 7) provide several videos for the general public, charging just shipping costs. *Jason and Robin's Awesome Hurricane Adventure Companion Video* is the companion video for the twelve-page, four-color workbook titled *Jason and Robin's Awesome Hurricane Adventure,* for children in third to fifth grades, about hurricane facts, hazard avoidance, planning, supplies, and what to do when watches and warnings are issued. *Against the Wind: Protecting Your Home From Hurricane Wind Damage* demonstrates how to inspect and make simple changes within homes to mitigate the potential devastating effects of hurricane wind damage. *Home Preparedness for Hurricanes* and *Hurricane Information Guide for Coastal Residents* are similar videos. *Before the Wind Blows* provides basic hurricane information, planning decisions, evacuation information, and how the Red Cross responds to hurricanes. An interactive CD-ROM, *Hurricane Strike,* is also available. It is designed for middle schoolers but is educational for all ages and provides background information on hurricanes and mitigation approaches as well as games and worksheets that incorporates math, science, and geography.

Insight Media, Inc., provides a variety of videos and CD-ROMS on most meteorology and oceanography topics. A few of the many include *The Weather Edu-Tutor Bundle,* four CD-ROMs on forecasting, weather phenomena (including hurricanes), data collection, and remote satellite sensing; *Understanding the Weather,* which addresses air masses, fronts, clouds, storms, and predictions, providing a basic understanding of weather phenomena and featuring a blend of live action, animation, and graphics; and *Pressure and Winds: Weather,* which explains what causes air pressure, explores the measurement of air pressure, and shows the effects of temperature and moisture content on air pressure. The address is:

Insight Media, Inc.
2162 Broadway
New York, NY 10024-0621
Phone: 800-233-9910
Fax: 212-799-5309
URL: http://www.insight-media.com

Films for the Humanities and Sciences offers the video *Inside a Hurricane*. This video provides background on deadly Hurricane Mitch (1998) and costly Hurricane Andrew (1992). It includes footage from Mitch, Andrew, and other hurricanes and follows scientists as they trace the growth of a tropical storm into a full-blown hurricane. It also provides footage of a flight into a hurricane's eye. The program documents the tragic story of the schooner *Fantome*, which tried to outrun Mitch and failed, losing all on board. This organization also offers a general meteorology video called *The World's Weather*. The address is:

Films for the Humanities and Sciences
P.O. Box 2053
Princeton, NJ 08543-2053
Phone: 800-257-5126
Fax: 609-671-0266
URL: http://www.films.com

Software Tracking

A variety of free, shareware, and commercial packages are available for tracking hurricanes. A list, with detailed descriptions, is available on Chris Landsea's FAQ page at http://www.aoml.noaa.gov/hrd/tcfaq/tcfaqHED.html. A few of the better packages include "Jstrack," a freeware hurricane tracking program for Unix and Windows 9x/ME/NT/2k/XP; "Eye of the Storm," a commercial package with 3D capabilities; "Global Tracks," which provides both historical and current hurricane tracks; and "Hurricane Watch 2000," which contains zooming capability.

Emergency Preparedness

The most popular emergency preparedness software is FEMA's (see chapter 7) HAZards U.S. MultiHazard (HAZUS-MH). HAZUS-MH is a risk assessment software program for analyzing potential losses from floods, hurricane winds, and earthquakes. It is a direct result of natural disasters from the early 1990s, particularly the Northridge earthquake (1994), Hurricane Andrew (1992), and the Midwest floods (1993), when FEMA decided to include prevention and risk assessment into emergency management.

Current scientific and engineering knowledge is coupled with the latest Geographic Information Systems (GIS) technology to produce estimates of hazard-related damage before, or after, a disaster occurs. HAZUS-MH takes into account various impacts of a hazard event such as:

- Physical damage to residential and commercial buildings, schools, critical facilities, and infrastructure.
- Economic loss, such as lost jobs, business interruptions, repair and reconstruction costs.
- Social impacts to people, including requirements for shelters and medical aid.

HAZUS-MH is designed to produce loss estimates for use by state, regional, and local governments in planning for earthquake loss mitigation, emergency preparedness, and response and recovery. HAZUS-MH helps states, communities, and businesses prepare for, mitigate the effects of, respond to, and recover from a hazard event. It has several interfaces for general, advanced, and expert use.

Federal, state, and local government agencies and the private sector can order HAZUS-MH free of charge from the FEMA Distribution Center. Details are provided at http://www.fema.gov/hazus/. Training courses are also listed on this website.

The Internet is also a fluid environment, but over the years certain websites have maintained a valuable and stable presence in the hurricane community. These informative websites are shown at the end of this section.

Hurricane and Weather Internet Resources

American Meteorological Society. Boston, Massachusetts. http://www.ametsoc.org.

The American Meteorological Society (AMS) is a nonprofit, professional society for meteorologists and related fields. This sight contains career, educational, publications, and conferences related to meteorology. Conference papers and presentations can be downloaded, while journal articles are available online for a fee (abstracts are free).

American Red Cross. Washington, D.C. http://www.redcross.org.

The American Red Cross is a humanitarian organization, led by volunteers, that provides relief to victims of disasters and helps prevent, prepare for, and respond to emergencies. This website contains information on hurricane preparedness and hurricane relief in disaster areas. Many articles can be downloaded in pdf format

Chris Landsea's Hurricanes, Typhoons, and Tropical Cyclones FAQ. NOAA/AOML/HRD, Miami, Florida. http://www.aoml.noaa.gov/hrd/tcfaq/tcfaqHED.html.

This up-to-date, informative website contains hurricane definitions, answers to specific hurricane questions, global statistics, links to other hurricane websites, and references for more information. Its is a popular and highly recommended website. One can also read about the Hurricane Research Division (http://www.aoml.noaa.gov/hrd/) on this site.

Cooperative Institute for Meteorological Satellite Studies, Tropical Cyclone Page. Madison, Wisconsin. http://cimss.ssec.wisc.edu/tropic/tropic.html.

This Web page displays state-of-the-art satellite imagery of hurricanes and the tropics around the world, with a research focus on winds derived from cloud motions. Archived hurricane imagery for the past few years as well as NHC and JTWC forecasts are also available on this website.

Federal Emergency Management Agency. Washington, D.C. http://www.fema.gov.

The Federal Emergency Management Agency (FEMA) is an independent agency of the federal government. Its mission is to reduce life and property losses through mitigation and preparedness programs. FEMA is also called in to help when the president declares a region a disaster area, such as coastal regions impacted by a hurricane. This website provides extensive documentation on hurricane preparedness and hurricane recovery issues. Current weather statements are also available from the National Hurricane Center.

Dr. William M. Gray, Atlantic Seasonal Hurricane Activity Forecasts. Colorado State University, Department of Atmospheric Science, Fort Collins, Colorado. http://tropical.atmos.colostate.edu.

Every December, April, June, August, and September, Dr. Bill Gray and colleagues issue forecasts on Atlantic hurricane activity at this website. These highly anticipated statements discuss in detail the forecast methodology and how different forecast parameters (such as El Niño and African rainfall) are expected to affect the upcoming hurricane season. Archives of previous forecasts as well as verifications are also available.

Hurricane Hunters. Keesler Air Force Base, Biloxi, Mississippi. http://www.hurricanehunters.org.

The 53rd Weather Reconnaissance Squadron, also known as the Hurricane Hunters, flies into Atlantic tropical disturbances, tropical depressions, tropical storms, and hurricanes on a routine basis and relays this information to the National Hurricane Center. This website includes aircraft reports, pictures of flights inside hurricanes, informative information about the Hurricane Hunters, and educational links.

Intellicast. Billerica, Massachusetts. http://www.intellicast.com.

Intellicast is a product of Weather Service International and provides extensive specialized weather information to help plan all outdoor and weather-sensitive activities. This site provides excellent satellite and radar images. A tropical weather section is also included.

Joint Typhoon Warning Center. Pearl Harbor, Hawaii. http://www.npmoc.navy.mil/jtwc.html.

This website contains the latest JTWC forecast statements for tropical cyclones around the world. You can also read the Annual Tropical Cyclone Report (beginning in 1996) or download tropical cyclone historical data sets containing positions and intensities for western Pacific and Australian storms.

National Data Buoy Center. Stennis Space Center, Mississippi. http://www.ndbc.noaa.gov.

This is an excellent source for obtaining real-time surface weather information over the ocean. It posts real-time measurements of wind, air temperature, water temperature, pressure, wave, and other data from all buoy and Coastal-Marine Automated Network (CMAN) stations. Current buoy data can also be obtained via a computer-generated voice by calling 228-688-1948 (it works fastest if you remember the five-character code of the buoy of interest) and from the National Data Buoy Center's ftp site. Plots of measurements for the last three days are available on the website to ascertain trends and recent weather history. Monitoring a buoy near a hurricane provides important information (although some buoys do not survive hurricane conditions and stop transmitting data). This site also provides archived buoy data.

National Hurricane Center. Miami, Florida. http://www.nhc.noaa.gov.

This website contains the latest NHC forecast statements as well as satellite imagery, reconnaissance reports, historical data, educational material, general information about the forecast procedures, and a description of the forecast facilities.

National Ocean Service (NOS) Center for Operational Oceanographic Products and Services (CO-OPS). Silver Spring, MD. http://co-ops.nos.noaa.gov/.

CO-OPS collects and distributes observations and predictions of water levels and currents. This website provides water level information, predictions based on tides, and any influences from the hurricane storm surge. A related website is the Tides Online at http://tidesonline.nos.noaa.gov, which provides users with immediate graphical and tabular water level and meteorological data from NOS water level stations located along the projected path of severe storms such as hurricanes.

National Oceanographic and Atmospheric Administration. Washington, D.C. http://www.noaa.gov.

This website is a good starting point for obtaining extensive weather facts, educational material, weather observations, and forecasts. It contains many links to NOAA's other branches, such as the National Weather Service.

National Weather Service. Washington, D.C. http://weather.gov.

Contains links to all National Weather Service forecasts and observations as well as current observations, radar, and satellite images.

Naval Research Laboratory, Tropical Cyclone Page. Monterey, California. http://www.nrlmry.navy.mil/tc_pages/tc_home.html.

This Web page displays state-of-the-art satellite imagery of hurricanes and the tropics around the world, including scatterometer winds, winds derived from cloud motions, and estimated rainfall rates.

NOAA Aircraft Operations Center. MacDill Air Force Base, Tampa, Florida. http://www.nc.noaa.gov/aoc.html.

The Aircraft Operations Center (AOC) provides the aircraft equipped with the scientific instruments required for hurricane research flights by the Hurricane Research Division. This site describes in detail the aircraft specifications. It also discusses the NOAA Corps, the smallest of the seven uniformed services of the United States. The AOC pilots are members of the NOAA Corps.

NOAA's Coastal Service Center's Historical Hurricane Database. Charleston, SC. http://hurricane.csc.noaa.gov/hurricanes/index.htm.

This website allows the search and display of detailed tropical storm and hurricane data and population trends. Searches can be made using a storm name, zip code, state, country, latitude, or longitude, back to 1851. This site also provides a searchable database of population changes from 1900 to 2000 for U.S. coastal counties. Reports of individual storms are also available.

Official U.S. Time. The National Institute of Standards and Technology (NIST), U.S. Naval Observatory (USNO). http://www.time.gov.

This site provides times accurate to 0.0000001 of a second for each time zone. Most helpful to the user is conversion information from local time to Coordinates Universal Time (UTC). UTC is the standard global time to avoid time zone confusion, and is based on local time in Greenwich, England, where 0° longitude passes through. For this reason, UTC is also called Greenwich Mean Time (GMT). GMT and UTC are also called Zulu (Z) time.

Unisys Weather. Kennett Square, Pennsylvania. http://weather.unisys.com.

This weather site from Unisys Corporation provides a complete source of graphical weather information. It is intended to satisfy the needs of the weather professional but can be a tool for the casual user as well. The graphics and data are displayed as a meteorologist would expect to see. There are detailed explanation pages to guide the novice user through the various plots, charts, and images. This site also includes a thorough archive of Atlantic, East Pacific, and West Pacific hurricane track graphics, historical track and intensity data, and some satellite imagery of well-known Atlantic storms.

The Weather Channel. Atlanta, Georgia. http://www.weather.com.

This cable media outlet provides excellent, understandable weather information and graphics on its website. In addition, the Weather Channel's meteorologists provide updated reports on tropical activity. Hurricane graphics and satellite imagery of the tropics are shown, and educational hurricane information is available. Short educational videos can also be downloaded.

The Weather Underground. Ann Arbor, Michigan. http://www.wunderground.com.

An excellent source of quick, accessible National Weather Service observations and forecasts, the Weather Underground also provides the latest information from the National Hurricane Center, the Central Pacific Hurricane Center, and the Joint Typhoon Warning Center. A similar site is http://cirrus.sprl.umich.edu/wxnet/, which is the University of Michigan Weather Website. The latter site also contains a list of more than 300 other weather-related websites and another list of vendors who sell weather software.

Epilogue

Hurricane Katrina

Probably the worst natural disaster in U.S. history, Hurricane Katrina occurred near the end of the publication deadline and could not be fully incorporated into the book. This epilogue provides the latest information on Katrina as of September 2005.

Hurricane Katrina first made landfall in south Florida on August 25, 2005, as a category 1 hurricane. Landfall occurred between Hallandale Beach and North Miami Beach, Florida, with wind speeds of approximately 80 mph and gusts to 90 mph. As the storm moved southwest across the tip of the Florida peninsula, Katrina's winds decreased slightly before regaining hurricane strength in the Gulf of Mexico. The storm caused moderate damage in the region and claimed eleven lives, providing a hint of its devastating force. Given that Katrina spent only seven hours over land, its strength was not significantly diminished, and it quickly reintensified shortly after moving over the warm waters of the Gulf of Mexico.

Atmospheric and oceanic conditions were conducive to rapid intensification, which lead to Katrina attaining major hurricane status on the afternoon of the August 26. Continuing to strengthen and move northward during the next forty-eight hours, Katrina reached maximum wind speeds of 172 mph (category 5) on the morning of Sunday, August 28, and its minimum central pressure dropped that afternoon to 902 mb--the fourth lowest on record for an Atlantic storm. Although Katrina was comparable to Hurricane Camille (1969), it was a significantly larger storm. Katrina's hurricane-force winds extended 120 miles from the storm center, with tropical storm-force winds extending 230 miles outward. Katrina also maintained a large eye, thereby providing large areal coverage of its most fierce winds. Finally, Katrina moved slower than Camille, thereby increasing the storm surge potential and time of wind exposure. All these ingredients resulted in catastrophic destruction and fatalities that dwarf the previous benchmark hurricanes of Camille and Betsy (1965) in southeast Louisiana, Mississippi, and Alabama.

Katrina made landfall on the morning of August 29 in Buras, Louisiana, as a strong category 4 hurricane with sustained winds of 140 mph and a central pressure of 923 mb--the fourth lowest on record for a land-falling Atlantic storm in the United States and the fifth highest of land-falling hurricanes in U.S. history. But, the size of the hurricane caused mass destruction. The resulting storm surge is the worst in U.S. history, peaking at 35 feet near the eastern eye wall in Buras and later during its second landfall just west of Waveland, Mississippi. However, 20-25 feet of water covered much of southeast Louisiana, the entire Mississippi coast, and parts of Alabama and penetrated an astounding 5 miles inland in the eye wall's path. Structures near the coast become scoured cement slabs, while inland all buildings suffered wind damage, flooding, or both.

New Orleans' levees, only built for a fast-moving category 3 hurricane, were toppled in the eastern section of the city, while waters in Lake Pontchartrain and floating objects such as barges precariously pounded the city's northern levees. Throughout August 29 and 30, levee breaches began to occur, flooding 80 percent of the city with up to 20 feet of water in some places. Because New Orleans is below sea level, water has to be removed from the city by pumps, which stop working due to wind damage and flooding of power equipment. Although most New Orleans residents evacuate ahead of the storm, tens of thousands more are stranded in the city, with some 20,000 in the Superdome sports facility, 20,000 more in the city's convention center, and many more in their homes. Local, state, and federal governments are widely criticized for both their lack of preparation and generally slow response in helping stranded citizens, while law and order breaks down into massive looting throughout the city.

The damage to the region, covering an area greater than the size of Britain, is truly mind-boggling. The national recovery effort may cost taxpayers $150 billion to $200 billion, with another $40–60 billion in insurance losses, obliterating the record damage of $26.5 billion caused by Hurricane Andrew (1992). In Mississippi, about 68,000 homes were destroyed, and another 65,000 suffered major damage. In Louisiana, about 250,000 homes were damaged or destroyed. It is the largest permanent displacement of people in history. The homeless are being sheltered in a variety of ways, including in the Houston Astrodome and on cruise ships.

At this writing, the death toll is trending toward 1,500, with about 70 percent of those deaths in Louisiana. This places Hurricane Katrina third behind the Lake Okeechobee Hurricane (1928) and the 1900 Galveston Hurricane. In Mississippi, the number of fatalities has already exceeded Camille.

The economic impact is enormous. In southeast Louisiana, the agriculture, oil, fishing, and tourism industries have been decimated, and commerce in New Orleans will be substantially reduced for the next year. Offshore facilities seem to have fared better, but oil pipelines have probably been broken and will take months to repair. Mississippi's tourist, agriculture, timber, and poultry industries also suffered immense losses. Mississippi's thriving water-bound casino industry, which generates $500,000 a day in tax revenue, has been heavily impacted, prompting lawmakers to consider moving them to land-based locations.

Environmental issues will also be of concern. The impact of oil spills, chemical seepage, and sewage pollution, both in the soil and in the surrounding prolific wildlife estuaries, could be devastating. The hurricane eroded more of the already rapidly vanishing wetlands. These wetlands also serve as a buffer against storm surge and hurricane intensity (which require water for their energy).

With regard to chapter 6, Katrina ranks:

1. Number 3 in U.S. hurricane fatalities (see table 6.1). Number 2 is still possible as of this writing.
2. The costliest hurricane in U.S. history (see tables 6.2 and 6.3).
3. The fourth most intense at landfall (see table 6.6).
4. The sixteenth category 4 hurricane to make landfall (see table 6.9).
5. The twenty-sixth category 5 hurricane (one day before landfall) in U.S. history (table 6.10).

6. The most extreme storm surge in U.S. history (see chapter 6's "World Record" section).

Sadly, Katrina also sets another precedent. Hurricane storm surge fatalities have not been a major issue since Hurricane Camille in 1969 (see tables 6.14–6.17). This catastrophe will focus renewed efforts on evacuation, mitigation, and public education issues.

Glossary

Acronyms

AMS	American Meteorological Society—Boston, MA
AOC	NOAA Aircraft Operations Center—Tampa, FL
CIMSS	Cooperative Institute for Meteorological Satellite Studies—University of Wisconsin-Madison
FEMA	Federal Emergency Management Agency—Washington, D.C.
GFDL	Geophysical Fluid Dynamics Laboratory—Princeton, NJ
HRD	Hurricane Research Division—Miami, FL
JTWC	Joint Typhoon Warning Center—Pearl Harbor, HI
McIDAS	Man-computer Interactive Data Access System
NASA	National Aeronautics and Space Administration—Washington, D.C.
NCDC	National Climatic Data Center—Asheville, NC
NCEP	National Centers for Environmental Prediction—Silver Springs, MD
NESDIS	National Environmental Satellite, Data and Information Service—Silver Spring, MD
NHC	National Hurricane Center—Miami, FL
NASA	National Aeronautics and Space Administration
NOAA	National Oceanic and Atmospheric Administration—Silver Spring, MD

NWA	National Weather Association—Montgomery, AL
NWS	National Weather Service—Silver Spring, MD
QBO	Quasi-Biennial Oscillation
WMO	World Meteorological Organization

Abbreviations for Cited Journals

Ann. N.Y. Acad. Sci.	*Annals of the New York Academy of Sciences*
Bull. Amer. Meteor. Soc.	*Bulletin of the American Meteorological Society*
Earth Obs. Mag.	*Earth Observation Magazine*
Geo. Res. Letters	*Geophysical Research Letters*
J. Appl. Meteor.	*Journal of Applied Meteorology*
J. Atmos. Sci.	*Journal of the Atmospheric Sciences*
J. Climate	*Journal of Climate*
J. Phys. Oceanogr.	*Journal of Physical Oceanography*
J. Meteor.	*Journal of Meteorology*
J. Meteor. Soc. Japan	*Journal of Meteorological Research, Japan*
Mar. Wea. Log	*Mariner's Weather Log*
Mon. Wea. Rev.	*Monthly Weather Review*
Quart. J. Roy. Meteor. Soc.	*Quarterly Journal of the Royal Meteorological Society*
Sci. Amer.	*Scientific American*
Wea. Forecasting	*Weather and Forecasting*

Terms

aerosonde   A small robotic aircraft that measures pressure, temperature, moisture, and wind. It is a light aircraft that is extremely fuel efficient and capable of flying long distances but possibly sturdy enough to survive severe weather such as hurricanes.

annular hurricane   A steady-state, intense hurricane with a circular eye surrounded by a nearly uniform ring of thunderstorms.

Arctic hurricane   *See* polar low.

beta drift A poleward and westward drift of a hurricane due to the vortex's interaction with the earth's rotation. Small storms drift 1–2 mph, while large storms drift 3–4 mph.

centrifugal force An outward-directed force in rotating flow that occurs because an object in motion wants to remain in a straight line. The sharper the curvature of the flow and/or the faster the rotation, the stronger is the centrifugal force.

cloud seeding The addition of agents that will alter the phase and size distributions of cloud particles, with the intent of influencing precipitation or cloud growth. These agents are usually granulated carbon dioxide (dry ice) or silver iodide with a molecular structure similar to ice nuclei. They are dropped into supercooled clouds, thereby converting the supercooled water to ice and promoting cloud development through the release of latent heat or freezing.

computer model A computer program that ingests current weather observations and approximates solutions to complicated equations so as to predict future atmospheric values such as wind, temperature, and moisture. Based on the field of *numerical weather prediction*.

consensus forecast The average of several different *computer model* forecasts.

concentric eyewall cycle A natural (but temporary) weakening process in which a new eyewall forms outside the original eyewall. The outer eyewall "chokes off" inflow to the inner eyewall, causing it to dissipate. The outer eyewall then propagates inward, replacing the original eyewall. This cycle typically takes one day to complete and is accompanied by fluctuations in central pressure and maximum wind speed.

convergence A net inflow of air. When convergence occurs at the surface, air must ascend.

Coriolis force An apparent deflection of air motion due to the earth's rotation since the coordinate system is changing relative to the earth (known as a noninertial, or relative, coordinate system). This effect becomes increasingly influential away from the equator where the earth's curvature increases. To someone living on earth, it appears that the wind is deflected to the right in the Northern Hemisphere and to the left in the Southern Hemisphere.

Doppler effect A shift in wavelength of radiation emitted or reflected from an object moving toward or away from the observer. Doppler radar translates the motions of air particles, cloud droplets, and raindrops into wind speed and wind direction measurements.

downburst A strong downdraft that exits the base of a thunderstorm and spreads out at the earth's surface as strong and gusty horizontal winds that can cause property damage.

Dvorak technique A methodology use to estimate intensity of a depression, tropical storm, or hurricane solely based on satellite-observed cloud organization and cloud height.

dynamic instability The breakdown of wind flow into a tropical wave under certain flow configurations and/or temperature patterns. This instability results because a perturbation acquires kinetic energy (the energy of motion) from another source and grows with time. This source may be from the kinetic energy of the large-scale wind flow or from potential energy (stored energy) due to certain temperature patterns.

easterly wave *See* tropical wave.

El Niño A 12–18-month period during which anomalously warm sea-surface temperatures occur in the eastern half of the equatorial Pacific Ocean. El Niño events occur irregularly, about once every three to seven years. Atlantic hurricane activity is usually suppressed during El Niño seasons. The opposite condition is called a *La Niña*.

ensemble forecast A set of different forecasts using the same *computer model*, valid at the same time, based on slightly different observation values. Generally, the mean of the ensemble forecast represents the most accurate forecast. Furthermore, the forecast spread indicates the possible forecast error and forecast uncertainty.

equatorial trough A *trough* that forms when air converges from both hemispheres near the equator, often resulting in towering thunderstorms. Also called the *InterTropical Convergence Zone* (ITCZ).

extratropical cyclone Any low-pressure system that derives its energy from horizontal temperature contrasts. Fifty percent of hurricanes undergo an extratropical transition, losing their tropical characteristics. Half of these will gradually decay, but about half reintensify, with a few experiencing rapid development with winds up to hurricane strength. In either situation (decay or reintensification), substantial rain can occur in hurricanes undergoing extratropical transition, resulting in severe flooding.

eye A region in the center of a hurricane (and tropical storms near hurricane strength) where the winds are light and skies are clear to partly cloudy. The eye is either completely or partially surrounded by the eyewall.

eyewall A wall of dense thunderstorms that surrounds at least half of the eye of a hurricane.

front A boundary between two air masses of different temperature and/or moisture properties.

Fujiwhara effect Interaction of two hurricanes (located less than 850 miles from each other) that orbit cyclonically about a midpoint between them; named after the pioneering experiments of Fujiwhara in 1921.

geosynchronous satellites A satellite that travels east at the same speed as the rotating earth, enabling the satellite to remain over the same location and provide continuous coverage of that region.

global warming A theoretical enhancement of the *greenhouse effect* due to an increase in carbon dioxide by fossil fuel emissions (such as the output from cars running on gasoline).

graupel An ice particle that has experienced heavy *riming* and no longer resembles the original ice crystal.

greenhouse effect The warming of the atmosphere by its absorbing and reemitting *infrared radiation* emitted by the earth (at electromagnetic wavelengths invisible to the human eye) back to the surface while allowing *solar radiation* emitted by the sun (including visible light) to pass through the atmosphere and reach the surface. The result is that the earth's surface is much warmer due to combined warming of shortwave and reemitted infrared radiation than if there was no atmosphere. Infrared radiation is absorbed primarily by two gases: water vapor and carbon dioxide. It's important to realize that the atmosphere's greenhouse effect is a *natural process,* and without it the earth would be a much colder, unlivable planet with an average surface temperature of 0°F.

gust factor The ratio of peak two-second winds to the sustained wind. In general, the rougher the terrain, the lower the sustained wind but the larger the gust factor due to turbulent eddies. Engineers frequently use the gust factor to determine wind-loading conditions.

hurricane A large mass of organized, oceanic thunderstorms with a complete cyclonic circulation and maximum sustained winds of at least 74 mph somewhere in the storm. Also called *typhoons* in the Northwest Pacific across 180°E, chubasco in the Philippines, and very severe cyclonic storms in India.

hurricane warning A warning given when it is likely that an area will experience sustained surface winds of hurricane force within twenty-four hours.

hurricane watch A hurricane watch indicates that hurricane conditions (sustained winds of 74 mph or greater) pose a threat to an area within thirty-six hours, and residents of the watch area should be prepared for hurricane conditions.

ice nuclei A floating aerosol with a molecular structure similar to ice.

infrared radiation The portion of the electromagnetic spectrum lying between visible radiation and microwave radiation, with wavelengths between 720 and 1 millimeter. This is the spectrum the earth typically emits radiation and is invisible to the human eye (as is most radiation).

intense hurricane *See* major hurricane.

InterTropical Convergence Zone (ITCZ) *See* equatorial trough.

inverted barometer effect The uplift of water in the center of a hurricane as an adjustment to the low air pressure there, corresponding to 3.9 inches sea level rise for every 10-mb drop in sea-level pressure.

La Niña A twelve- to eighteen-month period during which anomalously cool sea-surface temperatures occur in the eastern half of the equatorial Pacific Ocean. La Niña events occur irregularly, about once every three to seven years. Atlantic hurricane activity is usually enhanced during La Niña seasons. The opposite condition is called an *El Niño*.

latent heat Energy transfer conveyed through phase changes of matter. For example, the latent heat of condensation is the heat energy released when water vapor (a gaseous state) condenses to a liquid state; latent heat of evaporation is the heat energy absorbed by water vapor during the evaporation process; and latent heat of fusion is the heat energy released when water freezes.

longshore current An ocean current that develops when waves strike a coast at an angle, resulting in a net flow parallel to the coast.

Madden-Julian Oscillation A sinking/ascending undulation that travels eastward from India to North America. This oscillation, with a repeat time of thirty to sixty days, tends to favor genesis when the ascending branch occurs in the presence of a *monsoon trough* or *tropical wave*. This phenomenon forms in the near-equatorial Indian Ocean and propagates eastward. Its influence is strongest in the Indian Ocean and western Pacific Ocean.

major hurricane A hurricane that reaches a maximum sustained wind of at least 111 mph. This constitutes a category 3 hurricane or higher on the Saffir-Simpson Hurricane Scale. Also called an intense hurricane.

mesoscale vortices Whirling vortices that form at the boundary of the eyewall and eye where there is a tremendous change in wind speed. The updrafts in the eyewall stretch the vortices vertically, making them spin faster with winds up to 200 mph that are very destructive. Also called mesovortices.

midget A small hurricane with gale-force winds (35 mph) extending only 40 miles or less from the storm center.

monsoon depressions Weak cyclonic disturbances that form in the Bay of Bengal with no inner-core structure and track northwestward into the Indian subcontinent, bringing moderate winds and heavy rains.

monsoon gyre A reconfiguration of the *monsoon* trough into a very large, nearly circular vortex in the western North Pacific that rotates counterclockwise for two weeks with a cloud band rimming the southern and eastern periphery. A series of small tropical cyclones may emerge from the leading edge of the cloud band.

monsoon trough A *trough* that forms when Southern Hemisphere and Northern Hemisphere air converge at 10–20 degrees latitude. This type of trough occurs when air temperature increases away from the equator. The *Coriolis force* induces a partial cyclonic spin favorable for the genesis of hurricanes or the formation of tropical waves through *dynamic instability*. The vast majority of genesis cases are associated with the monsoon trough.

multidecadal changes Climate changes with ten-, twenty-, or thirty-year cycles. Atlantic hurricane activity, especially *major hurricanes*, tends to follow twenty- to thirty-year cycles with busy decades followed by inactive decades.

numerical weather prediction (NWP) Forecasting the weather based upon the solutions of mathematical equations by high-speed computers.

polar low A cyclone that forms in the arctic region, poleward of zones of strong temperature contrasts, ranges in sizes from 120 to 700 miles and has winds near or above gale force. Polar lows can have banding features with an eye similar to a hurricane, so some refer to them as *arctic hurricanes*. However, polar lows form over cold water, have a short life cycle less than thirty-six hours, have weaker but still significant winds, and are smaller in size.

polar-orbiting satellites A satellite following a north-south orbit around the earth's poles, providing pictures and data centered on different longitudes each hour as the earth rotates underneath them.

pressure The force per unit area exerted by air molecules on a surface. Conversely, the "weight" of the air above a given area of the earth's surface. Its standard unit of measurement is the millibar (mb), although it is also popularly measured as the height of a column of mercury supported by the atmosphere's weight using an instrument called a barometer. Sea-level pressure is normally close to 1,016 mb, or 30 inches of mercury.

Project STORMFURY A government-sponsored attempt in the 1960s to weaken hurricanes by *cloud seeding* just outside the eyewall to stimulate cloud growth, thereby depriving inflow to the eyewall.

Quasi-Biennial Oscillation (QBO) An oscillation of equatorial winds between 13 and 15 miles aloft. These winds change direction between westerly and easterly every twelve to sixteen months. Westerly winds are associated with more Atlantic hurricanes than easterly winds, especially when the wind is westerly at both 13 and 15 miles aloft.

radiosonde An instrument attached to a balloon that measures the vertical profile of temperature, moisture, pressure, and wind up to 40,000 feet from the ground.

reconnaissance planes Planes that fly into a hurricane's eye and take critical meteorological measurements. During the hurricane penetration, information about the horizontal wind and temperature structure is transmitted to the NHC. Once the plane enters the eye, a tube of instruments (called a dropsonde) is deployed that parachutes downward from flight level to the sea, sending valuable intensity measurements back to the NHC.

reinsurance The process whereby insurance companies buy insurance for themselves. A reinsurance program is divided into layers with different reinsurers assuming the risk of different layers.

rime A white or milky and opaque granular deposit of ice formed by the rapid freezing of *supercooled water* as it impinges on another object, typically another ice crystal in a cloud. Crystals that exhibit frozen droplets on their surfaces are referred to as rimed. Crystals that experience heavy riming are called *graupel*.

rip current A "rip" in the *longshore current* causing a narrow and strong offshore current, responsible for 80 percent of lifeguard rescues. Rip currents appear as a plume of dirty water (from stirred-up sediment) moving away from shore—often carrying foam, seaweed, or debris—with a break in incoming waves. They are mistakenly called an *undertow* or *riptide*.

riptide Another name for rip current, but misleading since there is no tidal component.

Saffir-Simpson Hurricane Scale A scale relating a hurricane's central pressure, maximum sustained winds, and storm surge to the possible damage it is capable of inflicting. The scale contains five categories increasing numerically with damage, with a category 1 being a minimal hurricane and a category 5 being a catastrophic hurricane.

silver iodide A compound whose molecules consist of one atom of silver and one atom of iodide. Its structure resembles ice crystals, and therefore silver iodide is used in *cloud seeding*.

solar radiation Radiation emitted by the sun, half of which is visible to the human eye.

spiral bands Curved thunderstorm bands that propagate around the hurricane circulation.

static instability A condition in which saturated air forced upward is less dense than surrounding unsaturated air that accelerates upward, forming towering, puffy clouds.

steering current Atmospheric current, generally somewhere between 5,000 and 15,000 feet above the surface, whose direction best relates to the motion of a hurricane.

steric effect A change in water level due to density changes, caused either by reduced salinity or warmer temperatures, or both.

storm surge An abnormal rise of the sea along a shore due to a meteorological influence, especially a hurricane. It is officially defined as the difference between the actual water level under the hurricane's influence and the level due to the astronomical tide and *wave setup*.

storm tide The actual sea level as influenced by the storm surge, astronomical tide, and *wave setup*. In practice, water level observations during posthurricane surveys are always storm tides.

subtropical cyclone A hybrid system that forms over the subtropical waters in the Atlantic and Pacific Oceans and contains a mixture of tropical and polar characteristics. Subtropical cyclones derive some of their energy from horizontal temperature contrasts (whereas tropical storms and hurricanes do not), as well as heat flux from the sea (similar to tropical storms and hurricanes).

supercooled water Water that remains in the liquid phase at temperatures colder than 32°F.

sustained winds The average wind speed over a period of time at roughly 30 feet above the ground. In the Atlantic and northern Pacific Oceans, this averaging is performed over a one-minute period, and in other ocean basins it is performed over a ten-minute period.

tornado A rapidly rotating column of air that protrudes from a cumulonimbus cloud in contact with the ground, often (but not always) visible in the shape of a funnel or a rope. The right front quadrant of a hurricane often produces tornadoes.

tornado warning A warning issued by the National Weather Service when a tornado has been observed by people trained as weather spotters or inferred by an instrument called Doppler radar. These warnings are issued with information on the tornado's current location and what communities are in the anticipated path.

tornado watch Issued by the Storm Prediction Center to alert the public that conditions are favorable for tornado development. These watches are issued with information on the location of the watch area and the length of time it is in effect, usually several hours.

trade winds Persistent tropical winds having a prevailing direction from the northeast in the Northern Hemisphere and from the southeast in the Southern Hemisphere.

trochoidal motion Short-term, oscillatory motion of a cyclone center about a mean path.

tropical cyclone The internationally designated general term for all large, cyclonically rotating thunderstorm complexes over tropical oceans. It includes depressions, tropical storms, and hurricanes in addition to other large, tropical, cyclonically rotating thunderstorm complexes that contain distinctly different temperature and organization characteristics, such as *monsoon depressions* and *subtropical cyclones*.

tropical depression A large mass of organized, oceanic thunderstorms with a complete cyclonic circulation and sustained winds everywhere less than 39 mph.

tropical disturbance A large mass of organized, oceanic thunderstorms originating in the tropics or subtropics generally with a diameter of 100–300 miles and persisting for twenty-four hours. Sometimes partial rotation is observed, but this is not required for a system to be designated a tropical disturbance.

tropical storm A large mass of organized, oceanic thunderstorms with a complete cyclonic circulation and maximum sustained winds between 39 and 73 mph somewhere in the storm. A storm is first given a name at this stage.

tropical waves A westward moving trough, shaped like an upside-down "V" and similar to a wave, embedded in northeasterly winds in the tropics. About fifty-five to seventy-five tropical waves are observed in the Atlantic each year, and 10–25 percent develop into a tropical depression or beyond. They are also called easterly waves.

trough An elongated area of low pressure.

typhoon A hurricane that forms over the western Pacific Ocean west of 180°E.

undertow The seaward pull of receding waves after they break on a shore. The definition is largely mythical, since the backwash will be replaced by another incoming wave and is therefore periodic. Often mistakenly called a *rip current*.

upwelling The ascending of subsurface water in association with the net outflow of surface water.

vertical wind shear The difference between wind speed and direction at 40,000 feet and the earth's surface. Hurricane formation and intensification are favored in regions where the wind speed and direction remain the same with height (weak vertical wind shear), thereby allowing the vertical orientation of thunderstorms to remain intact.

vortex Rossby waves A special class of wave solutions due to the curved flow and rapid variation in wind speed away from the storm center, which explains hurricane spiral bands.

vorticity The rate an imaginary paddle wheel placed in fluid would rotate due to curved flow and/or horizontally sheared flow. The radial variation of vorticity causes *vortex Rossby waves*, which explains the presence of hurricane spiral bands.

wave setup An increase in the water level on a beach due to the effects of waves running up at the beach with excessive incoming water not balanced by outgoing water, such as in hurricane conditions. This effect is more pronounced when deep water is near the beach.

wind gusts A sudden, brief increase in speed of the wind. According to U.S. weather-observing practices, gusts are reported when the peak wind speed reaches at least 18 mph and the variation in wind speed between the peaks and lulls is at least 10 mph. The duration of a gust is usually less than twenty seconds.

Conversion Tables

Adapted from Holland (1993b), Ahrens (1994), and author's lecture notes.

Length

1 kilometer (km) = 1000 m
   = 3281 ft
   = 0.621 mi
   = 0.54 nm

1 statute mile (mi) = 5280 ft
   = 1609 m
   = 1.609 km
   = 0.87 nm

1 nautical mile (nm) = 6080 ft
   = 1853 m
   = 1.853 km
   = 1.15 mi

1 meter (m) = 100 cm
   = 3.28 ft
   = 39.37 in

1 foot (ft) = 12 in
= 30.48 cm
= 0.305 m

1 centimeter (cm) = 0.39 in
= 0.01 m
= 10 mm

1 millimeter (mm) = 0.1 cm
= 0.001 m
= 0.039 in

1 inch (in) = 2.54 cm
= 0.08 ft

Approximate Conversion of Latitude, Longitude, and Length (Use with Some Caution)

1 degree latitude ≈ 111.137 km
≈ 60 nm
≈ 69 mi

1 degree longitude ≈ 111.137 km × cosine (latitude)
≈ 60 nm × cosine (latitude)
≈ 69 mi × cosine (latitude)

More Precise Distance Calculations between Two Latitude and Two Longitude Points

Let point 1 be latitude ϕ_1 and longitude θ_1, and let point 2 be latitude ϕ_2 and longitude θ_2. To accurately compute the distance between these two locations, one computes the distance along a "great circle" using the following steps:

Step 1: Compute C

$$C = \sin \phi_1 \times \sin \phi_2 + \cos \phi_1 \times \cos \phi_2 \times \cos (\theta_2 - \theta_1)$$

Step 2: Compute A (make sure the calculator is in radians mode)

$$A = \cos^{-1} C$$

Step 3: Compute distance

> distance = 6378 km × A
> distance = 3444 nm × A
> distance = 3963 mi × A

These computations assume that the earth is a perfect sphere, which introduces a small error of about 1 percent.

Area

1 square centimeter (cm^2) = 0.15 in^2
1 square inch (in^2) = 6.45 cm^2
1 square meter (m^2) = 10.76 ft^2
1 square foot (ft^2) = 0.093 m^2

Volume

1 cubic centimeter (cm^3) = 0.06 in^3
1 cubic inch (in^3) = 16.39 cm^3
1 liter (l) = 1000 cm^3
        = 0.264 gallons

Time

1 minute (min) = 60 s
1 hour (hr) = 60 min
       = 3600 s
1 day (d) = 24 hr
       = 1440 min
       = 86400 s

Time Zone Conversions

Coordinates Universal Time (UTC) is the standard global time to avoid time zone confusion and is based on local time in Greenwich, England, where 0° longitude (the international date line) passes through. For this reason, UTC is also called Greenwich Mean Time (GMT). GMT and UTC are also sometimes called Zulu (Z) time, universal time, and world time. The time notation ranges from 0000 to 2359. For example, 1 A.M. in Greenwich is 0100, pronounced "zero one hundred," and 8:35 P.M. in Greenwich is 2035. To compute local time, hours have to be added or subtracted from it based on the local time zone. Since the United States is west of the date line, hours are subtracted.

Daylight Saving Time, enacted to reduce energy use during the winter months, will also affect these conversions. Daylight Saving Time currently begins on the first Sunday of April at 2 A.M. local time, when an hour is added ("spring forward"). Time reverts to standard time on the last Sunday of October at 2 A.M. local time, when an hour is subtracted ("fall back"). Daylight Saving Time is not observed in Hawaii, American Samoa, Guam, Puerto Rico, the Virgin Islands, most of the Eastern Time Zone portion of Indiana, and Arizona (except for the Navajo Indian Reservation, which does observe it). To further reduce energy use, in August 2005 Congress passed an energy bill that extends Daylight Saving Time by about a month. Beginning in 2007, Daylight Savings Time will start on the second Sunday of March and end the first Sunday of November.

UTC conversions for the continental United States are shown below:

Eastern Standard Time (EST) = UTC minus 5 hours
Central Standard Time (CST) = UTC minus 6 hours
Mountain Standard Time (MST) = UTC minus 7 hours
Pacific Standard Time (PST) = UTC minus 6 hours

Eastern Daylight Time (EDT) = UTC minus 4 hours
Central Daylight Time (CDT) = UTC minus 5 hours
Mountain Daylight Time (MDT) = UTC minus 6 hours
Pacific Daylight Time (PDT) = UTC minus 7 hours

Speed

1 knot (kt) = 1 nautical mi/hr
    = 1.15 statute mi/hr, also written as 1.15 mph
    = 0.513 m/s
    = 1.85 km/hr

1 mile per hour (mi/hr, or mph) = 0.87 kt
    = 0.45 m/s
    = 1.61 km/hr

1 kilometer per hour (km/hr) = 0.54 kt
    = 0.62 mi/hr, or 0.62 mph
    = 0.28 m/s

1 meter per second (m/s) = 1.94 kt
 = 2.24 mi/hr, or 2.24 mph
 = 3.60 km/hr

Mass

1 gram (g) = 0.035 ounce
 = 0.0022 lb

1 kilogram (kg) = 1000 g
 = 2.2 lb

Pressure

1 millibar (mb) = 1000 dynes/cm^2
 = 0.75 millimeters of mercury (mm Hg)
 = 0.02953 inch of mercury (in Hg)
 = 0.01450 pound per square inch (lb/in^2)
 = 100 Pascals (Pa)

1 inch of mercury = 33.865 mb

1 millimeter of mercury = 1.3332 mb

1 Pascal = 0.01 mb

1 hectopascal (hPa) = 1 mb

1 kilopascal (kPa) = 10 mb

1 standard atmosphere = 1013.25 mb
 = 760 mm Hg
 = 29.92 in Hg
 = 14.7 lb/in^2

Converting °C to °F

°F = ⅘°C + 32
°C = ⅚(°F − 32)

If one can't remember the equations, then memorize that with each 10°C-temperature change there is an 18°F change. At freezing, 0°C = 32°F. Then, each time 10°C is added or subtracted, add or subtract another 18°F. The result is the following table:

−10°C = 14°F
 0°C = 32°F
 10°C = 50°F
 20°C = 68°F
 30°C = 86°F
 40°C = 104°F

After memorizing these key values, one can interpolate the values in between.

Converting °K to °C

°C = °K − 273.15
°K = °C + 273.15

Computing Dewpoint from Relative Humidity

Relative humidity (RH), frequently cited in the newspaper and TV, is the actual amount of water vapor in the air compared to the maximum possible water vapor amount at which air would saturate at that temperature. Expressed in percent, it indicates the amount of evaporation possible. At RH = 100%, the air is saturated and no evaporation can occur. Because RH will change with temperature even though water vapor content is the same, it is not a good moisture value. Dewpoint temperature is a better indication of water vapor content. The dewpoint temperature (T_D) is the temperature air must be cooled to at constant pressure and constant water vapor content in order for saturation to occur. However, T_D is complicated to compute. Two methods will now be shown.

If RH > 50%, T_D decreases by about 1°C for every 5% decrease in RH. When RH > 50%, the following simple equation can be used to compute T_D in °C (Lawrence 2005):

$$T_D = T - \frac{(100 - RH)}{5}$$

Note that T must be in °C.

The following complicated equation works for all RH (Parry 1969). T must be in °C:

$$a = 1 - \frac{RH}{100}$$

$$b = (14.55 + 0.114 \times T) \times a + ((2.5 + 0.007 \times T) \times a)^3 + (15.9 + 0.117 \times T) \times a^{14}$$

$$T_D = T - b$$

Converting One-Minute Maximum Sustained Winds Speed to Central Pressure in a Tropical Storm or Hurricane

Table G.1 gives empirical relationships that exist between the one-minute maximum sustained wind speeds and central pressure in tropical storms and hurricanes for four regions of the Atlantic and the western North Pacific. Atlantic hurricanes in the northern latitudes tend to have weaker winds for the same central pressure because they are moving over colder water and losing the thunderstorms that help maintain the tight wind structure. Note that Pacific storms have a lower central pressure for the same wind speed, because the environmental pressure surrounding the storms is also lower, and wind is the result of pressure differences outside and inside the hurricane. "NA" indicates not applicable, since tropical storms and hurricanes have never been observed for a region in that central pressure category. Caution is advised when using table G.1, as these relationships are only approximate.

One-Minute Sustained Winds to Ten-Minute Sustained Winds

In general, when averaging is applied over a longer period of time, sustained wind speeds decrease since wind gusts occur in short bursts. In other words, ten-minute sustained winds will be less than one-minute sustained winds under the same weather conditions. Since different countries use different definitions of sustained winds, comparing hurricane intensity statistics between oceans can be problematic. For example, the Atlantic Ocean and North Pacific Ocean basins use a one-minute averaging sequence, while other ocean basins use ten minutes.

The following conversion factors may be used to convert from one-minute sustained wind speeds to ten-minute sustained

TABLE G.1
Relationship of central pressure to hurricane winds in Atlantic and Pacific Oceans

Central pressure (mb)	Gulf of Mexico	1-minute maximum sustained wind speed (mph)			
		Atlantic, south of 25°N	Atlantic, 25–35°N	Atlantic, north of 35°N	Western North Pacific over warm water
1010	35	35	35	35	NA
1000	52	55	56	57	35
990	72	74	73	73	53
980	88	90	87	84	69
970	103	103	98	95	82
960	115	115	109	104	97
950	127	127	119	112	109
940	137	137	127	119	118
930	148	147	135	NA	129
920	158	156	143	NA	139
910	167	165	NA	NA	150
900	176	173	NA	NA	159
890	NA	181	NA	NA	170
880	NA	NA	NA	NA	179
870	NA	NA	NA	NA	187

Source: Adapted from Dvorak (1975) and Landsea et al. (2004).

wind speeds, although caution is advised when using this procedure since it is only approximate, and because it is possible for one-minute sustained winds to be less than ten-minute sustained winds (Holland 1993b).

10-minute winds = $0.871 \times$ 1-minute sustained winds
1-minute winds = $1.148 \times$ 10-minute sustained winds

Computing Wind Gusts from Sustained Wind Speed

While computing sustained winds over a given time period, short bursts of stronger winds known as *wind gusts* will occur. These wind gusts can cause isolated pockets of damage worse than in the surrounding area and therefore are useful to know. To compute these wind gusts, one can multiply by a *gust factor,* defined by the ratio of peak two-second winds to the sustained wind. The gust factors in table G.2 can be used for various exposures at 10-meter height (about 33 feet). In general, the rougher

TABLE G.2
Gust factors for different surface types

	Ocean	Flat Grassland	Woods/City
1-minute sustained winds	1.25	1.35	1.65
10-minute sustained winds	1.41	1.56	2.14

Source: Adapted from Holland (1993b).

the terrain, the lower the sustained wind but the larger the gust factor due to turbulent eddies. Engineers frequently use the gust factor to determine wind loading conditions.

The Modified Beaufort Scale

In 1831, Rear Admiral Sir Francis Beaufort developed a scale numbering from 0 to 12 based on the sea's impact on a ship (Kinsman 1990, 1991). From this, a common name for wind strength was assigned (i.e., "gentle breeze," "fresh gale," etc.). Since then, the scale has been modified several times to estimate wind speed by land or sea observations (table G.3). In addition, one can also use the table to determine wave height offshore for a given wind speed. The wind transfers some of its energy to the water through friction between the air molecules and the water molecules. Light wind generates ripples, and strong wind causes waves. Once initiated, waves can grow because the wind causes a pressure difference along the wave profile. As a wave grows, the pressure differences increase, which increases the wave height more, enhancing the pressure difference even more, and a positive feedback occurs. The waves also begin to interact among themselves to produce a spectrum of waves of different heights and lengths. The longer waves will actually propagate faster than the local wind that generated them, traveling far distances from the origination region. Eventually, these long-distance waves will reach an equilibrium height with the wind, and the sea becomes less chaotic, resulting in swells.

The typical time between successive passages of swell crests (known as the wave period) is six to eight seconds. Hurricanes far away produce very long waves, known as forerunners, with longer periods. In fact, before weather observations were available, seeing forerunners with wave periods of nine to fifteen sec-

TABLE G.3
The Modified Beaufort Scale

Beaufort number	Description	Mph	Wind speed in knots	Km/hr	Mean wind pressure (lb/ft$_2$)	Wind effects observed on land	Possible wave height in deep water offshore (use for rough estimate only)	Wind effects observed at sea
0	Calm	Less than 1	Less than 1	Less than 2	0	Calm; smoke rises vertically		Sea like mirror
1	Light air	1–3	1–3	2–6	0.01	Direction of wind shown by smoke drift but not by vanes	Less than ½ foot Less than ½ foot	Slight ripples; no foam crests
2	Light breeze	4–7	4–6	7–11	0.08	Wind felt on face; leaves rustle; vanes moved by wind; flags stir	½ to 1 feet	Small wavelets, still short but more pronounced; crests have a glassy appearance and do not break
3	Gentle breeze	8–12	7–10	12–19	0.28	Leaves, small twigs in constant motion; wind extends light flag	2–3 feet	Large wavelets; crests breaking; foam glassy; scattered whitecaps
4	Moderate breeze	13–18	11–16	20–29	0.67	Raises dust, loose paper; small branches move; flags flap	3½–5 feet	Small waves become larger; fairly frequent whitecaps
5	Fresh breeze	19–24	17–21	30–39	1.3	Small trees with leaves begin to sway; flags ripple; crested wavelets form on inland waters	6–8½ feet	Moderate waves form with many whitecaps; chance of some spray

(continues)

TABLE G.3
(Continued)

Beaufort number	Description	Mph	Wind speed in knots	Km/hr	Mean wind pressure (lb/ft$_2$)	Wind effects observed on land	Possible wave height in deep water offshore (use for rough estimate only)	Wind effects observed at sea
6	Strong breeze	25–31	22–27	40–50	2.3	Large branches in motion; whistling heard in utility wires; umbrellas used with difficulty	9½–12 feet	Large waves form; foam crests more extensive; some spray
7	Near gale	32–38	28–33	51–61	3.6	Whole trees in motion; inconvenient walking against wind; flags extend	13½–19 feet	Sea heaps up; some foam from waves blows in streaks
8	Gale	39–46	34–40	62–74	5.4	Twigs break off trees; walking is difficult	18–25 feet	Moderately high waves of greater length; well-marked streaks of foam
9	Strong gale	47–54	41–47	75–87	7.7	Slight structural damage occurs; signs and antennas blown down	23–32 feet	High waves; dense foam streaks; crests of waves begin to topple, tumble, and roll over; spray may affect visibility
10	Storm (whole gale)	55–63	48–55	88–101	10.5	Trees uprooted; considerable structural damage occurs	29–41 feet	Very high waves with long overhanging crests; foam blown in white dense streaks; on the

(continues)

11	Violent storm	64–73	56–63	102–119	14.0	Widespread damage	37–52 feet	whole, the sea takes a white appearance; visibility affected
								Exceptionally high waves (small and medium-sized ships might be lost to view behind the waves); sea completely covered with long white patches of foam; edges of waves blow into froth everywhere; visibility affected
12	Hurricane	74 and over	64 and over	120 and over	Over 17.0	Extensive damage	45 feet or greater	Air filled with foam and spray; sea completely white with driving spray; visibility very seriously affected

Source: Adapted from Holland (1993b) and NWS (1991).

onds often was the only clue that a hurricane was brewing somewhere, and trained mariners watched for them!

Because wave generation is a complicated process, wave heights in the modified Beaufort scale should be used for rough estimates only. Wave height depends on the magnitude and duration of the wind speed, how large an area the wind is affecting, and whether the waves have been generated locally or from some distant windy region. Typically, wave heights are reported as "significant wave height," defined as the average of the top one-third wave heights. However, in reality a spectrum of waves exists with different wave lengths and heights. Breaking waves will also occur when the upper segment of the wave moves faster than the lower segment. If waves are generated locally, the significant wave height (in feet) can be estimated based on wind speed (in mph) from the equation: $(0.45*wind)^2/13.58$. Wave heights propagating into the area from a windy region far away can be computed from complicated graphs based on wind speed, wind duration, and how big this generation area is. The National Centers for Environmental Prediction also runs wave models to predict height.

Caution should also be shown using the Beaufort scale outside U.S. waters, since the U.S. uses one-minute average winds and other countries use ten-minute averages; using ten-minute averages will result in smaller wind speeds.

The Beaufort table cannot be used to estimate wave heights near the coast in shallow water. As waves enter water less than 65 feet deep, they begin to "feel" the ocean bottom. In shallow water, the top part moves faster than the lower part and breaks into whitecaps. New waves—generally smaller—will form as they propagate inland. However, as the water depth continues to decrease, the wave slows down with a shorter wave length, and the wave height will grow again to conserve its energy; this process is called *shoaling*. When it reaches a height 75 percent of the depth, it will break as rolling whitecaps. If the coast has a shallow slope, the white foam just slides down the front; this is known as a *spilling breaker*. If the coast has a moderate slope, the break curls over in a *plunging breaker* (the best for surfing). On very steep coastlines, the wave may not break but just roll up the beach without plunging; this is known as a *surging breaker*, and such a situation is dangerous in strong storms because these type of waves are very powerful. Topography also plays a key role, as waves may be diffracted or concentrated by points and inlets. In regions where sharp, vertical cliffs occur near deep water, waves do not break but "reflect" off the cliff; when combined with the lifting mechanism of a storm surge, the waves can overtop the cliff, causing considerable destruction.

Tracking Hurricanes and Understanding National Hurricane Center Forecasts

Tracking a hurricane is an interesting pastime for many people as they plot where a hurricane has been and postulate where it might go. It is also an educational exercise for many students since it integrates geography, mapping skills, and science. When a tropical depression, tropical storm, or hurricane exists in the Atlantic or East Pacific, the National Hurricane Center (NHC) will issue forecast statements. You can plot the storm's position with a marker, following the storm's path. You can also plot the storm's projected paths based on NHC forecasts. Hurricane tracking charts can be ordered from the National Weather Service or downloaded from the FEMA, NOAA, National Weather Service, or NHC websites. Usually, coastal residents can also find tracking charts, often associated with a regional TV station, at many local stores.

The NHC issues numerous text products during the hurricane season. All these products are available on the NHC website, http://www.nhc.noaa.gov, or by automated email from the NHC. In addition, some are also distributed by radiofax, teletype, NOAA weather radio, and the Coast Guard using very high frequency (VHF), high frequency (HF), and medium frequency (MB) broadcasts.

The most useful to the general public is the "Tropical Weather Outlook" issued every six hours at 5:30 A.M., 11:30 A.M., 5:30 P.M., and 11:30 P.M. EST. This outlook briefly describes significant areas of disturbed weather and tropical disturbances. Additionally, this outlook discusses the potential for future development of these features out to forty-eight hours in the future. The outlook also lists any currently active tropical depressions, tropical storms, or hurricanes. The outlook is a valuable aid in maintaining tropical weather awareness and potential hurricane activity.

When a tropical depression, tropical storm, or hurricane exists in the Atlantic basin, the NHC issues advisories every six hours at 5 A.M., 11 A.M., 5 P.M., and 11 P.M. EST. Hurricane center positions are given by latitude (e.g., 28.4 degrees North) and by longitude (e.g., 88.7 degrees West). The current intensity and the predicted track and intensity are also issued. As a hurricane nears landfall, intermediate advisories may also be issued in three-hour increments between the forecast advisories. Special advisories are also issued whenever unanticipated significant changes occur or when tropical storm or hurricane watches or warnings must be issued immediately.

There are four kinds of advisory statements issued by the NHC every six hours: (1) a forecast advisory, (2) a public advisory, (3) discussion, and (4) strike probability. They include forecasts out to five days. Examples for Hurricane Georges as it approached the Louisiana and Mississippi coast are shown below (note that only three-day forecasts were issued in 1998).

Forecast Advisory

Forecast advisories contain detailed information about the hurricane's current location, intensity, and wind distribution. The wind distribution (in nautical miles) is given in "radius of 35 knot" winds and "radius of 50 knot" winds. Mariners generally avoid winds greater than 35 knots, and the U.S. Navy is required to avoid winds of these magnitudes. Forecast advisories also predict these parameters out to five days. When landfall is expected, watches and warnings are also included in the forecast advisory.

HURRICANE GEORGES FORECAST/ADVISORY NUMBER 49
NATIONAL WEATHER SERVICE MIAMI FL AL0798
1500Z SUN SEP 27 1998

A HURRICANE WARNING IS IN EFFECT FROM MORGAN CITY LOUISIANA TO PANAMA CITY FLORIDA. A HURRICANE WARNING MEANS THAT HURRICANE CONDITIONS ARE EXPECTED IN THE WARNED AREA WITHIN 24 HOURS. PREPARATIONS TO PROTECT LIFE AND PROPERTY SHOULD BE RUSHED TO COMPLETION.

A TROPICAL STORM WARNING AND A HURRICANE WATCH ARE IN EFFECT FROM EAST OF PANAMA CITY FLORIDA TO ST. MARKS FLORIDA. A HURRICANE WATCH IS IN EFFECT FROM WEST OF MORGAN CITY TO INTRACOASTAL CITY LOUISIANA.

HURRICANE CENTER LOCATED NEAR 28.4N 88.0W AT
27/1500Z POSITION ACCURATE WITHIN 25 NM

PRESENT MOVEMENT TOWARD THE NORTHWEST OR 315
DEGREES AT 7 KT

ESTIMATED MINIMUM CENTRAL PRESSURE 963 MB MAX
SUSTAINED WINDS 95 KT WITH GUSTS TO 115 KT
64 KT 80NE 40SE 20SW 60NW
50 KT 100NE 90SE 30SW 75NW
34 KT 130NE 150SE 90SW 100NW
12 FT SEAS.130NE 150SE 90SW 100NW
ALL QUADRANT RADII IN NAUTICAL MILES

REPEAT . . . CENTER LOCATED NEAR 28.4N 88.0W AT
27/1500Z
AT 27/1200Z CENTER WAS LOCATED NEAR 28.2N 87.8W

FORECAST VALID 28/0000Z 28.9N 88.7W
MAX WIND 100 KT . . . GUSTS 120 KT
64 KT . . . 90NE 40SE 20SW 60NW
50 KT . . . 100NE 90SE 30SW 75NW
34 KT . . . 130NE 150SE 90SW 100NW

FORECAST VALID 28/1200Z 29.6N 89.5W
MAX WIND 100 KT . . . GUSTS 120 KT
64 KT . . . 80NE 40SE 20SW 60NW
50 KT . . . 100NE 90SE 30SW 75NW
34 KT . . . 130NE 150SE 90SW 100NW

FORECAST VALID 29/0000Z 30.3N 89.9W . . . INLAND
MAX WIND 80 KT . . . GUSTS 95 KT
64 KT . . . 80NE 40SE 20SW 60NW
50 KT . . . 100NE 90SE 30SW 75NW
34 KT . . . 130NE 150SE 90SW 100NW

STORM SURGE FLOODING OF 10 TO 15 FEET . . . LOCALLY UP
TO 17 FEET AT THE HEADS OF BAYS . . . ABOVE NORMAL
TIDE LEVELS IS POSSIBLE IN THE WARNED AREA AND WILL
BE ACCOMPANIED BY LARGE AND DANGEROUS
BATTERING WAVES.

SMALL CRAFT FROM INTRACOASTAL CITY LOUISIANA
WESTWARD AND SOUTHWARD ALONG THE COAST OF
TEXAS SHOULD REMAIN IN PORT. SMALL CRAFT ALONG

THE WEST COAST OF THE FLORIDA PENINSULA SHOULD
REMAIN IN PORT UNTIL WINDS AND SEAS SUBSIDE.

REQUEST FOR 3 HOURLY SHIP REPORTS WITHIN 300 MILES
OF 28.4N 88.0W

EXTENDED OUTLOOK . . . USE FOR GUIDANCE ONLY . . .
ERRORS MAY BE LARGE

OUTLOOK VALID 29/1200Z 30.8N 89.9W . . . INLAND
MAX WIND 65 KT . . . GUSTS 80 KT
50 KT . . . 50NE 75SE 25SW 25NW

OUTLOOK VALID 30/1200Z 31.5N 89.5W . . . INLAND
MAX WIND 50 KT . . . GUSTS 60 KT
50 KT . . . 50NE 75SE 25SW 25NW

NEXT ADVISORY AT 27/2100Z

GUINEY

Public Advisory

A public advisory contains the hurricane's current position and intensity as well as general commentary regarding watches, warnings, future track and intensity, wind distribution, tornadoes, potential storm surge, and potential rainfall when landfall is possible or likely.

BULLETIN
HURRICANE GEORGES ADVISORY NUMBER 49
NATIONAL WEATHER SERVICE MIAMI FL
10 AM CDT SUN SEP 27 1998

. . . DANGEROUS HURRICANE GEORGES CLOSING IN ON
THE CENTRAL GULF COAST . . .

A HURRICANE WARNING IS IN EFFECT FROM MORGAN
CITY LOUISIANA TO PANAMA CITY FLORIDA. A
HURRICANE WARNING MEANS THAT HURRICANE
CONDITIONS ARE EXPECTED IN THE WARNED AREA
WITHIN 24 HOURS. PREPARATIONS TO PROTECT LIFE AND

PROPERTY SHOULD BE RUSHED TO COMPLETION . . . AND ADVICE FROM LOCAL EMERGENCY MANAGEMENT OFFICIALS SHOULD BE CLOSELY FOLLOWED

A TROPICAL STORM WARNING AND A HURRICANE WATCH ARE IN EFFECT FROM EAST OF PANAMA CITY FLORIDA TO ST. MARKS FLORIDA. A HURRICANE WATCH IS IN EFFECT FROM WEST OF MORGAN CITY TO INTRACOASTAL CITY LOUISIANA.

AT 10 AM CDT . . . 1500Z . . . THE CENTER OF HURRICANE GEORGES WAS LOCATED NEAR LATITUDE 28.4 NORTH . . . LONGITUDE 88.0 WEST. THIS POSITION IS ABOUT 80 MILES SOUTHEAST OF THE MOUTH OF THE MISSISSIPPI RIVER AND ABOUT 175 MILES SOUTHEAST OF NEW ORLEANS LOUISIANA.

GEORGES IS MOVING TOWARD THE NORTHWEST NEAR 8 MPH AND THIS GENERAL MOTION IS EXPECTED TO CONTINUE WITH A GRADUAL DECREASE IN FORWARD SPEED. THIS WOULD BRING THE CORE OF THE HURRICANE NEAR THE MOUTH OF THE MISSISSIPPI RIVER LATER TODAY. DO NOT FOCUS ON THE PRECISE LOCATION AND TRACK OF THE CENTER. THE HURRICANES DESTRUCTIVE WINDS . . . RAIN . . . AND STORM SURGE COVER A WIDE SWATH.

MAXIMUM SUSTAINED WINDS ARE NEAR 110 MPH WITH HIGHER GUSTS. GEORGES IS STRONG CATEGORY TWO HURRICANE ON THE SAFFIR-SIMPSON HURRICANE SCALE AND COULD REACH CATEGORY THREE STATUS BEFORE LANDFALL.

HURRICANE FORCE WINDS EXTEND OUTWARD UP TO 90 MILES FROM THE CENTER . . . AND TROPICAL STORM FORCE WINDS EXTEND OUTWARD UP TO 175 MILES. RAINBANDS OF GEORGES ARE SPREADING ACROSS PORTIONS OF THE WARNING AREA AND HURRICANE FORCE WINDS SHOULD BEGIN TO AFFECT THE AREA LATER TODAY.

AN AIR FORCE RESERVE UNIT HURRICANE HUNTER PLANE

REPORTED A MINIMUM CENTRAL PRESSURE OF 963 MB . . . 28.44 INCHES.

STORM SURGE FLOODING OF 10 TO 15 FEET ABOVE NORMAL TIDE LEVELS . . . AND UP TO 17 FEET AT THE HEADS OF BAYS . . . IS POSSIBLE IN THE WARNED AREA AND WILL BE ACCOMPANIED BY LARGE AND DANGEROUS BATTERING WAVES.

FLOODING RAINS OF 15 TO 25 INCHES . . . WITH LOCALLY HIGHER AMOUNTS . . . ARE LIKELY IN ASSOCIATION WITH THIS SLOW-MOVING HURRICANE.

ISOLATED TORNADOES ARE POSSIBLE EAST AND NORTHEAST OF THE TRACK OF GEORGES.

SMALL CRAFT FROM INTRACOASTAL CITY LOUISIANA WESTWARD AND SOUTHWARD ALONG THE COAST OF TEXAS SHOULD REMAIN IN PORT. SMALL CRAFT ALONG THE WEST COAST OF THE FLORIDA PENINSULA SHOULD REMAIN IN PORT UNTIL WINDS AND SEAS SUBSIDE.

REPEATING THE 10 AM CDT POSITION . . . 28.4 N . . . 88.0 W. MOVEMENT TOWARD . . . NORTHWEST NEAR 8 MPH. MAXIMUM SUSTAINED WINDS . . . 110 MPH. MINIMUM CENTRAL PRESSURE . . . 963 MB.

AN INTERMEDIATE ADVISORY WILL BE ISSUED BY THE NATIONAL HURRICANE CENTER AT 1 PM CDT FOLLOWED BY THE NEXT COMPLETE ADVISORY AT 4 PM CDT.

GUINEY

STRIKE PROBABILITIES ASSOCIATED WITH THIS ADVISORY NUMBER CAN BE FOUND UNDER AFOS HEADER MIASPFAT2 AND WMO HEADER WTNT72 KNHC.

Discussion

The discussion explains the hurricane specialist's reasoning for the predicted track and intensity (in this case, the hurricane specialist's last name is Guiney). Typically included are the forecaster's interpretations of the computer models; current oceanic observations from ships, buoys, oil rigs; and the reconnaissance flights by the Hurricane Hunters and the Hurricane Research Division.

> HURRICANE GEORGES DISCUSSION NUMBER 49
> NATIONAL WEATHER SERVICE MIAMI FL
> 11 AM EDT SUN SEP 27 1998
>
> THE AIR FORCE RESERVE HURRICANE HUNTER AIRCRAFT RECENTLY REPORTED A MINIMUM CENTRAL PRESSURE OF 963 MB FROM A GPS DROPSONDE WITH MAXIMUM FLIGHT-LEVEL WINDS OF 86 KNOTS. BUOY 42040 REPORTED A 45 KT 8-MINUTE SUSTAINED WIND . . . APPROXIMATELY 55 KT 1-MINUTE WIND SPEED . . . AND 33 FOOT SEAS AT 12Z WHILE THE C-MAN BUOY NEAR THE MOUTH OF THE MISSISSIPPI RIVER WAS 39 KT. THE INITIAL INTENSITY WILL BE HELD AT 95 KNOTS. THIS KEEPS GEORGES ON THE HIGH END OF CATEGORY TWO STATUS. THE FORECAST STILL CALLS FOR GEORGES TO STRENGTHEN TO A CATEGORY THREE BEFORE LANDFALL.
>
> BASED ON THE WIND DATA ABOVE . . . RECON WIND PLOTS AND HURRICANE RESEARCH DIVISION . . . HRD . . . ANALYSES THE WIND RADII HAVE BEEN ADJUSTED. THE PRIMARY CHANGE IS THE ADDITION OF HURRICANE-FORCE WINDS IN THE SOUTHWEST QUADRANT.
>
> FIXES FROM RECON AND NATIONAL WEATHER SERVICE DOPPLER RADAR SUGGEST A SLIGHT SLOWING IN THE FORWARD SPEED OVER THE LAST 6 HOURS. THE INITIAL MOTION ESTIMATE IS 315/07. THE SYNOPTIC REASONING REGARDING THE TRACK REMAINS UNCHANGED FROM THE LAST SEVERAL ADVISORIES. A GRADUAL SLOWDOWN IS FORECAST AS STEERING CURRENTS WEAKEN. THUS . . . THE OFFICIAL FORECAST TRACK IS VERY SIMILAR TO THE PREVIOUS ADVISORY. THIS MEANS THAT THE HURRICANE COULD PRODUCE EXTREMELY LARGE RAINFALL AMOUNTS COMBINED WITH A LONG PERIOD OF ONSHORE WINDS AND STORM SURGE FLOODING. GEORGES IS A VERY SERIOUS THREAT AND IT COULD BE EVEN WORSE IF THERE IS FURTHER STRENGTHENING.
>
> GUINEY

FORECAST POSITIONS AND MAX WINDS

INITIAL	27/1500Z	28.4N	88.0W	95 KTS
12HR VT	28/0000Z	28.9N	88.7W	100 KTS
24HR VT	28/1200Z	29.6N	89.5W	100 KTS
36HR VT	29/0000Z	30.3N	89.9W	80 KTS...INLAND
48HR VT	29/1200Z	30.8N	89.9W	65 KTS...INLAND
72HR VT	30/1200Z	31.5N	89.5W	50 KTS...INLAND

Strike Probability

As a means to quantify the likelihood of a hurricane passing near a particular coastal location, the NHC issues a strike probability discussion every six hours. The interpretation of these numbers requires some detailed explanation, which will begin with an example.

Suppose a hurricane is located in the central Gulf of Mexico and is forecast to make landfall in southeast Louisiana in twenty-four hours. Now suppose we could find 100 past cases where a storm was located near this position and was forecast to move along the same track. Since most twenty-four-hour forecasts contain some error, most twenty-four-hour positions will be different than what was predicted. In some cases, the location may be near southeast Louisiana, but in most other cases the actual twenty-four-hour position may be in Texas, Mississippi, Alabama, Florida, or some other location.

Now suppose that of these 100 past cases, southeast Louisiana experienced landfall twelve times within twenty-four hours. Then, the twenty-four-hour strike probability for this hurricane (currently located in the central Gulf of Mexico) making landfall in southeast Louisiana is 12 percent. Likewise, suppose six previous hurricanes actually made landfall in Galveston, Texas, within twenty-four hours; the strike probability for Galveston is then 6 percent. In this manner, strike probabilities can be issued for the entire threatened coastline. Similar probabilities can be computed for later forecast times.

The procedure through which the NHC computes strike probabilities is somewhat more complicated, but this example conveys the general procedure. A strike probability is defined by the NHC as the chance of the hurricane center passing within 65 nautical miles of a particular marine location. These probabilities are based on NHC forecast error statistics since 1970 and are an expression of forecast uncertainty. For example, if the NHC forecasts a hurricane to be 65 nautical miles from Galveston within twenty-four hours, what is the probability of this actually happening based on past forecast performance of hurricanes located at the same initial location?

The NHC issues strike probabilities for the following successive time periods: (1) less than twenty-four hours, (2) twenty-four to thirty-six hours, (3) thirty-six to forty-eight hours, and (4) forty-eight to seventy-two hours. Since forecast errors increase for longer-term forecasts, strike probabilities tend to "cluster" around a particular region for forecasts less than twenty-four hours, then tend to spread out with incrementally longer forecast time periods. These forecasts are issued in tabular form as shown below. The first column gives the strike probability within twenty-four hours. The next column gives the added increment to the probability within thirty-six hours, the next column the added increment within forty-eight hours, etc. For example, the strike probability below shows that Mobile's strike probability within twenty-four hours is 23 percent, within thirty-six hours is an added increment of 2 percent (or 25 percent total probability within thirty-six hours), within forty-eight hours is an added increment of 1 percent (or 26 percent total probability within forty-eight hours), and within seventy-two hours an added increment of 1 percent (or 27 percent total probability within seventy-two hours). The last column shows the total strike probability within seventy-two hours. It is important to remember that columns 2 through 4 cannot be utilized alone but are added increments to previous columns.

ZCZC MIASPFAT2 ALL
TTAA00 KNHC DDHHMM
HURRICANE GEORGES PROBABILITIES NUMBER 49
NATIONAL WEATHER SERVICE MIAMI FL
10 AM CDT SUN SEP 27 1998

PROBABILITIES FOR GUIDANCE IN HURRICANE
PROTECTION PLANNING BY GOVERNMENT AND DISASTER
OFFICIALS

AT 10 AM CDT . . . 1500Z . . . THE CENTER OF GEORGES WAS
LOCATED NEAR LATITUDE 28.4 NORTH . . . LONGITUDE 88.0

WEST

CHANCES OF CENTER OF THE HURRICANE PASSING WITHIN 65 NAUTICAL MILES OF LISTED LOCATIONS THROUGH 7AM CDT WED SEP 30 1998

LOCATION	A	B	C	D	E
29.6N 89.5W	42	X	X	X	42
30.3N 89.9W	31	1	1	X	33
30.8N 89.9W	24	3	1	1	29
MUAN 219N 850W	X	X	X	2	2
JACKSONVILLE FL	X	X	X	2	2
MARCO ISLAND FL	X	X	X	2	2
FT MYERS FL	X	X	X	2	2
VENICE FL	X	X	X	3	3
TAMPA FL	X	X	X	3	3
CEDAR KEY FL	X	X	1	3	4
ST MARKS FL	X	1	2	5	8
APALACHICOLA FL	3	2	2	5	12
PANAMA CITY FL	6	3	2	4	15
PENSACOLA FL	18	2	1	2	23
MOBILE AL	23	2	1	1	27

LOCATION	A	B	C	D	E
GULFPORT MS	31	1	X	1	33
BURAS LA	42	X	X	1	43
NEW ORLEANS LA	29	1	1	1	32
NEW IBERIA LA	13	5	2	3	23
PORT ARTHUR TX	1	3	4	5	13
GALVESTON TX	X	1	1	6	8
FREEPORT TX	X	X	1	5	6
PORT O CONNOR TX	X	X	X	3	3
GULF 29N 85W	3	1	3	5	12
GULF 29N 87W	65	X	X	X	65
GULF 28N 89W	99	X	X	X	99
GULF 28N 91W	5	3	3	5	16
GULF 28N 93W	X	1	2	7	10
GULF 28N 95W	X	X	X	5	5
GULF 27N 96W	X	X	X	2	2

```
COLUMN DEFINITION—PROBABILITIES IN PERCENT

A IS PROBABILITY FROM NOW TO 7AM MON
FOLLOWING ARE ADDITIONAL PROBABILITIES
B FROM 7AM MON TO 7PM MON
C FROM 7PM MON TO 7AM TUE
D FROM 7AM TUE TO 7AM WED
E IS TOTAL PROBABILITY FROM NOW TO 7AM WED
X MEANS LESS THAN ONE PERCENT

GUINEY
```

So, how should these probabilities be used for coastal residents and emergency preparedness officials? Ultimately, this is a personal decision, but general guidelines follow. First, one must decide the time window within which action must be initiated. This depends on the time it will take to complete evacuation procedures and on the intensity of the storm, since gale-force winds could arrive well before landfall in strong hurricanes. For populated coastal regions, decisions must be made with smaller probabilities than for those who can afford to wait for more precise forecasts. Second, the person has to decide on a probability threshold value at which action must be taken. This threshold value depends on the risk the decision-maker is willing to take for a given probability. A key factor in this decision would be some determination of the cost of not taking action and then being hit by the storm. Obviously, the decision-maker should err on the side of caution.

Final Comments on National Hurricane Center Statements

Because hurricanes change direction quickly, it is important to focus on the predicted path and not extrapolate from the past track. At the same time, you should realize that predicted paths may contain large errors, and typically even average forecasts beyond two days contain errors of several hundred miles. Furthermore, you should not concentrate on the landfall of the center alone, as strong winds, high surf, and torrential rains may extend far from the center.

Aircraft Reconnaissance Information

The National Hurricane Center website, http://www.nhc.noaa.gov, contains current aircraft reconnaissance information. Four products are available: the reconnaissance plan of the day, vortex data messages, airplane observations, and dropsonde observations. The reconnaissance plan of the day is a coded description of planned reconnaissance flights into Atlantic or Central Pacific tropical cyclones, subtropical cyclones, or suspect areas on a given day. This includes operational missions by the U.S. Air Force Reserve Hurricane Hunters and research missions by the Hurricane Research Division. It also includes a second-day outlook of potential flights.

When flights occur, the planes take a variety of observations. Reconnaissance observations are coded reports detailing the pressure, wind, temperature, and dew point along a reconnaissance aircraft's flight track. The reconnaissance observations are transmitted during passage to and from the storm and at turning points. During these flights, dropsondes are also ejected from the plane to take vertical profiles of pressure, wind, temperature, and dew point during the flight and in the storm center.

The most anticipated observation is the vortex data message—coded reports issued whenever a reconnaissance aircraft penetrates the center of a tropical or subtropical cyclone. These reports give the position of the cyclone as well as the time of the fix. They include information on winds, temperatures, pressure, and dew points encountered during penetration. If an eye is present, they also include information on its size, shape, and status. There are three types of vortex messages. Detailed Vortex Data Messages are the standard coded message issued when an aircraft penetrates a cyclone. Abbreviated Vortex Data Messages use the same format as the detailed messages, but they leave out some of the groups. Supplementary Vortex Data Messages give wind, temperature, pressure, and dew point data at 15-nm (nautical mile) intervals during the penetration into the storm and subsequent departure. Usually four detailed and four supplementary vortex messages are transmitted.

The first line of the report gives the mission identifier, which consists of (1) agency, (2) aircraft number, (3) number of missions in this storm system, (4) depression number, and (5) storm name. "AF554 WX OB 03 KNHC" means this mission is flown by the U.S. Air Force, with aircraft number 554, observation number 3, reported to NHC. The second line gives the name of the type of report, either ABBREVIATED VORTEX DATA MESSAGE or DETAILED VORTEX DATA MESSAGE.

The rest of the message is decoded as:

A. Date and time of fix: 06/1634Z means the report is from the sixth day of the month, at 1634Z.
B. Latitude of the vortex fix in degrees and minutes: for example, 26 DEG 00 MIN N. Longitude of the vortex fix in degrees and minutes: for example, 88 DEG 00 MIN W.
C. Minimum height of a standard pressure level, given in meters: for example, "700 MB 3150 M" means the lowest height of the 700 mb level was found to be 3150 meters above sea level.
D. Estimate of maximum surface wind observed in knots. For example, "30 KT" means the highest estimated surface wind is 30 knots with this particular storm system.
E. Bearing and range from center of the maximum surface wind, given in degrees and nautical miles. For example, "180 DEG 18 NM" means the 30-knot wind mentioned in D above is 18 nm south of the center of the storm.
F. Maximum flight level wind near storm center with direction from center given in degrees and speed in knots: for example, "110 DEG 45 KT" means that the wind is from 110 degrees at 45 knots.
G. Bearing and range from center of maximum flight level wind, given in degrees and nautical miles from the storm center: for example, "180 DEG 15 NM" means the maximum wind given in F above was found 15 nm south of the storm center.
H. Minimum sea-level pressure computed from dropsonde or extrapolation from within 1,500 feet of the sea surface, given in millibars (mb): for example, "1005 MB DROPSONDE" means that the lowest pressure found was 1,005 mb and was determined from a dropsonde.
I. Maximum flight-level temperature in Celsius/pressure altitude in meters, *outside* the eye: for example, "09 C / 3082 M" means that at the flight level of 700 mb, the highest temperature outside the eye is 9°C at a pressure altitude of 3,082 meters.
J. Maximum flight level temperature in Celsius/pressure altitude in meters, *inside* the eye: for example, "10 C / 3040 M" means that at the flight level of 700 mb, the

highest temperature inside the eye is 10°C at a pressure altitude of 3,040 meters.
K. Dewpoint temperature in Celsius/sea-surface temperature in Celsius inside the eye. For example, "08 C / 26 C" means that the dew point was 8°C inside the eye, and the temperature of the sea surface was 26°C.
L. Eye character: brief verbal description such as poorly defined, closed wall, open to NW, etc.
M. Eye shape orientation and diameter. Eye shapes are coded as follows: C (circular), CO (concentric), E (elliptical). Orientation of major axis of ellipse is transmitted in tens of degrees, and all diameters are transmitted in nautical miles. Examples: "E09/15/5" means elliptical eye oriented with major axis thru 90 degrees (and also 270 degrees), with length of major axis 15 nm, and length of minor axis 5 nm; "CO8-14" means concentric eye with inner eye diameter 8 miles, and outer diameter 14 miles.
N. Confirmation of latitude/longitude/time fix with format as in A and B above.
O. Fix determined and fix level. There are five means of determining fixes and nine means of indicating fix level. The fix determination will be a series of one to five numbers depending on how many items were used to determine the position of the storm center. The coding is as follows: 1, penetration; 2, radar; 3, wind; 4, pressure; and 5, temperature. The fix level will be either one or two numbers, depending on whether or not the surface and flight level centers were the same. The surface center will be given if visible; both the surface and flight level centers will be indicated only when they're the same. The coding is as follows: 0-surface, 1–1500 ft, 8–850 mb, 7–700 mb, 5–500 mb, 4–400 mb, 3–300 mb, 2–200 mb, 9-Other. For example, "1245/07" means that the fix was determined by four means: penetration, radar, temperature, and pressure. The fix level was both at the surface and at 700 mb.
P. Navigation fix accuracy in nautical miles and meteorological accuracy in nautical miles. For example, "5/10" means that the center is located within 5 nm of the latitude and longitude given for the center, with a meteorological accuracy to 10 nm.
Q. Remarks section.

Sample Report

AF554 WX OB 03 KNHC
DETAILED VORTEX DATA MESSAGE

A. 06/1634Z
B. 26 DEG 00 MIN N
88 DEG 00 MIN W
C. 700 MB 3150 M
D. 30 KT
E. 180 DEG 18 NM
F. 110 DEG 45 KT
G. 180 DEG 15 NM
H. 1005 MB DROPSONDE
I. 09 C/ 3082 M
J. 10 C/ 3040 M
K. 8 C/ 26 C
L. POORLY DEFINED
M. C08–14
N. 26 DEG 00 MIN N
88 DEG 00 MIN W
O. 1245/07
P. 5/10
Q. NONE

References

Abbey, R. F., Jr., L. M. Leslie, and G. J. Holland. 1995. Estimates of the inherent and practical limits of mean forecast errors of tropical cyclones. *Preprints, 21st Conference on Hurricanes and Tropical Meteorology.* Miami, FL, American Meteorological Society, 201–203.

Aberson, S. D. 2002. Two years of operational hurricane synoptic surveillance. *Wea. Forecasting,* **17,** 1101–1110.

———. 2003. Targeted observations to improve operational tropical cyclone track forecast guidance. *Mon. Wea. Rev.,* **131,** 1613–1628.

Adler, C. 2001. Neither wind nor rain. *Gambit Weekly,* May 15.

Ahrens, C. D. 1994. *Meteorology Today.* West Publishing, 591 pp.

Allaby, M. 1998. *A Chronology of Weather.* Facts on File, 154 pp.

Allen, E. S. 1976. *A Wind to Shake the World: The Story of the 1938 Hurricane.* Little, Brown, 370 pp.

Amato, I. 1997. *Pushing the Horizon: Seventy-Five Years of High Stakes Science and Technology at the Naval Research Laboratory.* Naval Research Laboratory, 409 pp. Also available at http://www.nrl.navy.mil.

Anon. 1996. The HOT ZONE. Hurricanes, Floods, and Blizzards: Blame Global Warming. *Newsweek,* January 21.

———. 2003. Hurricane Isabel recreational boat losses total over $150 million. *Saltwater Sportsman,* December, 38.

———. 2005. Record hurricane wave measured. *Stennis News,* March 31, pp. 24. (Available from Gulf Publishing Co., P.O. Box 4567, Biloxi, MS, 39535-4567, Ph: 228-896-2345.)

Anthes, R. A. 1982. *Tropical Cyclones: Their Evolution, Structure, and Effects.* American Meteorological Monographs, Vol. 19, American Meteorological Society, Boston, MA, 208 pp.

Atkinson, B. W. 1981a. Dynamical meteorology: Some milestones. In *Dynamical Meteorology: An Introductory Selection*, ed. B. W. Atkinson. Royal Meteorological Society, 116–129.

———. 1981b. Weather, meteorology, physics, mathematics. In *Dynamical Meteorology: An Introductory Selection*, ed. B. W. Atkinson. Royal Meteorological Society, 1–7.

Avila, L. A., and E. N. Rappaport. 1996. Atlantic hurricane season of 1994. *Mon. Wea. Rev.,* **124,** 1558–1578.

BAMS. 2004. Anti-gouging after Isabel. *Bull. Amer. Meteor. Soc.,* **85,** 156.

Barnes, J. 1998. *North Carolina's Hurricane History.* University of North Carolina Press, 206 pp.

Bazile, K. T. 2004. Sinking suspicions: Land-elevation data getting vital update. *Times Picayune,* May 14.

Bender, M. A., and I. Ginis. 2000. Real case simulations of hurricane-ocean interaction using a high-resolution coupled model: Effects on hurricane intensity. *Mon. Wea. Rev.,* **128,** 917–946.

Berdeklis, P., and R. List. 2001. The ice crystal-graupel charging mechanism of thunderstorm electrification. *J. Atmos. Sci.,* **58,** 2751–2770.

Bergeron, K. 1999. Hurricane Camille: 30 years later. *Sun Herald,* August 15.

———. 2000. Camille survivors want the truth to be heard. *Sun Herald,* August 17.

———. 2003. "Camille's wrath" rediscovered. *Sun Herald,* June 12.

Beven, J. L., S. R. Stewart, M. B. Lawrence, L. A. Avila, J. L. Franklin, and R. J. Pasch. 2003. Atlantic hurricane season of 2001. *Mon. Wea. Rev.,* **131,** 1454–1484.

Bister, M., and K. A. Emanuel. 1997. The genesis of Hurricane Guillermo: TEXMEX analyses and a modeling study. *Mon. Wea. Rev.,* **125,** 2662–2682.

Black, M. L., J. F. Gamache, F. D. Marks, C. E. Samsury, and H. E. Willoughby. 2002. Eastern Pacific Hurricanes Jimena of 1991 and Olivia of 1994: The effect of vertical shear on structure and intensity. *Mon. Wea. Rev.,* **130,** 2291–2312.

Black, P. G. 1983. Ocean temperature changes induced by tropical cyclones. Ph.D. dissertation, Pennsylvania State University, 278 pp.

———. 2004. An overview of CBLAST flights into Hurricanes Fabian and Isabel (2003). *Preprints, 26th Conference on Hurricanes and Tropical Meteorology.* Miami, FL, American Meteorological Society.

Black, P. G., and F. D. Marks. 1991. The structure of an eyewall mesovortex in Hurricane Hugo (1989). *Preprints, 19th Conference on*

Hurricanes and Tropical Meteorology. Miami, FL, American Meteorological Society, 579–582.
Black, R. A., and J. Hallett. 1986. Observations of the distribution of ice in hurricanes. *J. Atmos. Sci.,* **43**, 802–822.
Bowie, E. H. 1922. Formulation and movement of West Indian hurricanes. *Mon. Wea. Rev.,* **50**, 173–179.
Brouwer, G. 2003. The creeping storm. *Civil Engineering,* **73**, 46–55, 88.
Brown, S., and S. Cannizaro. 2002. Adios, Isidore: Soggy streets, cars, homes leave plenty of cleaning up; water invades St. Bernard homes. *Times Picayune,* September 27.
Browner, S. P., W. L. Woodley, and C. G. Griffith. 1977. Diurnal oscillation of the area of cloudiness associated with tropical storms. *Mon. Wea. Rev.,* **105**, 856–864.
Brueske, K. F., and C. S. Velden. 2003. Satellite-based tropical cyclone intensity estimation using the NOAA-KLM series Advanced Microwave Sounding Unit (AMSU). *Mon. Wea. Rev.,* **131**, 687–697.
Bunnemeyer, B. 1909. Climatological data for July, 1909. District No. 8, Texas and Rio Grand Valley. *Mon. Wea. Rev.,* **37**, 351–355.
Burpee, R. W. 1988. Grady Norton: Hurricane forecaster and communicator extraordinaire. *Wea. Forecasting,* **3**, 1050–1058.
———. 1995. Necrologies: Gordon E. Dunn. *Bull. Amer. Meteor. Soc.,* **76**, 260–261.
Burpee, R. W., J. L. Franklin, S. J. Lord, R. E. Tuleya, and S. D. Aberson. 1996. The impact of Omega dropwindsondes on operational hurricane track forecast models. *Bull. Amer. Meteor. Soc.,* **77**, 925–933.
Byers, H. R. 1944. *General Meteorology.* McGraw-Hill, 645 pp.
———. 1960. *Carl-Gustaf Arvid Rossby.* National Academy of Sciences, Biographical Memoirs, Vol. 34, 249–270.
———. 1974. History of weather modification. In *Weather and Climate Modification,* ed. W. N. Hess, 3–44. John Wiley and Sons.
Cabbage, M. 1998. Hurricane man. *Sun Sentinel,* May 31.
Carr, L. E., and R. L. Elsberry. 1997. Models of tropical cyclone wind distribution and Beta-effect propagation for application to tropical cyclone track forecasting. *Mon. Wea. Rev.,* **125**, 3190–3209.
Carter, T. M. 1983. *Probability of Hurricane/Tropical Storm Conditions: A User's Guide for Local Decision Makers.* National Oceanic and Atmospheric Administration, National Weather Service, 25 pp.
Case, B., and M. Mayfield. 1990. Atlantic hurricane season of 1989. *Mon. Wea. Rev.,* **118**, 1165–1177.

Case, R. A. 1986. Atlantic hurricane season of 1985. *Mon. Wea. Rev.*, **114**, 1390–1405.

Case, R. A., and H. P. Gerrish. 1984. Atlantic hurricane season of 1983. *Mon. Wea. Rev.*, **112**, 1083–1092.

Cecil, D. J., and E. J. Zipser. 2002. Reflectivity, ice scattering, and lightning characteristics of hurricane eyewalls and rainbands. Part II: Intercomparison of observations. *Mon. Wea. Rev.*, **130**, 785–801.

Chabert, J.-L. 1999. *A History of Algorithms from the Pebble to the Microchip*. Springer, 524 pp.

Chan, J. C. L. 2000. Tropical cyclone activity in the western North Pacific associated with El Niño and La Niña events. *J. Climate*, **13**, 997–1004.

Chan, J. C. L., and K. S. Liu. 2004. Global warming and Western North Pacific activity from an observational perspective. *J. Climate*, **23**, 4590–4602.

Chan, J. C. L., J. E. Shi, and K. S. Liu. 2001. Improvements in the seasonal forecasting of tropical cyclone activity over the western North Pacific. *Wea. Forecasting*, **16**, 491–498.

Chen, Y., G. Brunet, and M. K. Yau. 2003. Spiral bands in a simulated hurricane. Part II: Wave activity diagnostics. *J. Atmos. Sci.*, **60**, 1239–1256.

Chisholm, D. 2003. AMS introduces the Vonnegut-Schaefer undergraduate scholarship. *Bull. Amer. Meteor. Soc.*, **77**, 110–112.

Cione, J. J., and E. W. Uhlhorn. 2003. Sea surface temperature variability in hurricanes: Implications with respect to intensity change. *Mon. Wea. Rev.*, **131**, 1783–1796.

Cline, I. M. 1915. The tropical hurricane of September 29, 1915, in Louisiana. *Mon. Wea. Rev.*, **43**.

Colon, J. A. 1980. On the wind structure of the hurricane vortex. *Preprints, 13th Technical Conference on Hurricanes and Tropical Meteorology*. American Meteorological Society, 30–42.

Corbosiero, K. L., and J. Molinari. 2002. The effects of vertical wind shear on the distribution of convection in tropical cyclones. *Mon. Wea. Rev.*, **130**, 2110–2123.

Cressman, G. P. 1996. The origin and rise of numerical weather prediction. In *Historical Essays on Meteorology*, ed. J. R. Fleming. American Meteorological Society, 21–39.

Daintith, J., S. Mitchell, E. Tootill, and D. Gjertsen. 1994. *Biographical Encyclopedia of Scientists*, Vol. 2. Aylesbury, UK: Market House Books.

Danielson, E. W., J. Levin, and E. Abrams. 1998. *Meteorology*. WCB/McGraw-Hill, 462 pp.

Davis, W. R. 1954. Hurricanes of 1954. *Mon. Wea. Rev.*, **82**, 370–373.
DeAngelis, D. 1989. The hurricane priest. *Weatherwise*, **42**, 256–257.
DeMaria, M. 1996. A history of hurricane forecasting for the Atlantic Basin. In *Historical Essays on Meteorology*, ed. J. R. Fleming. American Meteorological Society, 263–305.
DeMaria, M., and J. Kaplan. 1994. Sea surface temperature and the maximum intensity of Atlantic tropical cyclones. *J. Climate*, **7**, 1324–1334.
DeMaria, M., J. A. Knaff, B. H. Connell. 2001. A tropical cyclone genesis parameter for the tropical Atlantic. *Wea. Forecasting*, **16**, 219–233.
Deppermann, C. E. 1946. Is there a ring of violent upward convection in hurricanes and typhoon? *Bull. Amer. Meteor. Soc.*, **27**, 6–8.
Dlugolecki, A. F., K. M. Clark, F. Knecht, D. McCauley, J. P. Palutikof, and W. Yambi. 1996. Financial services. In *Climate Change 1995: Impacts, Adaptations, and Mitigation of Climate Change: Scientific-Technical Analyses*, ed. R. T. Watson, M. C. Zinyowera, and R. H. Moss. Cambridge University Press, 539–560.
Dong, K., and C. J. Neumann. 1986. The relationship between tropical cyclone motion and environmental geostrophic flows. *Mon. Wea. Rev.*, **114**, 115–122.
Douglas, M. S. 1997. *The Everglades: River of Grass*. Pineapple Press, 480 pp.
Dunion, J. P., C. W. Landsea, S. H. Houston, and M. D. Powell. 2003. A reanalysis of the surface winds for Hurricane Donna of 1960. *Mon. Wea. Rev.*, **131**. 1992–2011.
Dunion, J. P., and C. S. Velden. 2002. Satellite applications for tropical wave/tropical cyclone tracking. Preprints, *25th Conf. On Hurricanes and Tropical Meteorology*, San Diego, CA, American Meteorological Society, 132–133.
———. 2004. The impact of the Saharan air layer on Atlantic tropical cyclone activity. *Bull. Amer. Meteor. Soc.*, **85**, 353–365.
Dunn, G. E. 1940a. Aerology in the hurricane warning service. *Mon. Wea. Rev.*, **68**, 303–315.
———. 1940b. Cyclogenesis in the tropical Atlantic. *Bull. Amer. Meteor. Soc.*, **21**, 215–229.
———. 1961. The hurricane season of 1960. *Mon. Wea. Rev.*, **89**, 99–108.
Dunn, G. E., W. R. Davis, and P. L. Moore. 1955. Hurricanes of 1955. *Mon. Wea. Rev.*, **83**, 315–326.
———. 1956. Hurricanes of 1956. *Mon. Wea. Rev.*, **84**, 436–443.
Dunn, G. E., and Staff. 1959. The hurricane season of 1959. *Mon. Wea. Rev.*, **87**, 441–450.

———. 1962. The hurricane season of 1961. *Mon. Wea. Rev.,* **90,** 107–119.

———. 1965. The hurricane season of 1964. *Mon. Wea. Rev.,* **93,** 175–187.

Durran, D. R. 1999. *Numerical Methods for Wave Equations in Geophysical Fluid Dynamics.* Springer, 465 pp.

Dvorak, V. F. 1975. Tropical cyclone intensity analysis and forecasting from satellite imagery. *Mon. Wea. Rev.,* **103,** 420–430.

———. 1984. *Tropical Cyclone Intensity Analysis Using Satellite Data.* NOAA Tech. Rep. NESDIS 11, 47 pp.

Elsberry, R. L. 1990. International experiments to study tropical cyclones in the western North Pacific. *Bull. Amer. Meteor. Soc.,* **71,** 1305–1316.

———. 2005. Achievement of USWRP hurricane landfall research goal. *Bull. Amer. Meteor. Soc.,* **86,** 643–645.

Emanuel, K. A. 1987. The dependence of hurricane intensity on climate. *Nature,* **326,** 483–485.

———. 2000. A statistical analysis of tropical cyclone intensity. *Mon. Wea. Rev.,* **128,** 1139–1152.

———. 2005. Increasing destructiveness of tropical cyclones over the past 30 years. *Nature,* **486,** 686–688.

ESSA (Environmental Science Services Administration). 1969. *Hurricane Camille, August 14–22 (Preliminary Report).* Weather Bureau, U.S. Department of Commerce, 58 pp.

Fassig, O. L. 1913. *Hurricanes of the West Indies.* U.S. Weather Bureau.

Fernandez, D. E., S. Frasier, J. Carswell, P. Chang, P. G. Black, and F. D. Marks Jr. 2004. 3-D atmospheric boundary layer wind fields from Hurricanes Fabian and Isabel. *26th Conference on Hurricanes and Tropical Meteorology.* Miami, FL, American Meteorological Society.

Fincher, L., and B. Read. 1999. *The 1943 "Surprise" Hurricane.* Available from the City of Houston Office of Emergency Management.

Fisher, S. M. 2003. From the archives. *Bull. Amer. Meteor. Soc.,* **84,** 942.

Fitzpatrick, P. J. 1997. Understanding and forecasting tropical cyclone intensity change with the Typhoon Intensity Prediction Scheme (TIPS). *Wea. Forecasting,* **12,** 826–846.

Fleming, J. R. 1990. *Meteorology in America, 1800–1870.* John Hopkins University Press, 264 pp.

Fletcher, R. D., J. R. Smith, and R. C. Bundgaard. 1961. Superior photographic reconnaissance of tropical cyclones. *Weatherwise,* **14,** 102–109.

Fox, William P. 1992. *Lunatic Wind: Surviving Out the Storm of the Century.* Algonquin Books of Chapel Hill, 197 pp.

Frankenfield, H. C. 1915. The tropical storm of August 10, 1915. *Mon. Wea. Rev.,* **43**, 405–412.

Franklin, J. L., M. L. Black, and K. Valde. 2003. GPS dropwindsonde wind profiles in hurricanes and their operational implications. *Wea. Forecasting,* **18**, 32–44.

Franklin, J. L., C. J. McAdie, and M. B. Lawrence. 2003. Trends in track forecasting for tropical cyclones threatening the United States, 1970–2001. *Bull. Amer. Meteor. Soc.,* **84**, 1197–1203.

Frederick, W. J. 2003. The rapid intensification and subsequent rapid weakening of Hurricane Lili as compared with historical hurricanes. *Wea. Forecasting,* **18**, 1295–1298.

Fujita, T. T. 1993. Wind fields of Andrew, Omar, and Iniki, 1992. *Preprint, 20th Conference on Hurricanes and Tropical Meteorology,* San Antonio, TX, American Meteorological Society, 46–49.

Garriot, E. B. 1900. *West Indian Hurricanes,* U.S. Weather Bureau.

———. 1909. Weather, forecasts, and warnings for the month. *Mon. Wea. Rev.,* **37**, 538–539.

Gillispie, C. C. 1975. *Dictionary of Scientific Biography,* Vol. 11. New York: Charles Scriber's Sons.

Goerss, J. S. 2000. Tropical cyclone forecasting using an ensemble of dynamical model. *Mon. Wea. Rev.,* **128**, 1187–1193.

Goldenberg, S., and L. Shapiro. 1996. Physical mechanisms for the association of El Niño and West African rainfall with Atlantic major hurricane activity. *J. Climate,* **9**, 1169–1187.

Gray, W. M. 1968. Global view of the origin of tropical disturbances and storms. *Mon. Wea. Rev.,* **96**, 669–700.

———. 1979. Hurricanes: Their formation, structure and likely role in the tropical circulation. In *Meteorology over the Tropical Oceans,* ed. D. B. Shaw. Royal Meteorological Society, 155–218.

———. 1984a. Atlantic seasonal hurricane frequency. Part I: El Niño and 30 mb quasi-biennial oscillation influences. *Mon. Wea. Rev.,* **112**, 1649–1668.

———. 1984b. Atlantic seasonal hurricane frequency. Part II: Forecasting its variability. *Mon. Wea. Rev.,* **112**, 1669–1683.

———. 1993. Seasonal forecasting. In *Global Guide to Tropical Cyclone Forecasting,* ed. G. J. Holland, chapter 5. World Meteorological Organization Technical Document, WMO/TD No. 560, Tropical Cyclone Programme, Report No. TCP-31, Geneva, Switzerland. Also available at http://www.wmo.int and http://www.bom.gov.au/bmrc (look for additional links under the mesoscale section to find this book)

Gray, W. M., and W. M. Frank. 1993. Hypothesis for hurricane intensity reduction from carbon black seeding. *Preprint, 20th Conference on Hurricanes and Tropical Meteorology,* San Antonio, TX, American Meteorological Society, 305–308.

Gray, W. M., and R. W. Jacobson Jr. 1977. Diurnal variation of deep cumulus convection. *Mon. Wea. Rev.,* **105,** 1171–1188.

Gray, W. M., and P. J. Klotzbach. 2004. *Summary of 2004 Atlantic Tropical Cyclone Activity and Verification of Author's Seasonal and Monthly Forecasts.* Fort Collins: Department of Atmospheric Science, Colorado State University, 39 pp. Available at http://hurricane.atmos.colostate.edu/Forecasts/2004/nov2004/.

Gray, W. M., C. W. Landsea, P. W. Mielke Jr., and K. J. Berry. 1998. Extended range forecast of Atlantic seasonal hurricane activity and U.S. landfall strike probability for 1999. Department of Atmospheric Science Paper, Colorado State University (also available at http://tropical.atmos.colostate.edu).

Gray, W. M., J. D. Sheaffer, and C. W. Landsea. 1996. Climate trends associated with multi-decadal variability of intense Atlantic hurricane activity. Chapter 2 in *Hurricanes, Climatic Change and Socioeconomical Impacts: A Current Perspective,* ed. H. F. Diaz and R. S. Pulwarty. Westview Press, 49 pp.

Guard, C., and M. A. Lander. 1993. A modified Saffir-Simpson hurricane scale for the tropical western Pacific Ocean. *Preprints, AMS 20th Conference on Hurricanes and Tropical Meteorology,* American Meteorological Society.

Haas, I. S., and R. Shapiro. 1982. The Nimbus satellite system: Remote sensing R&D platform of the 70's. In *Meteorological Satellites: Past, Present, and Future.* NASA Conference Publication 2227, 17–29.

Hall, J. D., A. J. Matthews, and D. J. Karoly. 2001. The modulation of tropical cyclone activity in the Australian region by the Madden-Julian Oscillation. *Mon. Wea. Rev.,* **129,** 2970–2982.

Harr, P. A., and R. L. Elsberry. 1996. Structure of a mesoscale convective system embedded in Typhoon Robyn during TCM-93. *Mon. Wea. Rev.,* **124,** 634–652.

Hart, R. E., and J. L. Evans 2001. A climatology of the extratropical transition of Atlantic tropical cyclones. *J. Climate,* **14,** 546–564.

Haurwitz, B. 1935. The height of tropical cyclones and of the "eye" of the storm. *Mon. Wea. Rev.,* **63,** 45–49.

Hawkins, J. D., T. F. Lee, J. Turk, C. Sampson, J. Kent, and K. Richardson. 2001. Real-time Internet distribution of satellite products for tropical cyclone reconnaissance. *Bull. Amer. Meteor. Soc.,* **82,** 567–578.

Hebert, P. J. 1976. Atlantic hurricane season of 1975. *Mon. Wea. Rev.*, **104**, 453–465.
———. 1980. Atlantic season of 1979. *Mon. Wea. Rev.*, **108**, 973–990.
Hebert, P. J., J. D. Jarrell, and M. Mayfield. 1997. *The Deadliest, Costliest, and Most Intense United States Hurricanes of this Century (and Other Frequently Requested Hurricane Facts)*. NOAA Technical Memorandum NWS TPC-1, National Oceanic and Atmospheric Administration, National Weather Service, National Hurricane Center, Miami, FL, 30 pp.; also available at http://www.nhc.noaa.gov.
Hemingway, E. 1935. Who murdered the Vets? A first-hand report on the Florida hurricane. *New Masses*, September 17, 9–10.
Henderson-Sellers, A., H. Zhang, G. Berz, K. Emanuel, W. Gray, C. Landsea, et al. 1998. Tropical cyclones and global climate change: A post-IPCC assessment. *Bull. Am. Meteor. Soc.*, **79**, 19–38.
Hock, T. F., and J. L. Franklin. 1999. The NCAR GPS dropwindsonde. *Bull. Amer. Meteor. Soc.*, **80**, 407–420.
Hoffman, R. N. 2004. Controlling hurricanes: Can hurricanes and other severe tropical storms be moderated or deflected? *Sci. Amer.*, **12**, 68–75.
Hogan, W. L. 1961. *Hurricane Carla: A Tribute to the News Media.* Leaman-Hogan, 192 pp.
Hohertz, L., and Coauthors. 1983. *Hurricane Alicia.* Barron Publications, 80 pp.
Holland, G. J. 1993a. *Global Guide to Tropical Cyclone Forecasting.* World Meteorological Organization Technical Document, WMO/TD No. 560, Tropical Cyclone Programme, Report No. TCP-31, Geneva, Switzerland. Also available at http://www.wmo.int and http://www.bom.gov.au/bmrc (look for additional links under the mesoscale section to find this book).
———. 1993b. Ready Reckoner. In *Global Guide to Tropical Cyclone Forecasting,* ed. G. J. Holland, chap. 9. World Meteorological Organization Technical Document, WMO/TD No. 560, Tropical Cyclone Programme, Report No. TCP-31, Geneva, Switzerland. Also available at http://www.wmo.int and http://www.bom.gov.au/bmrc (look for additional links under the mesoscale section to find this book).
———. 1995. Scale interaction in the western Pacific monsoon. *Meteor. Atmos. Phys.*, **56**, 57–80.
Hope, J. R. 1975. Atlantic season of 1974. *Mon. Wea. Rev.*, **103**, 285–293.
Houghton, J. T., L. G. Meira Filho, B. A. Callander, N. Harris, A. Kattenberg, and K. Maskell. 1996. *Climate Change 1995: The Science*

of Climate Change; Contribution of Working Group I to the Second Assessment of the Intergovernmental Panel on Climate Change. Cambridge University Press, 572 pp.

Huang, W., C. Wu, P. Lin, S. Aberson, and K. Hsu. 2004. The preliminary analysis of the dropsonde data from DOTSTAR and their impact on the typhoon track forecasts. *Preprints, 26th Conference on Hurricanes and Tropical Meteorology.* Miami, FL, American Meteorological Society.

Hughes, P. 1976. *American Weather Stories,* U.S. Department of Commerce, Washington, D.C., 116 pp.

———. 1987. Hurricanes haunt our history, *Weatherwise,* **40**, 134–140.

———. 1990. The Great Galveston Hurricane. *Weatherwise,* **43**, 190–198.

Illingworth, A. J. 1985. Charge separation in thunderstorms: Small scale processes. *J. Geophys. Res.,* **90**, 6026–6032.

Inoue, M., I. C. Handoh, and G. R. Bigg. 2002. Bimodal distribution of tropical cyclogenesis in the Caribbean: Characteristics and environmental factors. *J. Climate,* **15**, 2897–2905.

Jennings, G. 1970. *The Killer Storms: Hurricanes, Typhoons, and Tornadoes.* J. B. Lippincott, 207 pp.

Jones, D., S. R. Scheider, P. Wilczynski, and C. Nelson. 2004. NPOESS preparatory project: The bridge between research and operations. *Earth Obs. Mag.,* **13**, 12–22.

Jones, S. C., P. A. Harr, J. Abraham, L. F. Bosart, P. J. Bowyer, J. L. Evans, et al. 2003. The extratropical transition of tropical cyclones: Forecast challenges, current understanding, and future directions. *Wea. Forecasting,* **18**, 1052–1092.

JTWC (Joint Typhoon Warning Center). 1970. *Annual Typhoon Report.* U.S. Fleet Weather Central/Joint Typhoon Warning Center. Available at http://www.npmoc.navy.mil/jtwc/atcr/atcr_archive.html.

Jury, M. R., B. Pathack, and B. Parker. 1999. Climatic determinants and statistical prediction of tropical cyclone days in the Southwest Indian Ocean. *Journal of Climate,* **12**, 1738–1746.

Kaplan, J., and M. DeMaria. 1995. A simple empirical model for predicting the decay of tropical cyclone winds after landfall. *J. Appl. Meteor.,* **34**, 2499–2512.

Kareem, A., ed. 1985. *Hurricane Alicia: One Year Later.* American Society of Civil Engineers, 335 pp.

Kessler, E. 2004. *Historical overview of the National Severe Storm Laboratory.* Fortieth Anniversary Celebration, October 15, 2004. Available from http://www.nssl.noaa.gov.

Kinney, J. J. R. 1955. *Typhoon Forecasting Guide.* Tokyo Weather Central.

Kinsman, B. 1990. Who put the wind speeds in Admiral Beaufort's force scale? Part 1. The original scale. *Mar. Wea. Log,* **34,** 2–8.
———. 1991. Who put the wind speeds in Admiral Beaufort's force scale? Part 2: The new scale. *Mar. Wea. Log,* **35,** 12–18.
Klein, R., and R. A. Pielke Jr. 2002. Bad weather? Then sue the weatherman! Part I: Legal liability for public sector forecasts. *Bull. Amer. Meteor. Soc.,* **83,** 1791–1799.
Knaff, J. 1997. Implications of summertime sea level pressure anomalies in the tropical Atlantic region. *Mon. Wea. Rev.,* **10,** 789–804.
Knaff, J. A., J. P. Kossin, and M. DeMaria. 2003. Annular hurricanes. *Wea. Forecasting,* **18,** 204–223.
Knaff, J. A., S. A. Seseske, M. DeMaria, and J. L. Demuth. 2004. On the influences of vertical wind shear on symmetric tropical cyclone structure derived from AMSU. *Mon. Wea. Rev.,* **132,** 2503–2510.
Knox, J. L. 1955. The storm "Hazel": Synoptic resume of its development as it approached southern Ontario. *Bull. Amer. Meteor. Soc.,* **36,** 239–246.
Kossin, J. P., B. D. McNoldy, and W. H. Schubert. 2002. Vortical swirls in hurricane eye clouds. *Mon. Wea. Rev.,* **130,** 3144–3149.
Kossin, J. P., and W. H. Schubert. 2004. Mesovortices in Hurricane Isabel. *Bull. Amer. Meteor. Soc.,* **85,** 151–153.
Krupa, M. 2002. As floodwaters recede, the real work begins. *Times Picayune,* September 28.
Krupa, M., S. Staley, and P. Bartels. 2002. Isidore's tidal surge sucker punches parish. *Times Picayune,* September 27.
Kuettner, J. P., and D. E. Parker. 1976. GATE: Report of the field phase. *Bull. Amer. Meteor. Soc.,* **57,** 11–27.
Kummerow, C. J., J. Simpson, O. Thiele, W. Barnes, A. T. C. Chang, E. Stocker, et al. 2000. The status of the Tropical Rainfall Measuring Mission (TRMM) after two years in orbit. *J. Appl. Meteor.,* **39,** 1965–1982.
Kurihara, Y. M., R. E. Tuleya, and M. A. Bender. 1998. The GFDL hurricane prediction system and its performance in the 1995 hurricane season. *Mon. Wea. Rev.,* **126,** 1306–1322.
Kutzbach, G. 1996. *The Thermal Theory of Cyclones: A History of Meteorological Thought in the Nineteenth Century.* Boston: American Meteorological Society, 272 pp.
Lander, M. A. 1993. Comments on a "GCM simulation of the relationship between tropical storm formation and ENSO." *Mon. Wea. Rev.,* **121,** 2137–2143.

———. 1994. Description of a monsoon gyre and its effect on the tropical cyclones in the Western North Pacific during August 1991. *Wea. Forecasting*, **9**, 640–654.

———. 1999. A tropical cyclone with a very large eye. *Mon. Wea. Rev.*, **127**, 137–142.

Lander, M. A., and G. J. Holland. 1993. On the interaction of tropical cyclone-scale vortices. Part I: Observations. *Quart. J. Roy. Meteor. Soc.*, **119**, 1347–1361.

Landsea, C. W. 2000. Climate variability of tropical cyclones: Past, present, and future. In *Storms*, ed. R. A. Pielke Sr. and R. A. Pielke Jr., Routledge, 220–241.

———. 2005. Hurricanes, typhoons, and tropical cyclones: FAQ. Available at http://www.aoml.noaa.gov/hrd/tcfaq/tcfaqHED.html.

Landsea, C. W., C. Anderson, N. Charles, G. Clark, J. Dunion, J. Fernandez-Partagas, et al. 2004. The Atlantic hurricane database reanalysis project: Documentation for 1851–1910 alterations and additions to the HURDAT Database. In *Hurricanes and Typhoons: Past, Present, and Future*, ed. R. J. Murnane and K. Liu, chap. 7. Columbia University Press, 464 pp.

Landsea, C. W., N. Nicholls, W. M. Gray, and L. A. Avila. 1996. Downward trends in the frequency of intense Atlantic hurricanes during the past five decades. *Geo. Res. Letters*, **23**, 1697–1700.

Larson, E. 1999. *Isaac's Storm*. Crown Publishers, 323 pp.

Lawrence, M. G. 2005. The relationship between relative humidity and the dewpoint temperature in moist air. *Bull. Amer. Meteor. Soc.*, **86**, 225–234.

Lawrence, M. B., L. A. Avila, J. L. Beven, J. L. Franklin, J. L. Guiney, and R. J. Pasch. 2001. Atlantic hurricane season of 1999. *Mon. Wea. Rev.*, **129**, 3057–3084.

Lawrence, M. B., and J. M. Gross. 1989. Atlantic hurricane season of 1988. *Mon. Wea. Rev.*, **117**, 2248–2259.

Lawrence, M. B., B. M. Mayfield, L. A. Avila, R. J. Pasch, and E. N. Rappaport. 1998. Atlantic hurricane season of 1995. *Mon. Wea. Rev.*, **126**, 1124–1151.

Lawrence, M. B., and J. M. Pelissier. 1981. Atlantic hurricane season of 1980. *Mon. Wea. Rev.*, **109**, 1567–1582.

Lee, W.-C., F. D. Marks Jr., and R. E. Carbone. 1994. Velocity track display: A technique to extract real-time tropical cyclone circulations using a single airborne Doppler radar. *J. Atmos. Oceanic Technol.*, **11**, 572–578.

Lee, W.-C., F. D. Marks Jr., and C. Walther. 2003. Airborne Doppler radar data analysis workshop. *Bull. Amer. Meteor. Soc.*, **84**, 1063–1075.

Leggett, J., ed. 1994. *The Climate Time Bomb.* Greenpeace International, Amsterdam, Netherlands, 154 pp.

Leidner, M. S., L. Isaksen, and R. N. Hoffman. 2003. Impact of NSCAT winds on tropical cyclones in the ECMWF 4DVAR assimilation system. *Mon. Wea. Rev.,* **131**, 3–26.

LeMone, M. 1989. *Interview with Joanne Simpson.* National Center for Atmospheric Research. Available from the National Center for Atmospheric Research's archive at http://www.ucar.edu/archives.

LePore, F. 1996. Interview with *Earth and Sky* Radio Show.

Leslie, L. M., R. F. Abbey Jr., and G. J. Holland. 1998. Tropical cyclone track predictability. *Meteor. Atmos. Phys.,* **65**, 223–231.

Lewis, J. M. 2003. Ooishi's observation: Viewed in context of jet stream discovery. *Bull. Amer. Meteor. Soc.,* **84**, 357–369.

Lhermitte, R. M. 1971. Probing of atmospheric motion by airborne pulse-Doppler radar techniques. *J. Appl. Meteor.,* **10**, 234–246.

Lianshou, C., Z. Mingyu, and X. Xiangde. 2004. A tropical cyclone landfall research program (CLATEX) in China. *Preprints, 26th Conference on Hurricanes and Tropical Meteorology.* Miami, FL, American Meteorological Society.

List, R. 2004. Weather modification: A scenario for the future. *Bull. Amer. Meteor. Soc.,* **85**, 51–63.

Liu, K. S., and J. C. L. Chan. 1999. Size of tropical cyclones as inferred from ERS-1 and ERS-2 data. *Mon. Wea. Rev.,* **127**, 2992–3001.

———. 2002. Synoptic flow patterns associated with small and large tropical cyclones over the Western North Pacific. *Mon. Wea. Rev.,* **130**, 2134–2142.

Lorenz, E. N. 1963. Deterministic nonperiodic flow. *J. Atmos. Sci.,* **20**, 130–141.

Love, G. 1985. Cross-equatorial influence of winter hemisphere subtropical cold surges. *Mon. Wea. Rev.,* **113**, 1487–1498.

Ludlum, D. M. 1963. *Early American Hurricanes, 1492–1870.* American Meteorological Society, 198 pp.

———. 1989. *Early American Hurricanes: 1492–1870.* American Meteorological Society, 198 pp.

Lynch, P. 2002. Weather Forecasting: From Woolly Art to Solid Science. In *Meteorology at the Millenium,* ed. R. P. Pearce. Academic Press, 106–119.

Maloney, E. D., and D. L. Hartmann. 2000a. Modulation of eastern North Pacific hurricanes by the Madden-Julian Oscillation. *J. Climate,* **13**, 1451–1460.

———. 2000b. Modulation of hurricane activity in the Gulf of Mexico by the Madden-Julian Oscillation. *Science,* **287**, 2002–2004.

Marks, F. D., and P. G. Black. 1990. Close encounter with an intense mesoscale vortex within Hurricane Hugo (September 15, 1989). *Preprints, 4th Conference on Mesoscale Processes.* Boulder, CO, American Meteorological Society, 114–115.

Martin, C., and G. Parker. 2002. *The Spanish Armada: Revised Edition.* Manchester University Press, 320 pp.

Mattingly, G. 1974. *The Armada,* Mariners Books, 464 pp.

Mayfield, M., L. A. Avila, and E. N. Rappaport. 1994. Atlantic hurricane season of 1992. *Mon. Wea. Rev.,* **122**, 517–538.

McCarthy, J. 1969. *Hurricane!* American Heritage Publishing, 168 pp.

McDonald, W. F. 1935. The hurricane of August 31 to September 6, 1935. *Mon. Wea. Rev.,* **63**, 269–271.

McGeer, T., and J. Vagners. 1999. Historic crossing: An unmanned aircraft's Atlantic flight. *GPS World,* **10**, 24–30.

Merrill, R. T. 1984. A comparison of large and small tropical cyclones. *Mon. Wea. Rev.,* **112**, 1408–1418.

Minsinger, W. E. 1988. *The 1938 Hurricane.* American Meteorological Society, 128 pp.

Mitchell, C. L. 1924. West Indian hurricanes and other tropical cyclones of the north Atlantic ocean. *Mon. Wea. Rev.,* **52** (Suppl. 24), 47 pp.

———. 1926. The West Indian Hurricane of September 14–22, 1926. *Mon. Wea. Rev.,* **54**, 409–414.

Molinari, J., D. Knight, M. Dickinson, D. Vollaro, and S. Skubis. 1997. Potential vorticity, easterly waves, and tropical cyclogenesis. *Mon. Wea. Rev.,* **125**, 2699–2708.

Molinari, J., P. Moore, and V. Idone. 1999. Convective structure of hurricanes as revealed by lightning locations. *Mon. Wea. Rev.,* **127**, 520–534.

Montgomery, M. T., and R. J. Kallenbach. 1997. A theory for vortex Rossby waves and its application to spiral bands and intensity changes in hurricanes. *Quart. J. Roy. Meteor. Soc.,* **123**, 435–465.

Moore, P. L., and Staff. 1958. The hurricane season of 1957. *Mon. Wea. Rev.,* **86**, 401–408.

Morel, P., M. Desbois, and G. Szewach. 1978. A new insight into the troposphere with the water vapor channel of Meteosat. *Bull. Amer. Meteor. Soc.,* **59**, 711–714.

Murnane, R. J. 2004. Climate research and reinsurance. *Bull. Amer. Meteor. Soc.,* **85**, 697–707.

Mykle, R. 2002. *Killer 'Cane: The Deadly Hurricane of 1928.* Cooper Square, 235 pp.

Nebeker, F. 1995. *Calculating the Weather: Meteorology in the 20th Century.* Academic Press, 255 pp.

Nese, J. M., and L. M. Grenci. 1998. *A World of Weather: Fundamentals of Meteorology.* Kendall/Hunt, 539 pp.

Neumann, J. N. 1993. Global overview. In *Global Guide to Tropical Cyclone Forecasting,* ed. G. J. Holland, chapter 1. World Meteorological Organization Technical Document, WMO/TD No. 560, Tropical Cyclone Programme, Report No. TCP-31, Geneva, Switzerland. Also available at http://www.wmo.int and http://www.bom.gov.au/bmrc (look for additional links under the mesoscale section to find this book).

Neumann, C. J., B. R. Jarvinen, C. J. McAdie, and J. D. Elms. 1993. *Tropical Cyclones of the North Atlantic Ocean, 1871–1992.* National Environmental Satellite, Data, and Information Service, National Climatic Data Center, 193 pp.

Neumann, C. J., B. R. Jarvinen, C. J. McAdie, and G. R. Hammer. 1999. *Tropical Cyclones of the North Atlantic Ocean, 1871–1998.* National Climatic Data Center, 206 pp.

Newton, D. E. 1993. *Global Warming.* ABC-CLIO, 183 pp.

Newton, I., I. B. Cohen (trans.), and A. Whitman (trans.). 1999. *The Principia: Mathematical Principles of Natural Philosophy.* Preceded by "A Guide to Newton's Principia," by I. B. Cohen. University of California Press, 974 pp.

NHC. 2005. National Hurricane Center, http://www.nhc.noaa.gov (see link for NHC/TPC Glossary).

Nicholls, N. 1992. Recent performance of a method for seasonal forecasting of seasonal tropical cyclone activity. *Aust. Meteor. Mag.,* **21**, 105–110.

NOAA. 1982. *Some Devastating North Atlantic Hurricanes of the 20th Century.* U.S. Department of Commerce, National Oceanic and Atmospheric Administration, 16 pp.

———. 1991. *Centennial of American Weather Services: Past, Present and Future; U.S. History-making Weather, 1849–1990.* NOAA/National Weather Service, 12 pp.

———. 1993. *Memorable Gulf Coast Hurricanes of the 20th Century.* National Hurricane Center, National Oceanic and Atmospheric Administration, 11 pp.

———. 2001. *Signal Service Years: 1870–1891.* NOAA/National Weather Service. Available at http://www.history.noaa.gov.

Nolan, L. E., and J. M. Murphy. 2000. *Air Force Weather: A Brief History 1937–2000.* Available from the Air Force Weather Agency History Office, 27 pp.

Norton, G. 1951. Hurricanes of the 1950 season. *Mon. Wea. Rev.,* **79**, 8–15.

NOWCAST. 2002. Hurricane Andrew steps up. *Bull. Amer. Meteor. Soc.,* **83**, 1441.

NWS (National Weather Service). 1991. *Marine Surface Weather Observations.* National Weather Handbook, No. 1. U.S. Department of Commerce, National Oceanic and Atmospheric Administration, National Weather Service, Office of Systems Operations, Observing Systems Branch, Silver Spring, MD.

NWSTM. 1982. *Tropical Cyclones Report for the Central Pacific, 1982.* NOAA Technical Memorandum NWSTM PR-24. Also available at Central Pacific Hurricane Center, http://www.prh.noaa.gov/cphc/pages/hurrclimate.php.

———. 1992. *Tropical Cyclones Report for the Central Pacific, 1992.* NOAA Technical Memorandum NWSTM PR-34. Also available at Central Pacific Hurricane Center, http://www.prh.noaa.gov/cphc/pages/hurrclimate.php.

Ooyama, K. V. 1969. Numerical simulation of the life cycle of tropical cyclones. *J. Atmos. Sci.,* **26**, 3–40.

Owens, B. F., and C. W. Landsea. 2003. Assessing the skill of operational Atlantic seasonal tropical cyclone forecasts. *Wea. Forecasting,* **18**, 45–54.

Palmen, E. H. 1948. On the formation and structure of tropical cyclones. *Geophysica,* University of Helsinki, **3**, 26–38.

———. 1958. Vertical circulation and release of kinetic energy during the development of Hurricane Hazel into an extratropical storm. *Tellus,* **10**, 1–23.

Panofsky, H. A. 1981. Atmospheric hydrodynamics. In *Dynamical Meteorology: An Introductory Selection,* ed. B. W. Atkinson. Royal Meteorological Society, 8–20.

Parry, H. D. 1969. *The Semiautomatic Computation of Rawinsondes.* Technical memorandum WBTM EDL 10, U.S. Department of Commerce, Environmental Science Services Administration, Weather Bureau.

Pasch, R. J., and L. A. Avila. 1992. Atlantic hurricane season of 1991. *Mon. Wea. Rev.,* **120**, 2671–2687.

———. 1999. Atlantic hurricane season of 1996. *Mon. Wea. Rev.,* **127**, 581–610.

Pasch, R. J., L. A. Avila, and J. L. Guiney. 2001. Atlantic hurricane season of 1998. *Mon. Wea. Rev.,* **129**, 3085–3123.

Pasch, R. J., M. B. Lawrence, L. A. Avila, J. L. Beven, J. L. Franklin, and S. R. Stewart. 2004. Atlantic hurricane season of 2002. *Mon. Wea. Rev.* **132**, 1829–1859.

Pasch, R. J., and E. N. Rappaport. 1995. Atlantic hurricane season of 1993. *Mon. Wea. Rev.*, **123**, 871–886.

Penland, S., and Coauthors. 1999a. The impact of Hurricane Camille on the Chandeleur Islands in Southeast Louisiana. *Preprint, the Impact of Hurricane Camille: A Storm Impact Symposium to Mark the 30th Anniversary.* New Orleans, LA, August 17–18.

———. 1999b. The impact of Hurricane Georges on the Chandeleur Islands in Southeast Louisiana: A comparison to Hurricane Camille. *Preprint, the Impact of Hurricane Camille: A Storm Impact Symposium to Mark the 30th Anniversary.* New Orleans, LA, August 17–18.

Persson, A. 1998. How do we understand the Coriolis force? *Bull. Amer. Meteor. Soc.*, **79**, 1373–1385.

Pfost, R. L. 2003. Reassessing the impact of two historical Florida hurricanes. *Bull. Amer. Meteor. Soc.*, **84**, 1367–1372.

Pielke, R. A., Jr., and C. W. Landsea. 1998. Normalized Atlantic hurricane damage: 1925–95. *Wea. Forecasting*, **13**, 621–631.

———. 1999. La Niña, El Niño, and Atlantic hurricane damage in the United States. *Bull. Amer. Meteor. Soc.*, **80**, 2027–2033.

Pielke, R. A., Jr., and R. A. Pielke Sr. 1997. *Hurricanes: Their Nature and Impacts on Society.* John Wiley and Sons, 279 pp.

Pielke, R. A., J. Rubiera, C. Landsea, M. L. Fernandez, and R. Klein. 2003. Hurricane vulnerability in Latin America and the Caribbean: Normalized damage and loss potentials. *National Hazards Rev.*, 101–114.

Pielke, R. A., Jr., C. Simonpietri, and J. Oxelson. 1999. *Thirty Years after Hurricane Camille: Lessons Learned, Lessons Lost.* Available from Environmental and Societal Impacts Group, National Center for Atmospheric Research, Boulder, CO. Also available at http://sciencepolicy.colorado.edu.

Pierce, C. 1939. The meteorological history of the New England hurricane of 1938. *Mon. Wea. Rev.*, **67**, 237–288.

Platzman, G. W. 1979. The ENIAC computations of 1950: Gateway to numerical weather prediction. *Bull. Amer. Meteor. Soc.*, **60**, 302–312.

Posey, C. 1994. Hurricanes: Reaping the whirlwind. *Omni*, **16**, 34–47.

Purdom, J. F. W. 1995. Observations of thunderstorms and hurricanes using one-minute interval GOES-8 imagery. Abtracts, Week B, International Union of Geodesy and Geophysics, XXI General Assembly, Boulder, CO, Amer. Geophys. Union, Washington, D.C., B286.

Purdom, J. F. W., and W. P. Menzel. 1996. Evolution of satellite observations in the United States and their use in meteorology. In *Historical Essays on Meteorology*, ed. J. R. Fleming. American Meteorological Society, 99–155.

Ralston, A., and E. D. Reilly, eds. 1993. *Encyclopedia of Computer Science*. Van Nostrand Reinhold, 1558 pp.

Rappaport, E. N. 1999. Atlantic hurricane season of 1997. *Mon. Wea. Rev.*, **127**, 2012–2026.

———. 2000. Loss of life in the United States associated with recent Atlantic tropical cyclones. *Bull. Amer. Meteor. Soc.*, **81**, 2065–2073.

Rappaport, E. N., M. Fuchs, and M. Lorentson. 1998. *The Threat to Life in Inland Areas of the United States from Atlantic Tropical Cyclones*. Preprint from the 23rd Conference on Hurricanes and Tropical Meteorology, American Meteorological Society, 339–342.

Rappaport, E., and J. Partagas. 1995. History's deadly Atlantic hurricanes. *Mar. Wea. Log*, **39**, 44–47.

Rasmussen, E. A., and J. Turner. 2003. *Polar Lows: Mesoscale Weather Systems in the Polar Regions*. Cambridge University Press, 612 pp.

Reardon, L. F. 1986. *The Florida Hurricane and Disaster*. Lion and Thorne, 112 pp.

Riehl, H., and R. J. Shafer. 1944. The recurvature of tropical storms. *J. Meteor.*, **1**, 42–54.

Ritchie, E. A., and G. J. Holland. 1999. Large-scale patterns associated with tropical cyclogenesis in the western Pacific. *Mon. Wea. Rev.*, **127**, 2027–2043.

Roberts, N. 1969. *Extreme Hurricane Camille: August 14th through 22nd, 1969*. Nash Roberts Consultants, 135 pp.

Rogers, R. R., and P. L. Smith. 1996. A short history of radar meteorology. In *Historical Essays on Meteorology*, ed. J. R. Fleming. American Meteorological Society, 57–98.

Rosenfield, J. 1997. Storm surge! Hurricanes' most powerful and deadly force. *Weatherwise*, **50**, 18–24.

Roux, F. and F. D. Marks Jr. 1996. Extended velocity track display (EVTD): An improved processing method for Doppler radar observations of tropical cyclones. *J. Atmos. Oceanic Technol.*, **13**, 875–899.

Ryan, B. F., I. G. Watterson, and J. L. Evans. 1992. Tropical cyclone frequencies inferred from Gray's yearly genesis parameter: Validation of GCM tropical climates. *Geo. Res. Letters*, **24**, 1255–1258.

Sallee, R. 1993. Alicia: Streets of oaks, glass. *Houston Chronicle*, August 15.

Sanford, T. B., P. G. Black, J. Haustein, J. W. Fenney, G. Z. Forristall, and J. F. Price. 1987. Ocean response to hurricanes. Part I: Observations. *J. Phys. Oceanogr.*, **17**, 2065–2083.

Saunders, C. P. R. 1993. A review of thunderstorm electrification processes. *J. Appl. Meteor.*, **32**, 642–655.

Schmetz, J., P. Pili, S. Tjemkes, D. Just, J. Kerkmann, S. Rota, and A. Ratier. 2002. An introduction to METEOSAT Second Generation (MSG). *Bull. Amer. Meteor. Soc.*, **83**, 977–992.

Schnapf, A. 1982. The development of the TIROS global environmental satellite system. In *Meteorological Satellites: Past, Present, and Future.* NASA Conference Publication 2227, 7–16.

Sharp, R. J., M. A. Bourassa, and J. J. O'Brien. 2002. Early detection of tropical cyclones using seawinds-derived vorticity. *Bull. Amer. Meteor. Soc.*, **83**, 879–889.

Shay, L. K., P. G. Black, A. J. Mariano, J. D. Hawkins, and R. L. Elsberry. 1992. Upper ocean response to Hurricane Gilbert. *J. Geophys. Res.*, **97(12)**, 20, 227–220, 248.

Shay, L. K., A. J. Mariano, D. S. Jacob, and E. H. Ryan. 1998. Mean and near-inertial ocean current response to Hurricane Gilbert. *J. Phys. Oceanogr.*, **28**, 858–889.

Shea, E. L. 1987. *A History of NOAA, Being a Compilation of Facts and Figures regarding the Life and Times of the Original Whole Earth Agency.* NOAA, 44 pp. Also available in the history section at http://www.noaa.gov.

Sheets, B., and J. Williams. 2001. *Hurricane Watch.* New York: Vintage, 331 pp.

Sheets, R. C. 1984. *The National Weather Service Hurricane Probability Program.* NOAA Technical Report NWS 37, National Oceanic and Atmospheric Administration, National Weather Service, National Hurricane Center, Miami, FL.

———. 1990. The National Hurricane Center: Past, present, and future. *Wea. Forecasting*, **5**, 185–232.

Sheets, R. C., and J. Williams. 2001. *Hurricane Watch: Forecasting the Deadliest Storms on Earth.* Vintage, 331 pp.

Simpson, R. H. 1974. The hurricane disaster potential scale. *Weatherwise*, **27**, 169, 186.

Simpson, R. H., and P. J. Hebert. 1973. Atlantic hurricane season of 1972. *Mon. Wea. Rev.*, **101**, 323–333.

Simpson, R. H., and J. M. Pelissier. 1971. Atlantic season of 1970. *Mon. Wea. Rev.*, **99**, 269–277.

Simpson, R. H., A. L. Sugg, and Staff. 1970. The Atlantic hurricane season of 1969. *Mon. Wea. Rev.*, **98**, 293–306.

Singh, O. P., T. M. A. Khan, and M. S. Rahman. 2001. Tropical cyclone frequency in the north Indian Ocean in relation to the southern oscillation. *Mausam*, **52**, 511–514.

Smith, R. K., W. Ulrich, and G. Dietachmayer. 1990. A numerical study of tropical cyclone motion using a barotropic model. Part I: The role of vortex asymmetry. *Quart. J. Meteor. Soc.*, **116**, 337–362.

Smith, T. J., M. B. Robblee, H. R. Wanless, and T. W. Doyle. 1994. Mangroves, hurricanes, and lightning strikes: Assessment of Hurricane Andrew suggests an interaction across two differing scales of disturbance. *Bioscience*, **44**, 256–262.

Soden, B. J., C. S. Velden, and R. E. Tuleya. 2001. The impact of satellite winds on experimental GFDL hurricane model forecasts. *Mon. Wea. Rev.*, **129**, 835–852.

Sorbjan, Z. 1996. *Hands-on Meteorology: Stories, Theories, and Simple Experiments.* American Meteorological Society, 306 pp.

Stevenson, R. L. 1997. *A Footnote to History: Eight Years of Trouble in Samoa.* The World Wide School; originally published in 1899. Available at http://www.worldwideschool.org.

Stewart, G. R. 1941. *Storm.* Heyday Books, 352 pp.

Stull, R. B. 2000. *Meteorology for Scientists and Engineers.* Brooks/Cole, 502 pp.

Sugg, A. L. 1966. The hurricane season of 1965. *Mon. Wea. Rev.*, **94**, 183–191.

Sugg, A. L., and J. M. Pelissier. 1968. The hurricane season of 1967. *Mon. Wea. Rev.*, **96**, 242–250.

Sullivan, C. L. 1987. *Hurricanes of the Mississippi Gulf Coast, 1717 to Present.* Gulf Publishing, 139 pp.

Sumner, H. C. 1946. North Atlantic hurricanes and tropical disturbances of 1945. *Mon. Wea. Rev.*, **47**, 1–5.

———. 1947. North Atlantic hurricanes and tropical disturbances of 1947. *Mon. Wea. Rev.*, **48**, 251–256.

———. 1948. North Atlantic hurricanes and tropical disturbances of 1948. *Mon. Wea. Rev.*, **49**, 277–280.

Tannehill, I. R. 1955. *The Hurricanes Hunters.* Dodd, Mead, 271 pp.

———. 1956. *Hurricanes: Their Nature and History.* Princeton University Press, 308 pp.

Thorncroft, C. D., D. J. Parker, R. R. Burton, M. Diop, J. H. Ayers, H. Barjat, et al. 2003. The JET2000 Project: Aircraft observations of the African easterly jet and African easterly waves. *Bull. Amer. Meteor. Soc.*, **84**, 337–351.

Tomas, R. A., and P. J. Webster. 1997. The role of inertial instability in determining the location and strength of near-equatorial convection. *Quart. J. Roy. Meteor. Soc.*, **123**, 1445–1482.

Toth, Z., and E. Kalnay. 1993. Ensemble forecasting at NMC: The generation of perturbations. *Bull. Amer. Meteor. Soc.*, **74**, 2317–2330.
———. 1997. Ensemble forecasting at NCEP and the breeding method. *Mon. Wea. Rev.*, **125**, 3297–3319.
Tuleya, R. E., and T. R. Knutson. 2002. Impact of climate change on tropical cyclones. In *Atmosphere-Ocean Interactions*, Vol. 1. Southampton, UK: WIT Press, 293–312.
Turnipseed, P., G. L. Giese, J. L. Pearman, G. S. Farris, M. D. Krohn, and A. H. Sallenger. 1998. *Hurricane Georges: Headwater Flooding, Storm Surge, Beach Erosion, and Habitat Destruction on the Central Gulf Coast*. USGS Water Resources Investigations Report 98-4231.
U.S. Senate Bipartisan Task Force on Funding Disaster Relief. 1995. *Federal Disaster Assistance*, Document 104-4, 104th Congress, 1st Session, 250 pp.
Vaughan, W. W. 1982. Meteorological satellites: Some early history. In *Meteorological Satellites: Past, Present, and Future*. NASA Conference Publication 2227, 1–2.
Vaughan, W. W., and D. L. Johnson. 1994. Meteorological satellites: The very early years, prior to launch of TIROS-1. *Bull. Amer. Meteor. Soc.*, **75**, 2295–2302.
Velden, C., and L. M. Leslie. 1991. The basic relationship between tropical cyclone intensity and the depth of the environmental steering layer in the Australia region. *Wea. Forecasting*, **6**, 244–253.
Velden, C. S., T. L. Olander, and R. M. Zehr. 1998. Development of an objective scheme to estimate tropical cyclone intensity from digital geostationary satellite infrared imagery. *Mon. Wea. Rev.*, **126**, 1202–1218.
Vigh, J., S. Fulton, M. DeMaria, and W. H. Schubert. 2003. Evaluation of a multigrid tropical cyclone track model. *Mon. Wea. Rev.*, **131**, 1629–1636.
Von Baeyer, H. C. 1999. *Warmth Disperses and Time Passes: The History of Heat*. Modern Library, 240 pp.
Wakimoto, R. M., and P. G. Black. 1994. Damage survey of Hurricane Andrew and its relationship to the eyewall. *Bull. Amer. Meteor. Soc.*, **75**, 189–200.
Wang, Y., and G. J. Holland. 1996. The beta drift of baroclinic vortices. Part 1: Adiabatic vortices. *J. Atmos. Sci.*, **53**, 411–427.
Weber, H. C. 2001. Hurricane track prediction with a new barotropic model. *Mon. Wea. Rev.*, **129**, 1834–1858.
Weems, J. E. 1997. *A Weekend in September*. Texas A&M University Press, 192 pp.
Weik, M. H. 1961. The ENIAC story. *Ordnance, the Journal of the American Ordnance Association* (January–February): 3–7. Also available

in Ballistic Research Laboratory Report, Department of the Army Project No. 5B03-06-002.

Wexler, R. 1945. The structure of the September 1944 hurricane when off Cape Henry, Virginia. *Bull. Amer. Meteor. Soc.,* **26**, 156–159.

———. 1947. Structure of hurricanes as determined by radar. *Ann. N.Y. Acad. Sci.,* **48**, 821–844.

Wilemon, T. 1999. Information, money needed for memorial organizers seek to correct inaccuracies. *Sun Herald,* September 6.

Will, L. E. 1990. *Okeechobee Hurricane and the Hoover Dike.* Glades Historical Society, 204 pp.

Williams, J. 1992. *The Weather Book.* USA Today, 212 pp.

———. 1998. NASA chief kept her head in the clouds. *USA Today,* December 3.

Willoughby, H. E., J. A. Clos, and M. G. Shoreibah. 1982. Concentric eyewalls, secondary wind maxima, and the evolution of the hurricane vortex. *J. Atmos. Sci.,* **41**, 1169–1186.

Willoughby, H. E. 1988. Linear motion of a shallow-water barotropic vortex. *J. Atmos. Sci.,* **45**, 1906–1928.

Willoughby, H. E., and R. A. Black. 1996. Hurricane Andrew in Florida: Dynamics of a disaster. *Bull. Amer. Meteor. Soc.,* **77**, 543–549.

Willoughby, H. E., D. P. Jorgensen, R. A. Black, and S. L. Rosenthal. 1985. Project STORMFURY: A scientific chronicle 1962–1983. *Bull. Amer. Meteor. Soc.,* **66**, 505–514.

Willoughby, H. E., J. M. Masters, and C. W. Landsea. 1989. A record minimum sea level pressure observed in Hurricane Gilbert. *Mon. Wea. Rev.,* **117**, 2824–2828.

Wong, M. L. M., and J. C. L. Chan. 2004. Tropical cyclone intensity in vertical wind shear. *J. Atmos. Sci.,* **61**, 1859–1876.

Yamasaki, M. 1968. Numerical simulation of tropical cyclone development with the use of primitive equations. *J. Meteor. Soc. Japan,* **55**, 11–31.

Zehr, R. M. 1992. Tropical cyclogenesis in the western North Pacific. NOAA Tech. Rep. NESDIS 61, 181 pp.

Zhang, D.-L., Y. Liu, and M. K. Yau. 2001. A multiscale numerical study of Hurricane Andrew (1992). Part IV: Unbalanced flows. *Mon. Wea. Rev.,* **129**, 92–107.

Zhu, T., D. Zhang, and F. Weng. 2002. Impact of the Advanced Microwave Sounding Unit measurements on hurricane prediction. *Mon. Wea. Rev.,* **130**, 2416–2432.

Zipser, E. 1989. Interview with Robert H. Simpson, September 6 & 9, 1989. American Meteorological Society Tape Recorded Interview Project (TRIP). Available from the National Center for Atmospheric Research at http://www.ucar.edu/archives.

Zoch, R. T. 1949. North Atlantic hurricanes and tropical disturbances of 1949. *Mon. Wea. Rev.,* **50**, 339–341.

Index

Abbe, Cleveland, 111
Aberson, Sim, 193–194
Accumulated Cyclone Energy (ACI) index, 53–54
Adem, Julian, 127
ADvanced CIRCulation (ADCIRC) storm surge model, 37
Advanced Microwave Sounding Unit (AMSU), 148, 265
Aerosonde, 147
Airborne eXpendable BathyThermographs (AXBTs), 143, 288
Airborne eXpendable Current Profiler (AXCP), 141, 143, 288
Aircraft Satellite Data Link (ASDL), 143
American Meteorological Society (AMS), 132, 200, 259–261
 educational material and publications of, 261–262
 journals of, 259–260
Anthes, Richard, 135, 194–196
Arabian Sea, the, 103
Army Signal Corps, 111, 112
Atlantic hurricane season, 13–14
Atlantic Hurricanes (Dunn and Miller), 202
Atlantic Multidecadal Mode (AMM), 209
Atlantic Oceanic and Meteorological Laboratory (AOML), 272
Atmospheric pressure, 3

Australia, 85, 95
 Bureau of Meteorology and Research, 100
Automatic Picture Transmission (APT), 129
Azores high-pressure systems, 51

Bangladesh, 30–31
Barbados, 218
Bartie, Whitney, 131
Bartie v. United States, 131
Bay of Bengal, 93, 103
Beaufort, Francis, 108, 358. *See also* Modified Beaufort Scale
Bermuda, 189
Bjerknes, Vilhelm, 114, 220
Bowie, Edward H., 115
Brooks, Charles Franklin, 259
Burpee, Bob, 145, 148
Byers, Horace Robert, 120

Calculus, 105, 108
Caribbean Sea, 12–13, 43, 57, 117
 surface pressure of, 50
 and wind shear, 51
Centrifugal force, 18–19
Charney, Jules, 124
Cheng, Kenneth, 132
China, 31, 62, 84
China LAndfalling Typhoon EXperiment (CLATEX), 150
Clapeyron, Emile, 109

403

Clausius, Rudolf, 110
Cline, Isaac, 115–116, 196–199
Cline, Joseph, 197, 198
CLIPER (CLImatology and PERsistence) statistical equation, 136, 139
Cloud seeding, 61
Convergence, 3, 6, 9
Cooperative Institute for Meteorological Satellite Studies ([CIMSS], University of Wisconsin-Madison), 226, 264–266
  educational material and publications of, 266
Coriolis, Gaspard-Gustave de, 107, 109–110
Coriolis force, the, 3–4, 91, 97, 109–110
Coupled Boundary Layer Air Sea Transfer (CBLAST), 150, 151, 203
Cressman, George, 133
Cuba, 84, 191–192, 197
Currents, 34
  longshore, 34
  rip, 34
Cyclones, 110, 215, 217. *See also* Subtropical cyclones; Tropical cyclones
Cyclonic rotation, 1, 6, 8

DeMaria, Mark, 144, 199–201
Doppler, Christian, 140
Doppler effect, 140
Doppler radar, 140, 143, 145
  WSR-88 (NEXRAD) Doppler radar, 145
Douglas, Marjory Stoneman,
Downbursts, 25
Dropsonde Observations for Typhoon Surveillance near the Taiwan Region

(DOTSTAR), 151
Dropwindsondes, 137, 147, 149, 150
Duckworth, Joseph, 118–119
Dunn, Gordon, 117–118, 126, 134, 201–202

Dust storms, 7
Dvorak, V. F., 15–16, 136
Dvorak technique, 15–16, 42, 136, 226, 265
Dynamic instability, 5–6

Eckert, J. Presper, 120
El Niño-Southern Oscillation (ENSO), 49–50, 51, 52–53, 60, 100, 115, 141, 205, 252
Electronic Discrete Variable Automatic Computer (EDVAC), 122
Electronic Numerical Integrator and Computer (ENIAC), 120–122
  use of in weather forecasting, 124–125
Elsberry, Russell, 202–204
Elsner, Jim, 53
Environmental Science Service Administration (ESSA), 132, 133–134, 135
Everglades, The: River of Grass (Stoneman), 19–20
Euler, Leonhard, 107–108
Extratropical cyclone, 12–13

Facusse, Carlos Flores, 28
Farrar, John, 108
Fassig, O. L., 114
Federal Emergency Management Agency (FEMA), 64, 76–77, 80, 266–267, 273
  educational materials and publications of, 268
  services provided by, 267–268
Ferrel, William, 109
Fleet Numerical Meteorological and Oceanography Center (FNMOC), 125, 151
Florida, danger of hurricanes to, xvi–xvii
Fourth Convection and Moisture EXperiment (CAMEX-4), 149
France, 188, 189
Frank, Neil, 137, 143, 204–205
Franklin, Benjamin, 107

Front(s), 1–2
Fujiwhara effect, 40

Garriot, E. B., 113
Gay-Lussac, Joseph, 109
General Meteorology (Byers), 120
Geneva Convention (1864), 262–263
Geophysical Fluid Dynamics Laboratory (GFDL), 42, 125, 145–146, 149, 208, 269–270
  educational material and publications of, 270
Germany, 190–191
Gibbs, Willard, 110
Global Atmospheric Research Program (GARP), 137
Global Forecast System (GFS), 42
Global Learning and Observations to Benefit the Environment (GLOBE), 280–281
Global Positioning System (GPS), 146–147, 150
Global warming, 54, 56, 58. *See also* Greenhouse effect
  and the number and strength of hurricanes, 59–60
  and the number and strength of tropical storms, 56–59
Goddard, Robert, 116
Goldenberg, S., 50
Grant, Ulysses S., 111
Graupel, 23
Gray, William, xv—xvii, 49–53, 100, 141, 205–206, 209
Greenhouse effect, 54–56
Gulf of Mexico, 13–14, 38, 79, 190

Hadley, George, 106–107
Halley, Edmond, 105–106
Halsey, William ("Bull"), 119–120
Hamilton, Alexander, 253–254
Haurwitz, Bernhard, 117
Henry, Joseph, 110
Holland, Greg, 147, 206–207, 245
Home protection, 64–65. *See also* Insurance
  before hurricane season, 69–72
  during evacuation, 74
  during hurricane watch or warning, 72–74
  exterior doors and windows, 66–67
  post-hurricane, 74–77
  protecting pets, 77–78
  roof, 65–66
  storm shutters, 67–69
HURricane DATabase (HURDAT), 274, 292
Hurricane Hunters, the (53rd Weather Reconnaissance Squadron), 43–44, 48, 137, 142, 147, 270–272
  educational material and publications of, 272
Hurricane and Its Impact, The (R. Simpson and H. Riehl), 225
Hurricane modification attempts, 60–64
  chemical, 64
  hurricane seeding, 60–63
  microwave, 64
  nuclear, 63–64
Hurricane reconnaissance flights, 42–44, 119, 122, 135–136, 138, 141–142, 143, 374–376. *See also* Hurricane Hunters, the (53rd Weather Reconnaissance Squadron)
  sample report of, 377
  understanding reports of, 374–376
Hurricanes, 217. *See also* Hurricane modification attempts; Hurricane reconnaissance flights; Hurricanes, destruction caused by; Hurricanes, forecasting of; Hurricanes, formation of; Hurricanes, naming of; Hurricanes, structure of; Specific hurricanes
  annular, 21
  arctic, 92
  benefits of, 34–35

Hurricanes (continued)
  categories of, 31, 32–33 (table), 51–52, 238–23 9(table), 242 (table), 247(table)
    category 5, 31, 34, 57, 242 (table), 243
    worldwide, 93–97
  definition of, 1
  high-altitude detection of, 125–126, 129
  international impact of, 83–92
  and lightning, 22–24
  motion of, factors controlling, 39–40
    Fujiwhara effect, 40
    trochoidal motion, 40
  non-print resources concerning, 324–333
  preparation for, 46–49, 303. See also Home protection
    and multidecadal changes, 46
    and population increases, 46, 47
    and the use of drones and aerosondes, 48–49
  and rainfall amounts, 26, 244
  size of, 21–22
    major, 31
    midgets, 22
  worldwide naming conventions, 101–103
Hurricanes, destruction caused by, 25–26, 45–49, 46 (table), 83–84, 113, 237 (table). See also Saffir-Simpson Hurricane Scale
  deaths caused by, 232–233 (table), 240–241 (table), 248–249 (table)
  expense of, 47–48, 234–235 (table), 236 (table)
  eyewitness accounts of destruction, 255–258
  flooding, 26–28, 35–36
    deaths from, 26, 27–28
  international
    Caribbean, 84
    Latin America, 84
    in less-developed countries, 84–85
  ocean currents, 34
  storm surges and storm tides, 28–31, 45–46, 243–244
    deaths from, 31
  uprooted tree problems, 36
  waves, 29–30
Hurricanes, forecasting of, 35–37, 363–364. See also Hurricane reconnaissance flights; National Hurricane Center, understanding forecasts of; Viñes, Benito
  annual forecasting, 49–53
    factors affecting annual hurricane activity, 49–51
  and breeding cycles, 42
  computer models, 40–42
    numerical models, 135
  Congressional support for, 116–117, 126, 201, 213, 225
  consensus, 42
  ensemble, 41–42
  hurricane warnings, 44, 111, 112–113
  hurricane watches, 44
  observational platforms for, 42–44
  summary of forecasting procedures, 44
Hurricanes, formation of, 2. See also Tropical waves; Troughs
  extratropical transition, 2, 12–13
  intensification stage, 2, 8–10
  genesis stage, 2, 3–8, 9
    prerequisites for, 6–7
  and water temperature, 6, 9–10, 11, 13–14, 51, 124
  weakening stage, 10–12
Hurricanes, naming of, 14–15, 16 (table), 17 (table), 85
Hurricanes, structure of
  and cloud patterns, 15–18
    curved band, 16–17
    spiral bands, 17–18

Index **407**

and temperature, 20–21
types of patterns, 16
eye formation, 18–21
eyewall (s), 61, 62, 120
concentric eyewall cycle, 20

formation, 18–21
and temperature, 20–21
Hydrometeorological Prediction Center (HPC), 36

Ice nuclei, 60, 123
Imaging Wind and Rain Airborne Profiler (IWRAP), 151
Improved TIROS Operational System (ITOS), 133, 136
Improved Weather Reconnaissance System (IWRS), 142
India, 30, 31
Inducement, 23
Infrared spectrum, 55
Institute for Advanced Studies ([IAS] Princeton University), 122. *See also* Geophysical Fluid Dynamics Laboratory (GFDL) Meteorology Group, 124
Insurance, 64, 78–82. *See also* National Flood Insurance Program (NFIP); Reinsurance
flood, 79–80
hurricane, 78–79
Intergovernmental Panel on Climate Change (IPCC), report of, 251–252
International Workshop (s) on Tropical Cyclones (IWTC), 204
InterTropical Convergence Zone (ITCZ), 4, 106, 223

Jarrell, Jerry, 148
Jarvinen, Brian, 139
Jet stream, the, 123–124
JET2000, 149
Joint Numerical Weather Prediction Group (JNWPG), 124–125

Joint Typhoon Warning Center (JTWC), 95, 99, 127–128, 275–276
annual tropical cyclone summaries of, 321
educational material and publications of, 276
Joules, James Prescott, 110

Kamikaze ("divine wind"), 188
Kaplan, John, 144, 200
Kinetic energy, 5
Kosco, George F., 119

Landsea, Chris, 102, 208–210
Last Island (Isle Derniere) disaster, 256–258
Law of the Storms, The (Reid), 218
Leibniz, Gottfried, 105
Lhermitte, Roger, 135
Lightning, 22–24
intercloud, 23
intracloud, 23
Lorenz, Edward, 129–130

Madden-Julian Oscillation, 7–8, 100
Marconi, Guglielmo, 113
Marks, Frank, 210–211
Mauchly, John, 120
Maximum Envelop of Water (MEOW) maps, 37
McKinley, William, 112
Mesoscale vortices, 25
Mesovortices, 25, 129
Meteorograph. *See* radiosonde
Meteorological satellites, 127, 128–129. *See also* Television InfraRed Observation Satellite-1 (TIROS-1); TIROS Operational System (TOS)
Applications Technology Satellite (ATS-1), 133–134
Defense Meteorological Satellite Program (DMSP), 134
geostationary satellites, 133

408   Index

Meteorological satellites (continued)
  Geostationary Meteorological
    Satellite (GMS), 139
  Geostationary Operational
    Environmental Satellite
    (GOES), 138, 145
  Meteosat, 138–139
  Synchronous Meteorological
    Satellite-1 (SMS-1), 137–138
  Nimbus program, 132
  polar-orbiting, 128–129, 134
    Earth Observing System (EOS)
      satellites, 144–145, 148, 149
    Special Sensor Microwave/
      Imager (SSM/I), 142–143, 146
  Seasat satellite, 139
  Tropical Rainfall Measuring
    Mission (TRMM) satellite, 146
Miami Weather Bureau, 126. *See also*
  National Hurricane Center
  (NHC)
Microwave Imager (MI), 146
Miller, Banner, 202
Mitchell, C. L., 115
Modified Beaufort Scale, 358, 359–361
  (table), 362
Monsoon depressions, 93
Monsoon gyre, 104
Moore, William, 112, 197
Moore School of Electrical
  Engineering (University of
  Pennsylvania), 120, 121
National Aeronautics and Space
  Administration (NASA), 132,
  144, 147–148, 149, 276–277
  educational material and
    publications of, 277–278
  Goddard Space Flight Center, 129
National Center for Atmospheric
  Research (NCAR), 195–196
National Centers for Environmental
  Prediction (NCEP), 42, 125, 278,
  362
  educational material and
    publications of, 278–279

National Climate Data Center
  (NCDC), 279–280
  educational material and
    publications of, 280, 323–324
National Data Buoy Center, 136
National Environmental Satellite and
  Data Information Service
  (NESDIS). *See* National
  Oceanographic and Atmospheric
  Administration (NOAA)
National Flood Insurance Program
  (NFIP), 79, 80, 134
National Hurricane Center ([NHC],
  formerly Miami Weather
  Bureau), 15, 38, 43, 44, 48, 101,
  126, 202, 214, 225, 226, 290
  and the Man computer Interactive
    Data Access System (McIDAS),
    141
  summary of forecasting
    procedures, 44–45
  understanding forecasts of,
    363–364, 373
    forecast advisory, 364–366
    forecast discussion, 368–370
    public advisory, 366–368
    strike probability, 370–373
  use of computer models for
    forecasting, 40–42
National Hurricane Research
  Laboratory (NHRL), 126–127
National Lightning Detection
  Network, 24
National Oceanographic and
  Atmospheric Administration
  (NOAA), 43, 132, 135, 138, 140,
  143, 146, 147, 148, 149, 277,
  280–281
  Aircraft Operations Center (AOC),
    43, 286–289
    educational material and
      publications of, 289–290
  Climate Prediction Center (CPC), 53
    educational material and
      publications of, 281

Hurricane Research Division
(HRD), 127, 225, 272, 280
  annual summaries of, 318–321
  educational material and
    publications of, 274–275
  research topics of, 272–273
National Environmental Satellite
  and Data Information Service
  (NESDIS), 132,
  265, 279
National Weather Service (NWS),
  35–36, 37, 134, 280, 283
  educational material and
    publications of, 283–284,
    322–323
  Spaceflight Meteorology Group,
    211–212
National Weather Association (NWA),
  281–282
  educational material and
    publications of, 282
Naval Research Laboratory (NRL),
  284–285
  educational material and
    publications of, 285–286
Navier, Claude-Louis, 108–109
Navier-Stokes equations, 109
Neumann, Charles, 136, 139,
  211–212
Newton, Isaac, 105, 106
Norton, Grady, 126, 201, 212–214
"Numerical Simulation of the Life
  Cycle of Tropical Cyclones"
  (Ooyama), 214

Ooyama, Vic, 135, 200, 214–215

Parent, Robert, 128
Philippines, the, 28, 31, 102–103
Piddington, Henry, 110, 215–216,
  228
Pielke, R. A., 64, 84–85
Pielke, R.A., Sr., 64
Polar low, 92
Potential energy, 5

Project Cirrus, 61
Project STORMFURY, 62–64, 123, 129,
  141, 225, 229
  excerpts from report of, 249–251

Quasi-Biennial Oscillation (QBO), 50,
  100, 141

Racer's Storm (1837), 190
RAdio Detection And Range
  (RADAR), 118, 126. *See also*
  Doppler radar
Radiometer, 128, 149. *See also* Stepped
  Frequency Microwave
  Radiometer (SFMR)
Radiosonde, 117
Rasmussen, NAME, 92
Red Crescent Society, 84, 262–263
  educational material and
    poublications, 264
Red Cross, the, 77, 262–263
  educational material and
    publications of, 264
Redfield, William, 109, 216–217
Regional Specialized Meteorological
  Centers (RSMCs), 98–100
Reid, William, 111, 190, 215, 217–218
Reinsurance, 53
Richardson, Lewis Fry, 108, 114–115,
  218–219
Riehl, Herbert, 120, 127, 202, 205,
  219–220, 222
Riming, 23
Rodney, George, 255
Rossby, Carl-Gustave Arvid, 18,
  220–222, 260
Rossby wave motion, 18, 221–222

Saffir, Herbert, 31
Saffir-Simpson Hurricane Scale, 31,
  32–33 (table), 34, 85
  modified form of for tropical
    Pacific area, 86–90 (table)
Sailor's Horn Book, The (Piddington),
  215

410 Index

SANders' BARotropic (SANBAR) model, 135
Satellite imagery, 20
Satellites. *See* Meteorological satellites
Scatterometer, 139
Schafer, Robert, 120
Schaefer, Vincent, 123
Science reports, reliability of, 58–59
Sea, Lake, and Overland Surges from Hurricanes (SLOSH) computer model, 37
Sea Venture, 189
Shapiro, L., 50
Sheets, Robert, 143, 145
SHIFOR (Statistical Hurricane Intensity FORcast), 139–140
SHIPS (Statistical Hurricane Intensity Prediction Scheme), 144, 200
Silver iodide, 123
Simpson, Joanne, 222–223
Simpson, Robert, 31, 126, 134, 137, 224–226
Solar radiation. *See* Greenhouse effect
Spain, 188–189
Specific hurricanes
    Alicia (1983), 54, 313–314
    Allen (1980), 21, 54, 243
    Andrew (1992), 22, 24, 25, 35, 54, 57–58, 64, 129, 245, 315–316
    Audrey (1957), 131
    Bob (1991), 24, 314
    Bonnie (1998), 148
    Camille (1969), 22, 30, 37, 43, 45, 137, 243, 313
    Celia (1970), 129
    Charley (2004), xv, 45, 148
    Claudette (1979), 26, 34, 146, 151
    Danny (1997), 27
    Debbie (1969), 62
    Dennis (1999), 26
    Diana (1984), 24
    Donna (1960), 187
    Earl (1998), 148
    Easy (1950), 15
    Elena (1985), 24
    Erin (2001), 149
    Esther (1961), 61
    Fifi (1974), 27
    Flora (1963), 27
    Florence (1988), 24
    Floyd (1999), 26
    Frederic (1979), 313
    Galveston Hurricane (1900), 196, 197, 309–310
    Georges (1998), 31, 148, 187, 317, 369
    Gilbert (1988), 21, 28, 143, 243
    Ginger (1972), 62
    Gloria (1995), 19
    Gordon (1994), 28
    Great Hurricane (1780), 189–190, 254–255
    Hazel (1954), 126, 313–313
    Hugo (1989), 25, 43–44, 314
    Humberto (2001), 149
    Iniki (1992), 316
    Ivan, xvii
    Jeanne (2004), xv
    Katrina (2005), 335–338
    King (1950), 15
    Mitch (1998), 28, 84, 187
    Opal (1995), 38, 43, 45
    Pauline (1997), 28
    Privy Hurricane (1898), 192
    Roxanne (1995), 12
Specific supertyphoons
    Forrest, 243
    Tip (1979), 21, 103, 243, 244
Specific tropical cyclones
    Bangladesh Cyclone (1970), 245
    Tracy (1974), 22, 103, 244
Specific tropical storms
    Alberto (1994), 151, 316
    Allison (2001), 27, 151
    Chantal (2001), 149
    Debbie (1982), 140
    Gabrielle (2001), 149, 151
    Isabel (2003), 80
    Thelma (1991), 27

Specific typhoons
 Carmen (1960), 245
 Cobra (1944), 119
 Herb (1996), 27
 John (1944), 245
 Kelly (1981), 27
 Nina (1975), 27
 Omar, 43
 Peggy (1986), 28
 Thelma (1987), 28
Static instability, 3
Stepped Frequency Microwave
 Radiometer (SFMR), 148–149,
 151
Steric effect, 29
Stewart, George, 14–15
Stokes, George, 109
Storm (Stewart), 14
Storm Spotters, 142
Storm Trackers, 138
Storms, Floods, and Sunshine (I. Cline),
 199
Subtropical cyclones, 15, 93
Suomi, Verner, 128
Supertyphoons, 95. *See also* Specific
 supertyphoons
Synoptic Flow Experiment, 140

Taiwan, 150–151
Television InfraRed Observation
 Satellite-1 (TIROS-1), 128–129,
 132, 277
Thailand, 31
Thermodynamics, 110
Third Convection and Moisture
 EXperiment (CAMEX-3),
 147–148
TIROS Operational System (TOS),
 133
Tornadoes, 25
 tornado warning, 35
 tornado watch, 35
Tropical Cyclone Motion (TCM-90,
 TCM-92, TCM-93) Experiments,
 144, 203

Tropical Cyclone Warning Centers
 (TCWCs), 98–99
Tropical cyclones, 83–84, 93–94. *See
 also* Specific tropical cyclones
 categories of, 86–90(table)
 intensity of, 95, 97
 naming of, 85, 95, 96(table),
 101–103
 occurrence of, 91–92
 size variations of, 103–104
 worldwide monitoring and
 forecasting of, 97–100
Tropical Cyclones (I. Cline), 116
"Tropical Cyclones and Global
 Climate Change: A Post-IPCC
 Assessment" (Henderson-
 Sellers et al.), 252–253
Tropical depression(s), 2, 7, 8
Tropical disturbance, 2
Tropical EXperiment in MEXico
 (TEXMEX), 144
Tropical Prediction Center (TPC), 290.
 See also National Hurricane
 Center (NHC)
 educational material and
 publications of, 291–292
 Technical Support Branch (TSB),
 291
 Tropical Analysis and Forecast
 Branch (TAFB), 291
Tropical Rainfall Measuring Mission
 (TRMM), 146, 277
Tropical storms, 2, 9. *See also* Specific
 tropical storms
Troughs, 3–4
 equatorial, 4
 frontal, 6
 monsoon, 4–5
 surface, 6
Turner, NAME, 92

United States, 190–191
University Corporation for
 Atmospheric Research (UCAR),
 195

University of Hong Kong, 100, 102
Upwelling, 11
U.S. Air Force, 127, 138, 141–142
U.S. Air Force Reserve, 138, 142
U.S. Army Air Force, 118
U.S. Department of Defense (DOD), 99–100, 129
  Defense Meteorological Satellite Program (DMSP), 134
U.S. Navy, 127, 135, 138

U.S. Naval Western Oceanography Center, 99, 275
U.S. Weather Bureau, 112

Velden, Chris, 226
VICBAR hurricane track forecast model, 200
Viñes, Benito, 111–112, 227–228
von Mayer, Julius Robert, 110
von Neumann, John, 121, 122
Vonnegut, Bernard, 123
Vortex Rossby waves, 18
Vorticity, 221

Walker, Gilbert, 115
Wasaburo Ooishi, 124
water temperature, 9–10, 11, 13–14, 29, 51, 60, 124
Waves, 29–30, 244
  tropical, 5–6, 117
  wave setup, 29
Weather balloons, 117
Weather Channel, the, 292–293
  educational material and publications of, 293
Weather magazines, 302–303
Weather modification, 122–123
Weather prediction, 129–131
  errors in, 130

legal issues, 130–131
Weather Prediction by Numerical Procedures (Richardson), 218–219
West Indian Hurricanes (Garriot), 113
West Indian Hurricanes and Other Tropical Cyclones of the North Atlantic Ocean
  (Mitchell), 115
Wiggins, Allan, 119
Willoughby, Hugh E., 228–229
Wind(s), 25–26. *See also* Modified Beaufort Scale; Wind shear
  categories of, 96(table)
  cyclonic, 9, 217
  maximum, 2
  sustained, 2
  rotary (counterclockwise), 107, 109
  speed, 217, 356–358, 357(table)
  trade, 106
Wind shear, 6, 8, 9–10
  Caribbean, 51
  vertical, 10–11
Winthrop, John, 107
Wireless telegraphy, 113–114
World Bank, 84
World Meteorological Organization (WMO), 83, 97, 98, 293–295
  educational material and publications of, 295
  Tropical Cyclone Program (TCP) of, 294
World Weather Watch (WWW), 97–100
Wragge, Clement, 14
WSR-88 (NEXRAD) Doppler radar, 145

Yoshio Kurihara, 207–208